LABORATORY EXPERIMENTS IN
MICROBIOLOGY

ELEVENTH EDITION

TED R. JOHNSON

St. Olaf College

CHRISTINE L. CASE

Skyline College

ACQUISITIONS EDITOR: Kelsey Churchman
PROJECT MANAGER: Courtney Towson
PROGRAM MANAGER: Chriscelle Palaganas
EDITORIAL ASSISTANT: Ashley Williams
TEXT AND PHOTO PERMISSIONS PROJECT MANAGER: Rachel Youdelman
TEXT PERMISSIONS SPECIALIST: James Fortney, Lumina Datamatics
PROGRAM MANAGEMENT TEAM LEAD: Mike Early
PROJECT MANAGEMENT TEAM LEAD: Nancy Tabor

PRODUCTION MANAGEMENT: Thistle Hill Publishing Services, LLC
COMPOSITOR: Cenveo® Publishing Services
ART HOUSE: Precision Graphics
DESIGN MANAGER: Marilyn Perry
INTERIOR DESIGNER: Cenveo® Publishing Services
COVER DESIGNER: Tandem Creative
PHOTO RESEARCHER: Liz Kincaid, Lumina Datamatics
MANUFACTURING BUYER: Stacey Weinberger
EXECUTIVE MARKETING MANAGER: Lauren Harp

COVER PHOTO CREDITS: Background image TEK Image/Science Photo Library/Getty Images. *Serratia marcescens* growing on nutrient agar courtesy of Christine L. Case. *Serratia marcescens* is distinguished by its red pigment. In hospital situations, this bacterium can be found on catheters and in wash basins. *S. marcescens* is an important cause of healthcare-associated infections, especially in neonatal intensive care units and in post-surgical patients. *S. marcescens* may cause pneumonia, wound infections, urinary tract infections, endocarditis, bacteremia, and septicemia.

Library of Congress Cataloging-in-Publication Data
Johnson, Ted R., 1946–
  Laboratory experiments in microbiology / Ted R. Johnson, St. Olaf College,
Christine L. Case, Skyline College. — Eleventh edition.
    pages cm
  Includes bibliographical references and index.
  ISBN 978-0-321-99493-6 — ISBN 0-321-99493-0
1. Microbiology—Laboratory manuals. I. Case, Christine L., 1948– II. Title.
  QR63.J65 2016
  579.078—dc23
                                                    2014043308

ISBN 10: 0-321-99493-0
ISBN 13: 978-0-321-99493-6

www.pearsonhighered.com

2 3 4 5 6 7 8 9 10—V064—18 17 16 15

Contents

Preface

Now in its Eleventh Edition, *Laboratory Experiments in Microbiology* is valued by instructors and students for its comprehensive coverage of every area of microbiology, its user-friendly laboratory reports, and its clear, straightforward organization. Containing 57 thoroughly class-tested exercises, this manual provides basic microbiology techniques with applications for undergraduate students in diverse areas, including the biological sciences, the allied health sciences, agriculture, environmental science, nutrition, pharmacy, and various preprofessional programs. It is designed to supplement any nonmajors microbiology textbook. Coauthored by Christine L. Case, this manual is an ideal companion to *Microbiology: An Introduction,* Twelfth Edition, by Gerard J. Tortora, Berdell R. Funke, and Christine L. Case.

OUR APPROACH

Exercises have been designed to include the American Society for Microbiology laboratory core curriculum, which is considered essential to teach in every introductory microbiology laboratory, regardless of its emphasis.

This laboratory manual has two primary goals—teaching microbiology techniques and showing students the importance of microbes in our daily lives and their central roles in nature. Most of the exercises are investigative by design and require the students to evaluate their experimental results and draw conclusions. We hope in this way to promote analytical reasoning and provide students with a variety of opportunities to reinforce the technical skills they have learned. We also highlight practical uses of microbiology by including material with direct applications to procedures performed in clinical and commercial laboratories.

With a strong emphasis on laboratory safety, this laboratory manual encourages students not only to learn but also to practice safety techniques so that safety becomes part of their professional behavior. We have included a safety contract that students can hand in to their instructors to indicate that they understand safety requirements. (For specific safety suggestions, see the section titled Specific Hazards in the Laboratory on page 4 and the sections of the Introduction that follow it.) We also alert students with yellow safety boxes at key points in the exercises. These safety boxes are marked with either a biohazard icon ☣, a general safety icon ⚠, or a biosafety level 2 icon **BSL-2** indicating appropriate safety techniques.

NEW TO THIS EDITION

For the Eleventh Edition, the overall goal was to make this manual even more navigable and visually effective for students and instructors. The following changes have helped us fulfill that goal.

- **Each exercise has been updated** to reflect the American Society for Microbiology 2012 Guidelines for Biosafety in Teaching Laboratories.
- **Biosafety levels (BSLs)** are noted. Every effort has been made to use biosafety level 1 (BSL-1) organisms. BSL-2 organisms are required in a few exercises to demonstrate specific principles and processes.
- **Each part begins with a case study** to give students a real-world example of the applications of the material they are learning. The solution to the case study requires content from the laboratory exercises in that part.
- **Each exercise includes a Lab and Lecture: Putting it all together** activity available through MasteringMicrobiology.® These activities are designed to help students see how lab and lecture are integrated. Students will use their new information and incorporate their new knowledge into lab.

ORGANIZATION

This manual is divided into 14 parts. The introduction to each part explains the unifying theme for that part. Each Part begins with a Case Study that relates the exercises with clinical applications. Exercises in the first four parts provide sequential development of fundamental techniques. The remaining exercises are as independent as possible to allow instructors to select the most desirable sequence for their course. The exercises are organized as follows.

OBJECTIVES This introductory section defines the specific skills or concepts to be mastered in the exercise. The objectives can easily be used as the basis for assessment.

BACKGROUND This narrative section provides definitions and explanations for each exercise. Students should refer to their text for more detailed explanations of the concepts introduced in the laboratory exercises.

MATERIALS This comprehensive list includes supplies, media, and equipment needed for the exercise.

CULTURES This list identifies the living organisms required for the exercise.

TECHNIQUES REQUIRED This section provides a list of techniques from earlier exercises in the manual needed to complete the current exercise.

PROCEDURE At the core of each exercise, this section provides step-by-step instructions, stated as simply as possible and frequently supplemented with diagrams. Questions are occasionally asked in the Procedure section to remind the student of the rationale for a step.

LABORATORY REPORT Designed to help students learn to collect and present data in a systematic fashion, the laboratory report concludes each exercise. Students are first asked to write the *Purpose* of the exercise so they can relate their laboratory work to their learning. Students are asked to formulate their *Hypothesis* or write their *Expected Results* using information provided in the Background and their own experience. Tables are provided to record *Results*. The questions in *Conclusions* are designed to lead the student from a collection of data or observations to a conclusion. In most instances, the results for each student

team will be unique; they can be compared with the information given in the Background and other references but will not be identical to that information. The *Questions* reinforce the conclusions and ask the student to interpret results. The range of questions requires students to think about their results, recall facts, and then use this information to answer the questions. *Critical Thinking* questions are designed to help students use their new knowledge and practice analytical skills. *Clinical Application* questions have been collected from the literature. They are designed to encourage students to synthesize their new information to relate concepts and techniques to clinical applications. This lab manual and the course textbook should provide sufficient background to enable students to answer *Critical Thinking* and *Clinical Application* questions.

PREPARATION GUIDE

The comprehensive *Preparation Guide for Laboratory Experiments in Microbiology,* Eleventh Edition (ISBN 0-134-02449-4) provides all the information the instructor needs to set up and teach a laboratory course with this manual. It includes the following:

- General instructions for setting up the lab
- Information on obtaining and preparing cultures, media, and reagents, and expected results for each of the biochemical tests and cultures used
- A master table showing the techniques and biosafety level required for each exercise
- Cross-references for each exercise to specific pages in Tortora/Funke/Case *Microbiology: An Introduction,* Twelfth Edition.
- For each exercise: helpful suggestions, detailed lists of materials needed, and answers to all the questions in the student manual

To make *Laboratory Experiments in Microbiology,* Eleventh Edition, easy to use, we have carefully designed the experiments to use inexpensive, readily available, nonhazardous materials. Moreover, the exercises have been thoroughly tested in our classes in Minnesota and California by students with a wide variety of talents and interests. Our students have enjoyed their microbiology laboratory experiences; we hope yours will, too!

ACKNOWLEDGMENTS

We are most grateful to the following individuals for their time, talent, and interest in our work. Each person carefully read and edited critical parts of the manuscript.

Chuck Hoover of the University of California, San Francisco, for making us aware of the new techniques used in dental microbiology for Exercise 47.

Anne Jayne of the University of San Francisco for offering suggestions and a capsule stain.

Kylin Johnson of Skyline College for her invaluable assistance in preparing materials.

We are indebted to St. Olaf College and Skyline College for providing the facilities and resources in which innovative laboratory exercises can be developed. We are grateful for the wonderful students who have inspired us and have made teaching microbiology a joy.

We would like to commend the staff at Pearson Education for their support. In particular, we thank Kelsey Churchman, our Acquisitions Editor; Chriscelle Palaganas and Ashley Williams for their editorial skill and conscientiousness; and Nancy Tabor and Courtney Towson for expertly guiding this manual through the production process.

Special thanks go to Clifton Franklund of Ferris State University for his expert work on the pre-lab questions that accompany this manual.

Last, but not least, our gratitude goes to Michelle Johnson, who gave her professional insights and was a sustaining presence; and Don Biederman, who provided timely encouragement and support.

REVIEWERS

We are deeply grateful to the following reviewers who helped shape the direction of the Eleventh Edition:

Veronica Amaku, *Lone Star College–CyFair Campus*
Jennifer Bess, *Hillsborough Community College– Dale Mabry Campus*
Laurie Bradley, *Hudson Valley Community College*
Lynn B. DeSanto, *Lackawanna College*
Regina D. Foster, *Oklahoma State University Institute of Technology*
Jennifer Hatchel, *College of Coastal Georgia*
Suzanne Kempke, *St. Johns River State College*
Mustapha Lahrach, *Hillsborough Community College–SouthShore Campus*
Luis A. Materon, *University of Texas–Pan American*
Stacy Pfluger, *Angelina College*
Ines Rauschenbach, *Rutgers University* and *Union County College*
Jason Rothman, *California State Polytechnic University, Pomona*
Misty D. Wehling, *Southeast Community College*
Daniel Westholm, *The College of St. Scholastica*

A SPECIAL NOTE TO STUDENTS

This book is for you. The study of microbiology is dynamic because of the diversity of microbes and the variability inherent in every living organism. Outside the laboratory—on a forest walk or tasting a fine cheese—we experience the activities of microbes. We want to share our excitement for studying these small organisms. Enjoy!

Ted R. Johnson and Christine L. Case

Introduction

Welcome to microbiology! Microorganisms are all around us, and as Pasteur pointed out over a century ago, they play vital roles in the ecology of life on Earth. In addition, some microorganisms provide important commercial benefits through their use in the production of chemicals (including antibiotics) and certain foods. Microorganisms are also major tools in basic research in the biological sciences. Finally, as we all know, some microorganisms cause disease—in humans, other animals, and plants.

In this course, you will have firsthand experience with a variety of microorganisms. You will learn the techniques required to identify, study, and work with them. Before getting started, you will find it helpful to read through the suggestions on the next few pages.

SUGGESTIONS TO HELP YOU BEGIN

1. Science has a vocabulary all its own. New terms will be introduced in **boldface** throughout this manual. To develop a working vocabulary, make a list of these new terms and their definitions.
2. The microbes used in the exercises in this manual are referred to by their *scientific names.* Common names were never given to microbes because they are not visible to the human eye without a microscope. The word *microbe,* now in common use, was introduced in 1878 by Charles Sedillot. The scientific names will be unfamiliar at first, but do not let that deter you. Practice saying them aloud. Most scientific names are taken from Latin and Greek roots. If you become familiar with these roots, the names will be easier to remember.
3. Microbiology usually provides the first opportunity that undergraduate students have to experiment with *living organisms.* Microbes are relatively easy to grow and lend themselves to experimentation. Because there is variability in any population of living organisms, not all the experiments will "work" as the lab manual says. The following exercise will illustrate what we mean:

Write a description of *Homo sapiens* for a visitor from another planet: _____
_____
_____
_____
_____
_____
_____

After you have finished, look around you. Do all your classmates fit the description exactly? Probably not. Moreover, the more detailed you make your description, the less conformity you will observe. During lab, you will make a detailed description of an organism and probably find that this description does not match your reference exactly.

4. Microorganisms must be cultured or grown to complete most of the exercises in this manual. Cultures will be set up during one laboratory period and will be examined for growth in the next laboratory period. Accurate record keeping is therefore essential. Mark the steps in each exercise with a bright color or a bookmark so you can return to complete your Laboratory Report on that exercise. *Accurate records* and *good organization* of laboratory work will enhance your enjoyment and facilitate your learning.
5. *Observing* and *recording* your results carefully are the most important parts of each exercise. Ask yourself the following questions for each experiment:
   What did the results indicate?
   Are they what I expected? If not, what happened?
6. If you do not master a technique, try it again. In most instances, you will need to use the technique again later in the course.
7. Be sure you can answer the questions that are asked in the Procedure for each exercise. These questions are included to reinforce important points that will ensure a successful experiment.
8. Finally, carefully study the general procedures and safety precautions that follow.

GENERAL PROCEDURES IN MICROBIOLOGY

In many ways, working in a microbiology laboratory is like working in the kitchen. As some very famous chefs have said,

> Our years of teaching cookery have impressed upon us the fact that all too often a debutant cook will start in enthusiastically on a new dish without ever reading the recipe first. Suddenly an ingredient, or a process, or a time sequence will turn up, and there is astonishment, frustration, and even disaster. We therefore urge you, however much you have cooked, always to read the recipe first, even if the dish is familiar to you. . . . We have not given estimates for the time of preparation, as some people take half an hour to slice three pounds of mushrooms, while others take five minutes.*

1. Read the laboratory exercises *before* coming to class.
2. *Plan* your work so that you complete all experiments during the assigned laboratory period. A good laboratory student, like a good cook, is one who can do more than one procedure at a time—that is, one who is efficient.
3. Use only the *required* amounts of materials, so that everyone can do the experiment.
4. *Label* all of your experiments with your name, date, and lab section.
5. Even though you will do most exercises with another student, you must become familiar with *all* parts of each exercise.
6. Keep *accurate* notes and records of your procedures and results so that you can refer to them for future work and tests. Many experiments are set up during one laboratory period and observed for results in the next laboratory period. Your notes are essential to ensure that you perform all the necessary steps and observations.
7. *Demonstrations* will be included in some of the exercises. Study the demonstrations and learn the content.
8. If you are color-blind, let your instructor know; many techniques require discrimination of colors.
9. Keep your cultures current; discard old experiments.
10. *Clean up* your work area when you are finished. Leave the laboratory clean and organized for the next student. Remember:
    - Return stain and reagent bottles to their original locations.

*J. Child, L. Bertholle, and S. Beck. *Mastering the Art of French Cooking*, Vol. 1. New York: Knopf, 1961.

- Place slides in the appropriate disinfectant container as instructed.
- Remove all markings on glassware (such as Petri plates and test tubes) before putting glassware into the marked autoclave trays.
- Place glass Petri plates agar-side down in marked autoclave containers.
- Place swabs and pipettes in the appropriate disinfectant jars or biohazard containers.
- Place disposable plasticware in marked biohazard or autoclave containers.
- Discard used paper towels.

BIOSAFETY

The most important element for managing microorganisms is strict adherence to standard microbiological practices and techniques, which you will learn during this course. There are four biosafety levels (BSLs) for working with live microorganisms; each BSL consists of combinations of laboratory practices and techniques, safety equipment, and laboratory facilities. (See Table 1 on page 3.) Each combination is specifically appropriate for the operations performed, the documented or suspected routes of transmission of the microorganisms, and the laboratory function or activity.

Biosafety Level 1 represents a basic level of containment that relies on standard microbiological practices with no specific facilities other than a sink for handwashing. When standard laboratory practices are not sufficient to control the hazard associated with a particular microorganism, additional measures may be used. Gloves should be worn if skin on hands is broken or if a rash is present.

Biosafety Level 2 includes handwashing, and an autoclave for decontaminating laboratory wastes must be available. Precautions must be taken for handling and disposing of contaminated needles or sharp instruments. BSL-2 is appropriate for working with human body fluids. A lab coat should be worn. Gloves should be worn when hands may contact potentially hazardous materials.

Biosafety Level 3 is used in laboratories where work is done with pathogens that can be transmitted by the respiratory route. BSL-3 requires special facilities with self-closing, double doors and sealed windows.

Biosafety Level 4 is applicable for work with pathogens that may be transmitted via aerosols and for which there is no vaccine or treatment. The BSL-4 facility is generally a separate building with specialized ventilation and waste management systems to prevent release of live pathogens to the environment.

Which biosafety level is your lab? _____

TABLE 1 BIOSAFETY LEVELS

Biosafety level	Practices	Personal Protection (Primary Barriers)	Facilities (Secondary Barriers)
	BSL-3 plus • Separate building	BSL-3 plus full-body air-supplied, positive-pressure personnel suit	BSL-3 plus separate building and decontamination facility
BSL-4 / BSL-3	BSL-2 plus • Controlled access • Decontamination of clothing before laundering	BSL-2 plus protective lab clothing; enter and leave lab through clothing-changing and shower rooms	BSL-2 plus self-closing, double-door access
BSL-2	BSL-1 plus • Limited access • Biohazard warning signs • "Sharps" precautions • Safety manual of waste decontamination policies	Lab coat; goggles and gloves, as needed	BSL-1 plus autoclave
BSL-1	Standard microbiological practices	Lab coat; goggles and gloves, as needed	Open benchtop sink; autoclave recommended

Biosafety Practices

The lab exercises in this course involve the use of living organisms. Although the microorganisms we use are not considered to be highly virulent, all microorganisms should be treated as potential pathogens (organisms capable of causing disease).

The following rules must be observed at all times to prevent accidental injury to or infection of yourself and others and to minimize contamination of the lab environment. These guidelines are in agreement with *Guidelines for Biosafety in Teaching Laboratories* (American Society for Microbiology, 2012).

1. Never place books, backpacks, purses, or the like on benchtops. Always place these in the assigned cubicles.
2. Electronic devices should not be brought into the lab. This includes, but is not limited to, iPods, MP3 players, radios, cell phones, and calculators.
3. Clean your work area with disinfectant at the beginning AND end of each lab.
4. Wash your hands with soap and dry with paper towels when entering and leaving the lab.
5. Wear a lab coat at all times while working in the lab to prevent contamination or accidental staining of your clothing.
   a. Closed-toe shoes (no sandals) are to be worn in the lab.
   b. Tie back long hair to prevent exposure to flame and contamination of cultures.
   c. Gloves should be worn when staining microbes and handling hazardous chemicals.
   d. Wear safety goggles when performing a procedure (such as pipetting, spread plates, and so on) that may generate a splash hazard.

6. Do not place anything in your mouth or eyes while in the lab. This includes pencils, food, and fingers. Keep your hands away from your mouth and eyes.
   a. Eating (including gum, cough drops, and candy) and drinking are prohibited in the lab at all times. Do not bring water bottles into the lab.
   b. Do not apply cosmetics in the lab.
   c. Never pipette by mouth. Use a mechanical pipetting device.
7. Do not remove media, equipment, or bacterial cultures from the laboratory. This is absolutely prohibited and unnecessary.
8. Do not place contaminated instruments such as inoculating loops, needles, and pipettes on benchtops. Loops and needles should be sterilized by incineration, and pipettes should be disposed of in designated receptacles.
9. Carry cultures in a test-tube rack when moving around the lab and when keeping cultures on benchtops for use.
10. Immediately cover spilled cultures or broken culture tubes with paper towels and then saturate with disinfectant. Notify your instructor that there has been a spill. After 20 minutes, clean up the area and dispose of the towels and broken items as indicated by your instructor.
11. Report accidental cuts or burns to the instructor immediately.
12. At the end of each lab session, place all cultures and materials in the proper disposal area.
13. Persons who are immunocompromised (including those who are pregnant) and students living with or caring for an immunocompromised individual are advised to consult their physician to determine the appropriate level of participation in the lab.

SPECIFIC HAZARDS IN THE LABORATORY

Keep containers of alcohol away from open flames.

Glassware Not Contaminated with Microbial Cultures

1. If you break a glass object, sweep up the pieces with a broom and dustpan. Do not pick up pieces of broken glass with your bare hands.
2. Place broken glass in one of the containers marked for this purpose. The one exception to this rule concerns broken mercury thermometers; consult your instructor if you break a mercury thermometer.

Electrical Equipment

1. The basic rule to follow is this: Electricity and water don't mix. Do not allow water or any water-based solution to come into contact with electrical cords or electrical conductors. Make sure your hands are dry when you handle electrical connectors.
2. If your electrical equipment crackles, snaps, or begins to give off smoke, do not attempt to disconnect it. Call your instructor immediately.

Fire

1. If *gas burns* from a leak in the burner or tubing, turn off the gas.
2. If you have a *smoldering sleeve,* run water on the fabric.
3. If you have a *very small fire,* the best way to put it out is to smother it with a towel or book (not your hand). Smother the fire quickly.
4. If a *larger fire* occurs, such as in a wastebasket or sink, use one of the fire extinguishers in the lab to put it out. Your instructor will demonstrate the use of the fire extinguishers.
5. In case of a *large fire* involving the lab itself, evacuate the room and building according to the following procedure:
   a. Turn off all gas burners, and unplug electrical equipment.
   b. Leave the room and proceed to _____ _____
   c. It is imperative that you assemble in front of the building so that your instructor can take roll to determine whether anyone is still inside. Do not wander off.

Accidents and First Aid

1. Report all accidents immediately. Your instructor will administer first aid as required.
2. For spills in or near the eyes, use the eyewash immediately.
3. For large spills on your body, use the safety shower.
4. For heat burns, chill the affected part with ice as soon as possible. Contact your instructor.

5. Place a bandage on any cut or abrasion.

Earthquake

Turn off your gas jet and get under your lab desk during an earthquake. Your instructor will give any necessary evacuation instructions.

ORIENTATION WALKABOUT

Locate the following items in the lab:

Broom and dustpan	Instructor's desk
Eyewash	Reference books
Fire blanket	Safety shower
Fire extinguisher	To Be Autoclaved area
First-aid cabinet	Biohazard containers
Fume hood	

SPECIAL PRACTICES

Potential pathogens used in the exercises in this manual present a minimal hazard and require ordinary aseptic handling conditions (Biosafety Level 2). They are marked throughout this manual with the **BSL-2** icon. No special competence or containment is required. These organisms are the following:

- *Enterococcus faecalis*
- *Proteus* species
- *Pseudomonas aeruginosa*
- *Staphylococcus aureus*
- *Streptococcus pneumoniae*
- *Str. pyogenes*

BSL-1 labs will not be using the BSL-2 microbes but will observe demonstration cultures using BSL-2 organisms.

Treat all microorganisms subcultured from the environment as potential BSL-2 organisms, and subculture them only if you are in a BSL-2 laboratory.

LABORATORY FACILITIES

1. Interior surfaces of walls, floors, and ceilings are water resistant so that they can be easily cleaned.
2. Benchtops are impervious to water and resistant to acids, alkalis, organic solvents, and moderate heat.
3. Windows in the laboratory are closed and sealed.
4. An autoclave for decontaminating laboratory wastes is available, preferably within the laboratory.
5. Keep laboratory doors closed when experiments are in progress.
6. The instructor controls access to the laboratory and allows access only to people whose presence is required for program or support purposes.
7. Place contaminated materials that are to be decontaminated at a site away from the laboratory into a durable, leakproof container that is closed before being removed from the laboratory.
8. An insect- and rodent-control program is in effect.

Student Safety Contract

During your microbiology course, you will learn how to safely handle fluids containing microorganisms. Through practice you will be able to perform experiments in such a way that bacteria, fungi, and viruses remain in the desired containers, uncontaminated by microbes in the environment. These techniques, called **aseptic techniques,** will be a vital part of your work if you are going into health care or biotechnology.

1. Do not eat, drink, smoke, store food, or apply cosmetics in the laboratory.
2. Wear closed-toe shoes at all times in the laboratory.
3. Tie back long hair.
4. Disinfect work surfaces at the beginning and end of every lab period and after every spill. The disinfectant used in this laboratory is _____.
5. Wash your hands before and after every laboratory period. Because bar soaps may become contaminated, use liquid or powdered soaps.
6. Use mechanical pipetting devices; do not use mouth pipetting.
7. Wear safety goggles while pipetting.
8. Wash your hands immediately and thoroughly with soap and water if they become contaminated with microorganisms.
9. Cover spilled microbial cultures with paper towels, and saturate the towels with disinfectant. Leave covered for 20 minutes, and then clean up the spill and dispose of the towels.
10. Do not touch broken glassware with your hands; use a broom and dustpan. Place broken glassware *contaminated* with microbial cultures or body fluids in the To Be Autoclaved container. (See page 4 for what to do with broken glassware that is not contaminated.)
11. Place glassware and slides contaminated with blood, urine, and other body fluids in disinfectant.
12. To avoid transmitting disease, work only with your own body fluids and wastes in exercises that require saliva, urine, blood, or feces. The Centers for Disease Control and Prevention (CDC) states that "epidemiologic evidence has implicated only blood, semen, vaginal secretions, and breast milk in transmission of HIV" (*Biosafety in Microbiological and Biomedical Laboratories,* www.cdc.gov).
13. Do not perform unauthorized experiments.
14. Do not use equipment without instruction.
15. Do not engage in horseplay in the laboratory.
16. If you got this far in the instructions, you'll probably do well in lab. Enjoy lab, and make a new friend.

 Procedures marked with this biohazard icon should be performed carefully to minimize the risk of transmitting disease.

 Procedures marked with this safety icon should be performed carefully to minimize risk of exposure to chemicals or fire.

I have read the above laboratory safety rules and agree to abide by them when in the laboratory.

Name: _____ Date: _____

PART

1

Microscopy

EXERCISES

1 Use and Care of the Microscope

2 Examination of Living Microorganisms

Antoni van Leeuwenhoek is the first person known to have observed living microbes in suspension. Unfortunately, he was very protective of his homemade microscopes and left no description of how to make them (see the photograph on page 8). During his lifetime he kept "for himself alone" his microscopes and his method of observing "animalcules." Directions for making a replica of van Leeuwenhoek's microscope can be found in *American Biology Teacher*.* Fortunately, you will not have to make your own microscope.

The microscope is a very important tool for a microbiologist. Microscopes and microscopy (microscope technique) are introduced in Exercises 1 and 2, which are designed to help you become familiar with the compound light microscope and proficient in using it. This knowledge will be valuable in later exercises.

Beginning students frequently become impatient with the microscope and forgo this opportunity to practice and develop their observation skills. Simple observation is a critical part of any science. Making discoveries by observation requires *curiosity* and *patience*. We cannot provide procedures for observation, but we can offer this suggestion: Make *careful sketches* to enhance effective observation. You need not be an artist to draw what you see. In your drawings, pay special attention to

1. *Size relationships.* For example, how big are bacteria relative to protozoa?
2. *Spatial relationships.* For example, where is one bacterium in relation to the others? Are they all together in chains?
3. *Behavior.* For example, are individual cells moving, or are they all flowing in the liquid medium?
4. *Sequence of events.* For example, were cells active when you first observed them?

Looking at objects through a microscope is not easy at first, but with a little practice, you, like Antoni van Leeuwenhoek, will make discoveries in the microcosms of peppercorn infusions and raindrops. In 1684, van Leeuwenhoek wrote the following:

Tho my teeth are kept usually very clean, nevertheless when I view them in a Magnifying Glass, I find growing between them a little white matter as thick as wetted flour: In this substance tho I do not perceive any motion, I judged there might probably be living creatures.

---

*W. G. Walter and H. Via. "Making a Leeuwenhoek Microscope Replica." *American Biology Teacher* 30(6): 537–539, 1968.

I therefore took some of this flour and mixed it either with pure rain water wherein were no Animals; or else with some of my Spittle (having no air bubbles to cause a motion in it) and then to my great surprise perceived that the aforesaid matter contained very many small living Animals, which moved themselves very extravagantly. Their motion was strong and nimble, and they darted themselves thro the water or spittle, as a Jack or Pike does thro the water.*

---

*Quoted in E. B. Fred. "Antoni van Leeuwenhoek." *Journal of Bacteriology* 25: 1, 1933.

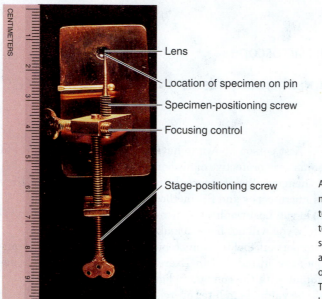

A replica of the simple microscope made by Antoni van Leeuwenhoek to observe living organisms too small to be seen with the naked eye. The specimen was placed on the tip of the adjustable point and viewed from the other side through the tiny round lens. The highest magnification with his lenses was about 300×.

CASE STUDY: Too Many Slides

Your first microbiology field trip is to a large university hospital lab. Urinary tract infections are quite common, so it is not surprising that urine specimens make up a large proportion of the samples submitted for routine laboratory diagnosis. Nevertheless, you are surprised to learn that the lab technicians may examine 200 microscope slides of urine every day. You are given the opportunity to look at some of the slides and are asked to describe any microorganisms that you see.

Use the following choices to indicate which type of microorganism the one described in each question is most likely to be.

Questions

   a. Alga
   b. Bacterium
   c. Fungus
   d. Protozoan

1. In a wet mount of urine, you observe flagellated, nucleated cells. Which type of microorganism is most likely?
2. In a fixed, stained slide, you don't see any cells until you use the oil immersion objective. Which type of microorganism is most likely?
3. In a wet mount of the scrapings, you observe long filaments composed of many cells. Which type of microorganism is most likely?

Use and Care of the Microscope

The most important DISCOVERIES *of the laws, methods and progress of nature have nearly always sprung from the* EXAMINATION *of the smallest objects which she contains.* – J E A N   B A P T I S T E   L A M A R C K

OBJECTIVES

After completing this exercise, you should be able to:

1. Demonstrate the correct use of a compound light microscope.
2. Name the major parts of a compound microscope.
3. Determine the relative sizes of different microbes.
4. Identify the three basic morphologies of bacteria.

BACKGROUND

Virtually all organisms studied in microbiology are invisible to the naked eye and require the use of optical systems for magnification. The microscope was invented shortly before 1600 by Zacharias Janssen of the Netherlands. The microscope was not used to examine microorganisms until the 1680s, when a clerk in a dry-goods store, Antoni van Leeuwenhoek, examined scrapings of his teeth and any other substances he could find. The early microscopes, called **simple microscopes,** consisted of biconvex lenses and were essentially magnifying glasses. (See the photograph on page 8.) To see microbes requires a compound microscope, which has two lenses between the eye and the object. This optical system magnifies the object, and an illumination system (sun and mirror or lamp) ensures that adequate light is available for viewing. A **brightfield compound microscope,** which shows dark objects in a bright field, is used most often.

The Microscope

You will be using a brightfield compound microscope similar to the one shown in **FIGURE 1.1a.** The basic frame of the microscope consists of a **base,** a **stage** to hold the slide, an **arm** for carrying the microscope, and a **body tube** for transmitting the magnified image. The stage may have two clips or a movable mechanical stage to hold the slide. The light source is in the base. Above the light source is a **condenser,** which consists of several lenses that concentrate light on the slide by focusing it into a cone, as shown in **FIGURE 1.1b.** The condenser has an **iris diaphragm,** which controls the

angle and size of the cone of light. This ability to control the *amount* of light ensures that optimal light will reach the slide. Above the stage, on one end of the body tube, is a revolving nosepiece holding three or four **objective lenses.** At the upper end of the tube is an **ocular** or **eyepiece lens** ($10\times$ to $12.5\times$). If a microscope has only one ocular lens, it is called a **monocular** microscope; a **binocular** microscope has two ocular lenses.

Focusing the Microscope

By moving the lens closer to the slide or the stage closer to the objective lens, using the coarse- or fine-adjustment knob, one can focus the image. The larger knob, the **coarse adjustment,** is used for focusing with the low-power objectives ($4\times$ and $10\times$), and the smaller knob, the **fine adjustment,** is used for focusing with the high-power and oil immersion lenses. The coarse-adjustment knob moves the lenses or the stage longer distances. The area seen through a microscope is called the **field of vision.**

The **magnification** of a microscope depends on the type of objective lens used with the ocular. Compound microscopes have three or four objective lenses mounted on a nosepiece: scanning ($4\times$), low-power ($10\times$), high-dry ($40\times$ to $45\times$), and oil immersion ($97\times$ to $100\times$). The magnification provided by each lens is stamped on the barrel. The total magnification of the object is calculated by multiplying the magnification of the ocular (usually $10\times$) by the magnification of the objective lens. The most important lens in microbiology is the **oil immersion lens;** it has the highest magnification ($97\times$ to $100\times$) and must be used with immersion oil. Optical systems could be built to magnify much more than the $1000\times$ magnification of your microscope, but the resolution would be poor.

The Light Source

Compound microscopes require a light source. The light may be reflected to the condenser by a mirror under the stage. If your microscope has a mirror, the sun or a lamp may be used as the light source. Most compound microscopes have a built-in illuminator in

Ocular lens
(eyepiece)
Remagnifies the
image formed by
the objective lens

Body tube Transmits
the image from the
objective lens to the
ocular lens

Arm

Objective lenses
Primary lenses that
magnify the specimen

Mechanical stage
Holds the microscope
slide

Condenser Focuses
light through specimen

Diaphragm Controls the amount
of light entering the condenser

Illuminator Light source

Base

Coarse focusing knob

Fine focusing knob
Used for focusing the specimen; turning the
knob changes the distance between the
objective lens and the specimen

Light intensity
Adjusts current
to lamp

(a) Principal parts and functions

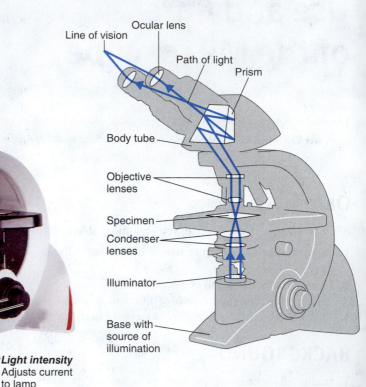

Line of vision
Ocular lens
Path of light
Prism
Body tube
Objective lenses
Specimen
Condenser lenses
Illuminator
Base with source of illumination

(b) The path of light (bottom to top)

FIGURE 1.1 **The compound light microscope. (a)** Its principal parts and their functions. **(b)** Blue lines from the light source through the ocular lens trace the path of light.

the base. The *intensity* of the light can be adjusted with a wheel that regulates the amount of current to the bulb. Higher magnification usually requires more light, which can be obtained by adjusting the iris diaphragm.

Resolution

Resolution, or **resolving power,** is the ability of a lens to reveal fine detail or two points distinctly separated. An example of resolution involves a car approaching you at night. At first only one light appears, but as the car nears, you can distinguish two headlights. The resolving power is a function of the wavelength of light used and a characteristic of the lens system called **numerical aperture.** Resolving power is best when two objects are seen as distinct even though they are very close together. Resolving power is expressed in units of length; the smaller the distance, the better the resolving power.

$$\text{Resolving power} = \frac{\text{Wavelength of light used}}{2 \times \text{numerical aperture}}$$

Smaller wavelengths of light improve resolving power. The effect of decreasing the wavelength can be seen in electron microscopes, which use electrons as a source of "light." The electrons have an extremely short wavelength and result in excellent resolving power. A

light microscope has a resolving power of about 200 nanometers (nm), whereas an electron microscope has a resolving power of less than 0.2 nm. The numerical aperture is engraved on the side of each objective lens (usually abbreviated N.A.). Increasing the numerical aperture—for example, from 0.65 to 1.25—improves the resolving power. The numerical aperture depends on the maximum angle of the light entering the objective lens and on the **refractive index** (the amount the light bends) of the material (usually air) between the objective lens and the slide. This relationship is defined by the formula:

N.A. = $N \sin \theta$, where
  N = Refractive index of medium
  θ = Angle between the most divergent light ray gathered by the lens and the center of the lens

As shown in **FIGURE 1.2**, light is refracted when it emerges from the slide because of the change in medium as the light passes from glass to air. When immersion oil is placed between the slide and the oil immersion lens, the light ray continues without refraction because immersion oil has the same refractive index ($N = 1.52$) as glass ($N = 1.52$). This can be seen easily. When you look through a bottle of immersion oil, you cannot see the glass rod in it because of

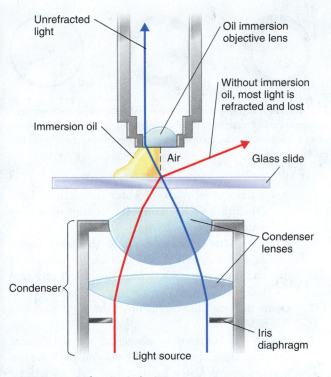

FIGURE 1.2 **Refractive index.** Because the glass microscope slide and immersion oil have the same refractive index, the oil keeps the light rays from refracting.

the identical *N* values of the glass and immersion oil. Using oil minimizes light loss, and the lens focuses very close to the slide.

As light rays pass through a lens, they are bent to converge at the **focal point,** where an image is formed (**FIGURE 1.3a**). When you bring the center of a microscope field into focus, the periphery may be fuzzy because of the curvature of the lens, resulting in multiple focal points. This is called **spherical aberration** (**FIGURE 1.3b**). Spherical aberrations can be minimized by using the iris diaphragm, which eliminates light rays to the periphery of the lens, or by a series of lenses resulting in essentially a flat optical system. Sometimes a multitude of colors, or **chromatic aberration,**

is seen in the field (**FIGURE 1.3c**). This is caused by the prismlike effect of the lens as various wavelengths of white light pass through to a different focal point for each wavelength. Chromatic aberrations can be minimized by using filters (usually blue); or by lens systems corrected for red and blue light, called *achromatic lenses;* or by lenses corrected for red, blue, and other wavelengths, called *apochromatic lenses.* The most logical, but most expensive, method of eliminating chromatic aberrations is to use a light source of one wavelength, or **monochromatic light.**

GENERAL GUIDELINES

The microscope is a very important tool in microbiology, and it must be used carefully and correctly. Follow these guidelines *every* time you use a microscope.

1. Carry the microscope with both hands: one hand beneath the base and one hand on the arm.
2. Do not tilt the microscope; instead, adjust your stool so you can comfortably use the instrument.
3. Observe the slide with both eyes open, to avoid eyestrain.
4. Always focus by moving the lens away from the slide.
5. Always focus slowly and carefully.
6. When using the low-power lens, keep the iris diaphragm barely open to achieve good contrast. Higher magnification requires more light.
7. Before using the oil immersion lens, have your slide in focus under high power. *Always focus with low power first.*
8. Keep the stage clean and free of oil. Keep all lenses except the oil immersion lens free of oil.
9. Keep all lenses clean. Use *only* lens paper to clean them. Wipe oil off the oil immersion lens before putting your microscope away. Do not touch the lenses with your hands.
10. Clean the ocular lens carefully with lens paper. If dust is present, it will rotate as you turn the lens. If needed, wet the lens paper with optical lens cleaner.

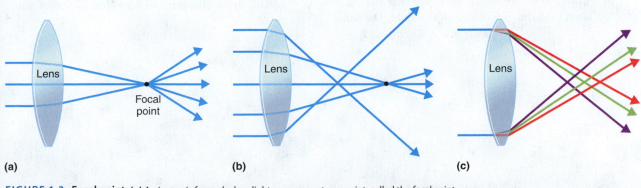

(a) **(b)** **(c)**

FIGURE 1.3 **Focal point. (a)** An image is formed when light converges at one point, called the focal point. **(b)** Spherical aberration. Because the lenses are curved, light passing through one region of the lens has a different focal point from light passing through another part of the lens. **(c)** Chromatic aberration. The lens may give each wavelength of light a different focal point.

11. After use, remove the slide, wipe oil off it, put the dust cover on the microscope, and return it to the designated area.
12. When a problem does arise with the microscope, obtain help from the instructor. Do not use another microscope unless yours is declared "out of action."

MATERIALS

Compound light microscope
Immersion oil
Lens paper and optical lens cleaner
Prepared slides of algae, fungi, protozoa, and bacteria

PROCEDURE

1. Place the microscope on the bench squarely in front of you.
2. Obtain a slide of algae, fungi, or protozoa, and place it in the clips on the mechanical stage.
3. Adjust the eyepieces on a binocular microscope to your own personal measurements.
   a. Look through the eyepieces and, using the thumb wheel, adjust the distance between the eyepieces until one circle of light appears.
   b. With the low-power (10×) objective in place, cover the left eyepiece with a small card and focus the microscope on the slide. When the right eyepiece has been focused, remove your hand from the focusing knobs and cover the right eyepiece. Looking through the microscope with your left eye, focus the left eyepiece by turning the eyepiece adjustment. Make a note of the number at which you focused the left eyepiece so you can adjust any binocular microscope for your eyes.
4. Raise the condenser up to the stage. On some microscopes, you can focus the condenser by the following procedure:
   a. Focus with the 10× objective.
   b. Close the iris diaphragm so only a minimum of light enters the objective lens.
   c. Lower the condenser until you see the light as a circle in the center of the field. On some

FIGURE 1.4 Focusing the condenser. (a) Using low power, lower the condenser until a distinct circle of light is visible. **(b)** Center the circle of light using the centering screws. **(c)** Open the iris diaphragm until the light just fills the field.

microscopes, you can center the circle of light (**FIGURE 1.4**) by using the centering screws found on the condenser.
   d. Raise the condenser up to the slide, lower it, and stop when the color on the periphery changes from pink to blue (usually 1 or 2 mm below the stage).
5. Open the iris diaphragm until the light just fills the field.
6. Adjust the contrast by changing the diaphragm opening. Diagram some of the cells on the slide under low power. Use a minimum of light by adjusting the _____.
7. When you have brought an image into focus with low power, rotate the turret to the next lens, and the subject will remain almost in focus. All of the objectives (with the possible exception of the 4×) are **parfocal;** that is, when a subject is in focus with one lens, it will be in focus with all of the lenses. However, the **working distance**—that is, the distance between the objective lens and the specimen—will change (Figure 1.2). In general, the working distance decreases as magnification increases. When you have completed your observations under low power, swing the high-dry objective into position and focus. Use the fine adjustment. Only a slight adjustment should be required. Why? _____

_____

More light is usually needed at a higher magnification. How can you increase the amount of light? __

_____

Again, draw the general size and shape of some cells.

(a) Move the high-dry lens out of position.

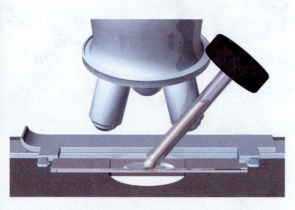

(b) Place a drop of immersion oil in the center of the slide.

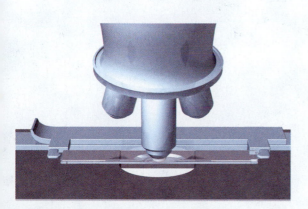

(c) Move the oil immersion lens into position.

FIGURE 1.5 **Using the oil immersion lens.**

8. Move the high-dry lens out of position, and place a drop of immersion oil on the area of the slide you are observing. Carefully click the oil immersion lens into position. It should now be immersed in the oil (**FIGURE 1.5**). Careful use of the fine-adjustment knob should bring the object into focus. Note the shape and size of the cells. Did the color of the cells change with the different lenses? _____ Did the size of the field change? _____

9. Record your observations, and note the magnifications. (Use the following figures as references: Algae: Figure 2.2a and b on page 21 and 34.3 on page 273; protozoa: Figure 2.2c on page 21 and 35.1 on page 279; fungi: Figure 33.2 on page 262.)

10. Repeat this procedure with all the available slides. When observing the bacteria, note the three different morphologies, or shapes, shown in **FIGURE 1.6** on the next page. When your observations are completed, move the turret to bring a low-power objective into position. *Do not* rotate the high-dry (40×) objective through the immersion oil. Remove the slide. Clean the oil off the objective lens with lens paper, and clean off the slide with tissue paper or a paper towel.

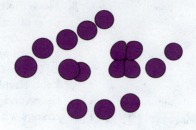

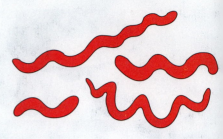

(a) Bacillus (plural: bacilli) or rod

(b) Coccus (plural: cocci)

(c) Spiral

FIGURE 1.6   Basic shapes of bacteria.

LABORATORY REPORT
Use and Care of the Microscope

PURPOSE _____
_____
_____

EXPECTED RESULTS

1. The high-dry lens will be optimal for observing multicellular fungi or algae. Agree/disagree

2. The oil immersion objective is necessary to determine the morphology of prokaryotes. Agree/disagree

RESULTS

Microscope number: _____ Monocular or binocular: _____

Eyepiece adjustment notes: _____

Draw a few representative cells from each slide, and show how they appeared at each magnification. Note the differences in size at each magnification.

Algae

Slide of _____

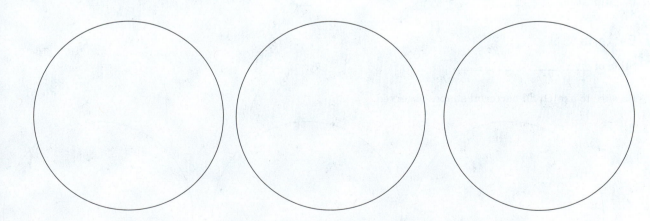

Total
magnification   ___×   ___×   ___×

Fungi

Slide of _____

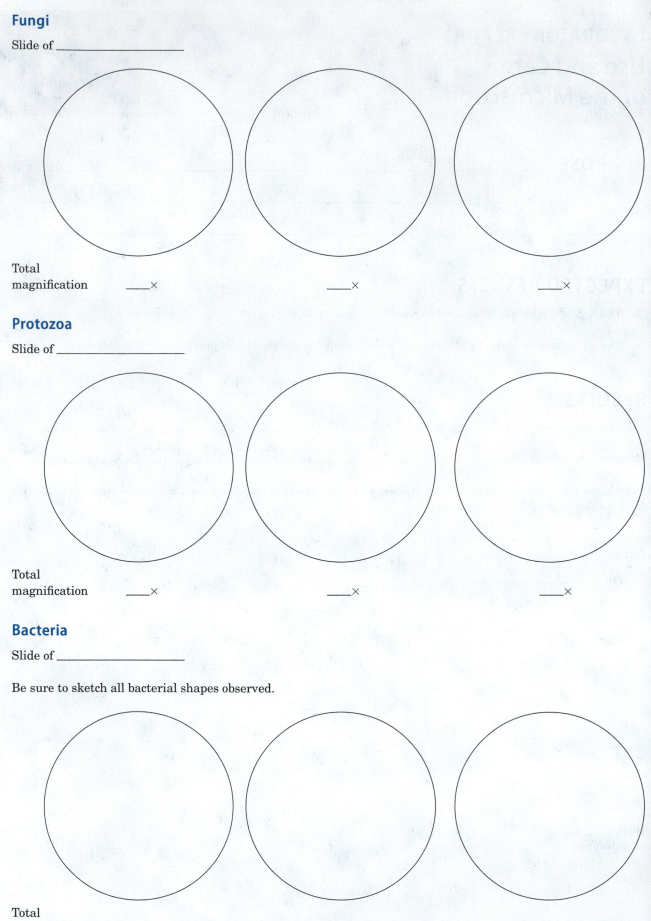

Total
magnification ____× ____× ____×

Protozoa

Slide of _____

Total
magnification ____× ____× ____×

Bacteria

Slide of _____

Be sure to sketch all bacterial shapes observed.

Total
magnification ____× ____× ____×

CONCLUSIONS

1. Do your results agree with your expected results? _____

2. What was the largest organism you observed? _____

   The smallest? _____

3. What three bacterial shapes did you observe? _____

   _____

4. How does increased magnification affect the field of vision? _____

QUESTIONS

1. Why is it desirable that microscope objectives be parfocal? _____

   _____

   _____

2. Which objective focuses closest to the slide when it is in focus? _____

   _____

3. Which controls on the microscope affect the amount of light reaching the ocular lens? _____

   _____

CRITICAL THINKING

1. Assume the diameter of the field of vision in your microscope is 2 mm under low power. If one *Bacillus* cell measures 2 μm long, how many *Bacillus* cells could fit end-to-end across the field? How many 10-μm yeast cells could fit across the field?

2. Name two ways in which you can enhance the resolving power.

3. What are the advantages of the low-power objective over the oil immersion objective for viewing fungi or algae?

4. What would occur if water were accidentally used in place of immersion oil?

CLINICAL APPLICATION

Assume you are looking for microorganisms in a tissue sample from a lung biopsy. The microbes become apparent when you switch to 1000×. What type of microbe is most likely?

Examination of Living Microorganisms

OBJECTIVES

After completing this exercise, you should be able to:

1. Prepare and observe wet-mount slides and hanging-drop slides.
2. Distinguish motility from Brownian movement.
3. Use a phase-contrast microscope.
4. Explain how phase-contrast microscopy differs from brightfield microscopy.

BACKGROUND

In **brightfield microscopy,** objects are dark and the field is light. Brightfield microscopy can be used to observe unstained microorganisms. However, because the optical properties of the organisms and of their aqueous environment are similar, very little contrast can be seen. Two other types of compound microscopes, however, are useful for observing living organisms: darkfield and phase-contrast microscopes. These microscopes optically increase the contrast between organism and background by using special condensers.

In **phase-contrast microscopy,** small differences in the refractive properties of the objects and the aqueous environment are transformed into corresponding variations of brightness. In phase-contrast microscopy, a ring of light passes through the object, and light rays are **diffracted** (retarded) and out of phase with the light rays that do not hit the object. The phase-contrast microscope enhances these phase differences so that the eye detects the difference as contrast (**FIGURE 2.1**) between the organisms and background and between structures within a cell. The organisms appear as degrees of brightness against a darker background. Phase-contrast microscopy allows study of structural detail within live cells.

Observing Living Microbes

In this exercise, you will examine, using wet-mount techniques, different environments to help you become aware of the numbers and varieties of microbes found in nature. The microbes will exhibit either Brownian movement or true motility. **Brownian movement** is not true motility but rather movement caused by the molecules in the liquid striking an object and causing the object to shake or bounce. In Brownian movement, the particles and microorganisms all vibrate at about the same rate and maintain their relative positions. Motile microorganisms move from one position to another. Their movement appears more directed than Brownian movement, and occasionally the cells may spin or roll.

Many kinds of microbes, such as protozoa, algae, fungi, and bacteria, can be found in pond water and in infusions of organic matter. Van Leeuwenhoek made some of his discoveries using a peppercorn infusion similar to the one you will see in this exercise. Direct examination of living microorganisms is very useful in determining size, shape, and movement. Using a wet mount is a fast way to observe bacteria, but motility is difficult to determine because of the movement of the water. Motility and larger microbes are more easily observed in the greater depth provided by a hanging drop. Using a petroleum jelly seal reduces evaporation of the suspended drop of fluid.

In this exercise, you will examine living microorganisms by brightfield microscopy and by phase-contrast microscopy.

MATERIALS

Slides
Coverslips
Hanging-drop (depression) slide
Petroleum jelly
Toothpick
Pasteur pipettes
Alcohol
Gram's iodine
Phase-contrast microscope and centering telescope
Inoculating loop

CULTURES

Hay infusion, incubated 1 week in light
Hay infusion, incubated 1 week in dark
Peppercorn infusion
18- to 24-hour-old broth culture of *Bacillus*

TECHNIQUES REQUIRED

Compound light microscopy (Exercise 1)

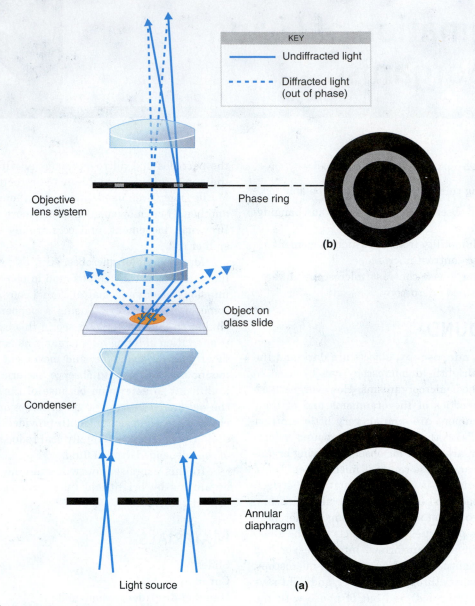

FIGURE 2.1 Phase-contrast microscopy. (a) A hollow cone of light is formed by the annular diaphragm. Diffracted rays are further retarded by the phase ring **(b)** in the objective lens.

PROCEDURE

Wet-Mount Technique

1. Suspend the infusions by stirring or shaking them carefully. Using a Pasteur pipette, transfer a very small drop of one hay infusion to a slide, or transfer a loopful using the inoculating loop, as demonstrated by your instructor.
2. Handle the coverslip carefully by its edges, and place it on the drop.
3. Gently press on the coverslip with the end of a pencil or loop handle.
4. Place the slide on the microscope stage and observe it with low power (10×). Adjust the iris diaphragm to admit a small amount of light. Concentrate your observations on the larger, more rapidly moving organisms. At this magnification, bacteria are barely discernible as tiny dots. (**FIGURE 2.2** and Figure 34.3 on page 273 may help you to identify some of the microorganisms.)
5. Examine the slide with the high-dry lens (40× to 45×); increase the light and focus carefully. Some microorganisms are motile, whereas others exhibit Brownian movement.
6. After recording your observations, examine the slide with the oil immersion lens. Bacteria should now be magnified sufficiently to be seen.
7. If you want to observe the motile organisms further, place a drop of alcohol or Gram's iodine at the edge of the coverslip, and allow it to run under and mix with the infusion. What does the alcohol or iodine do to these organisms? _____

_____

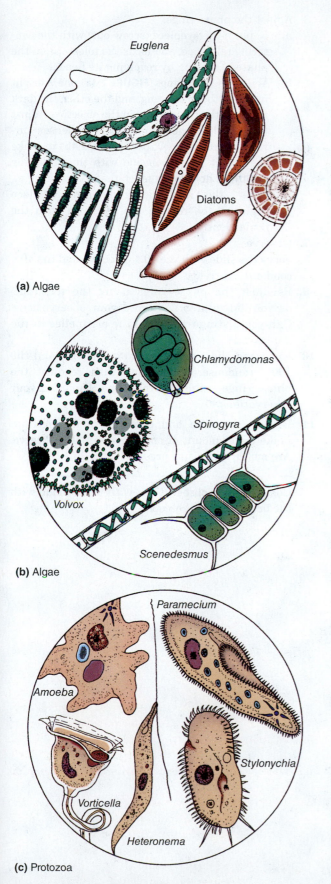

(a) Algae

(b) Algae

(c) Protozoa

FIGURE 2.2 **Eukaryotes found in pond water.** Some common algae and protozoa that can be found in infusions.

You can now observe them more carefully.

8. Record your observations, noting the relative sizes and the shapes of the organisms.
9. Make a wet mount from the other hay infusion, and observe it. Record your observations.
10. Discard slides and coverslips in the disinfectant jar. Wipe the oil from the objective lens with lens paper.

Hanging-Drop Procedure

1. Obtain a clean hanging-drop (depression) slide.
2. Pick up a small amount of petroleum jelly on a toothpick.
3. Pick up a coverslip (by its edges), and carefully touch the petroleum jelly with an edge of the coverslip to get a small rim of petroleum jelly. Repeat with the other three edges (**FIGURE 2.3a**), keeping the petroleum jelly on the same side of the coverslip.
4. Place the coverslip on a paper towel, with the petroleum jelly–side up.
5. Transfer a small drop or loopful of the peppercorn infusion to the coverslip (**FIGURE 2.3b**).
6. Place a depression slide (well-side down) over the drop, and quickly invert it so the drop is suspended in the well (**FIGURE 2.3c**). Why should the drop be hanging? _____
   _____
7. Examine the drop under low power (**FIGURE 2.3d**) by locating the edge of the drop and moving the slide so the edge of the drop crosses the center of the field.
8. Reduce the light with the iris diaphragm, and focus. Observe the different sizes, shapes, and types of movement.
9. Switch to high-dry and record your observations. Do not focus down. Why not? _____
   _____

   Do not attempt to use the oil immersion lens.
10. When finished, discard your slides and coverslips in disinfectant. Using a new coverslip, repeat the procedure with the culture of *Bacillus*. Record your observations.
11. Return your microscope to its proper location. Clean your slides well and return them. Discard the coverslips in the disinfectant jar.

Phase-Contrast Microscopy

1. Make a wet mount of any of the suspensions.
2. Place the slide on the stage and turn on the light.
3. Start with the 10× objective and move the condenser diaphragm to match setting "10."
4. Focus on an obvious clump of material with the coarse and fine adjustments.
5. Close the iris diaphragm, and move the condenser up and down until a light octagon comes into focus. Then open the diaphragm until light just fills the field.

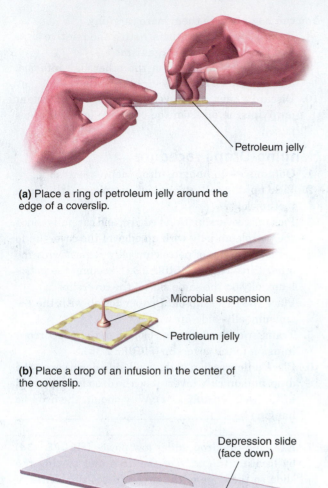

(a) Place a ring of petroleum jelly around the edge of a coverslip.

Petroleum jelly

Microbial suspension

Petroleum jelly

(b) Place a drop of an infusion in the center of the coverslip.

Depression slide (face down)

(c) Place the depression slide on the coverslip.

Microscope objective lens

Depression slide (face up)

Petroleum jelly

(d) Turn the slide over; place the slide, coverslip up, on the microscope stage; and observe it under the low and high-dry objectives.

FIGURE 2.3 **Hanging-drop preparation.**

6. Adjust the phase rings.
   a. Replace the eyepiece (screw out) with the centering telescope, and focus the telescope on the phase plate ring by revolving its head.
   b. You will see two rings (**FIGURE 2.4a**): one, a bright image of the phase ring, and the other, the dark image of the phase plate in the objective. Adjust the knurled wheel under the condenser turret with your fingertips (**FIGURE 2.4b**) to make the bright image coincide with the dark ring (**FIGURE 2.4c**). Do not touch the iris diaphragm.
   c. When the ring has been centered (see Figure 2.4c), replace the telescope with the ocular lens.
7. Observe.
8. Focus the slide with the 40× objective and the 40× condenser turret.
9. Readjust the phase rings using the telescope, as you did in step 6. Record your observations. Can you distinguish any of the organelles in the organisms?
10. Focus the slide with the 100× objective and the 100× condenser turret, and then readjust the phase rings using the telescope. Diagram your observations.
11. Discard the slide and coverslip in disinfectant. Make a wet mount of one other sample. Observe. Are motile organisms present?

Compare your observations with those made with the brightfield microscope.

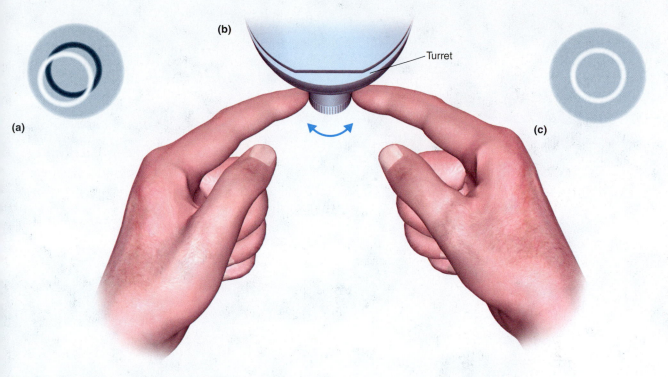

FIGURE 2.4 Adjustment of phase rings. (a) The two images seen before adjustment of the phase rings. Moving the wheel under the condenser turret **(b)** makes the images coincide **(c)**.

Name: _____   Date: _____   Lab Section: _____

LABORATORY REPORT
Examination of Living Microorganisms

PURPOSE _____

_____

_____

EXPECTED RESULT

What types of organisms will be more prevalent in the hay infusion incubated in the light than in the hay infusion incubated in the dark? _____

RESULTS

Wet-Mount Technique

Draw the types of protozoa, algae, fungi, and bacteria observed. Indicate their relative sizes and shapes. Record the magnification.

Sample: Hay infusion, light   Hay infusion, dark

Total
magnification: ____×   ____×

Compare the size and shape of organisms observed in the "light" and "dark" hay infusions.

_____

_____

_____

Hanging-Drop Procedure

Draw the types of microorganisms seen under high-dry magnification. Indicate their relative sizes and shapes.

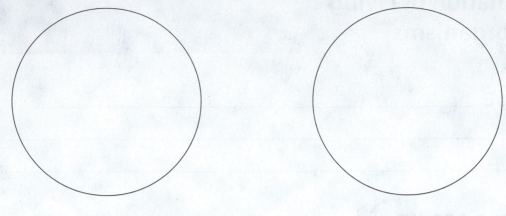

Sample: Peppercorn infusion   *Bacillus* culture

Total
magnification: ____×   ____×

Phase-Contrast Microscopy

Carefully draw the organisms and their internal structure.

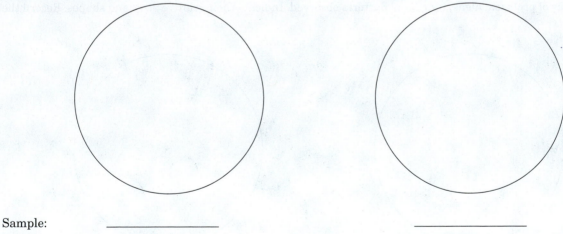

Sample: _____   _____

Total
magnification: ____×   ____×

Bacteria

In the following table, record the relative numbers of the bacterial shapes observed. Record your data as 4+ (most abundant), 3+, 2+, +, − (not seen).

Culture	Shape		
	Bacilli (rods)	Cocci	Spirals
Hay infusion, dark			
Hay infusion, light			
Peppercorn infusion			
Bacillus culture			

CONCLUSIONS

1. Did your results agree with your expected results? _____

2. How did you distinguish true motility from Brownian movement or motion of the fluid? _____

    _____

3. Compare the appearance of microorganisms you observed using phase-contrast microscopy with that of those

    you observed using brightfield microscopy. _____

    _____

    _____

4. Which infusion had the most bacteria? _____ Briefly explain why. _____

    _____

    _____

    _____

QUESTIONS

1. What, if any, practical value do these techniques have? _____

    _____

2. What is the advantage of the hanging-drop procedure over the wet mount? _____

    _____

3.   Why is petroleum jelly used in the hanging-drop procedure? _____

_____

_____

CRITICAL THINKING

1.   Why are microorganisms hard to see in wet preparations?

2.   Can you distinguish the prokaryotic organisms from the eukaryotic organisms? Explain.

3.   Why isn't the oil immersion lens used in the hanging-drop procedure?

4.   Where did the organisms in the infusions come from?

CLINICAL APPLICATION

A wet mount may be used for quick examination of fecal samples. Is this most useful for identifying protozoan or bacterial parasites? Briefly explain.

Handling Bacteria

EXERCISES

Robert Koch and his colleagues successfully transferred *Bacillus anthracis* and *Mycobacterium tuberculosis* between diseased animals and laboratory media. To do this without infecting themselves, they used aseptic techniques. **Aseptic techniques** are procedures used to avoid introducing unwanted microbes and to prevent spreading microbes where they are not wanted. Aseptic techniques are of utmost importance in the laboratory and in medical procedures.

Bacteria are transferred between laboratory media and microscope slides using an inoculating loop. An **inoculating loop** is a Nichrome wire held with an insulated handle. Before and after it is used, the inoculating loop is sterilized by heating or **flaming**—that is, the loop is held in the flame of a burner (photograph a on page 30) or electric incinerator (photograph b) until it is red-hot. Single-use plastic loops should not be heated. They are removed aseptically from the package.

The first bacterial cultures were grown in *broth* (liquid) media, such as infusions and blood. The first solid culture media, used in 1872, were made from potato starch, coagulated egg white, and meat. Robert Koch used gelatin. Gelatin had two disadvantages, however: The gelatin liquefied under standard incubation conditions, and some bacteria liquefied the gelatin. A solution came from Koch's colleague, the American-born Angelina Hess, who suggested

> the use of agar-agar which she had been using for years in her kitchen in
> the preparation of fruit and vegetable jellies. While yet in America she had
> received the recipe from her mother; her mother in turn had obtained the
> formula from some Dutch friends who had formerly lived in Java. In the East
> Indies, where the source of agar-agar, Japanese seaweed (Gelidium corneum),
> also abounds, this curious material has been used for generations as a jellify-
> ing agent and as a thickening for soups.*

---

*A. P. Hitchens and M. C. Leikind. "The Introduction of Agar-Agar into Bacteriology." *Journal of Bacteriology* 35(5): 485–493, May 1939.

The bacteria you culture in Exercise 3 will demonstrate the need for the aseptic techniques (Exercise 4) that you will use throughout this course.

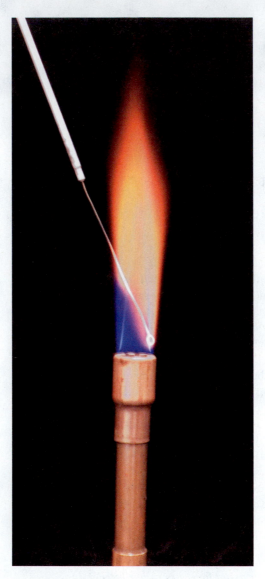

(a) Flaming a loop

(b) An electric incinerator

Sterilizing an inoculating loop. (a) Before and after use, sterilize the loop in a flame until it is red-hot. Place the wire in the hottest part of the flame, at the edge of the inner blue area. **(b)** The loop can be sterilized in an electric incinerator.

CASE STUDY: Spaghetti Scum

You notice that a white pellicle (scum) has formed on the leftover spaghetti you put in the refrigerator a week ago. You know that starch dissolved in the water during cooking can precipitate out at cold temperatures. Microscopic examination of the pellicle shows a myriad of small particles of various sizes.

Question

How could you determine whether the pellicle is just starch from the pasta or bacterial growth?

Microbes in the Environment

Whatever is WORTH doing at all is worth doing WELL.
–PHILIP DORMER STANHOPE

OBJECTIVES

After completing this exercise, you should be able to:

1. Describe why agar is used in culture media.
2. Prepare nutrient broth and nutrient agar.
3. Compare bacterial growth on solid and liquid culture media.
4. Describe colony morphology using accepted descriptive terms.

BACKGROUND

Microbes are almost everywhere; they are found in the water we drink, the air we breathe, and the earth on which we walk. They live in and on our bodies. Microbes occupy ecological niches on all forms of life and in most environments. In most situations, the ubiquitous microorganisms are harmless. However, in microbiology, work must be done carefully to avoid contaminating sterile media and materials with these microbes.

In this exercise, we will attempt to culture (grow) some microbes from the environment. When a medium is selected for culturing bacteria, macronutrients (carbon, nitrogen, phosphorus, and sulfur), an energy source, and any necessary growth factors must be provided. A medium whose exact chemical composition is known is called a **chemically defined medium** (TABLE 3.1).

Culture Media

Most chemoheterotrophic bacteria are routinely grown on **complex media**—that is, media for which the exact chemical composition varies slightly from batch to batch. Protein in the form of meat extracts and partially digested proteins called peptones usually supply organic carbon, energy, and nitrogen sources. **Nutrient broth** is a commonly used liquid complex medium. When agar is added, it becomes a solid medium called **nutrient agar** (TABLE 3.2).

Agar, an extract from marine red algae, has some unique properties that make it useful in culture media. Few microbes can degrade agar, so it remains solid during microbial growth. It liquefies at 100°C and remains

TABLE 3.1   GLUCOSE–MINIMAL SALTS BROTH

Ingredient	Amount/100 ml
Glucose	0.5 g
Sodium chloride (NaCl)	0.5 g
Ammonium dihydrogen phosphate ($NH_4H_2PO_4$)	0.1 g
Dipotassium phosphate (K_2HPO_4)	0.1 g
Magnesium sulfate ($MgSO_4$)	0.02 g
Distilled water	100 ml

TABLE 3.2   NUTRIENT AGAR

Ingredient	Amount/100 ml
Peptone	0.5 g
Beef extract	0.3 g
Sodium chloride (NaCl)	0.8 g
Agar	1.5 g
Distilled water	100 ml

in a liquid state until cooled to 40°C. Once the agar has solidified, it can be incubated at temperatures of up to 100°C and remain solid.

Media must be sterilized after preparation. The most common method of sterilizing culture media that are heat stable is **steam sterilization,** or **autoclaving,** which uses steam under pressure. During this process, material to be sterilized is placed in the autoclave and heated to 121°C at 15 pounds of pressure (15 psi) for 15 minutes.

Growing Bacteria

Culture media can be prepared in various forms, depending on the desired use. **Petri plates** containing solid media provide a large surface area for examination of colonies. The microbes will be **inoculated,** or intentionally introduced, onto nutrient agar or into

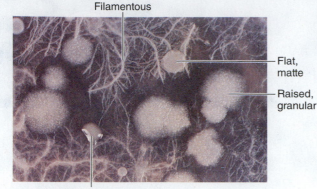

FIGURE 3.1 **Bacterial colonies.** Each different-appearing colony is a different bacterial species. In this sample, one species is producing an antibiotic that inhibits the filamentous *Bacillus mycoides* from growing near it. Some terms used to describe colonies are given.

nutrient broth. The bacteria that are inoculated into culture media increase in number during an **incubation period.** After suitable incubation, liquid media become **turbid,** or cloudy, as a result of bacterial growth. On solid media, colonies will be visible to the naked eye. A **colony** is a population of cells that arises from a single bacterial cell. A colony may arise from a group of the same microbes attached to one another, which is therefore called a **colony-forming unit.** Although many species of bacteria give rise to colonies that appear similar, each colony that appears different is usually a different species (**FIGURE 3.1**).

MATERIALS

250-ml Erlenmeyer flask with cap or plug
100-ml graduated cylinder
Distilled water
Nutrient broth powder
Agar
Glass stirring rod
5-ml pipette
Propipette or pipette bulb
Test tubes with caps (3)
Sterile Petri dishes (4)
Balance
Weighing paper or boat
Autoclave gloves
Hot plate
Tube containing sterile cotton swabs
Tube containing sterile water

DEMONSTRATION

Use of the autoclave

TECHNIQUES REQUIRED

Pipetting (Appendix A)

PROCEDURE First Period

1. Preparing culture media:
   a. Prepare 100 ml of nutrient broth in a 250-ml flask. Using the graduated cylinder, add 100 ml of distilled water to the flask. Read the preparation instructions on the nutrient broth bottle. Calculate the amount of nutrient broth powder needed for 100 ml. If the amount needed for 1000 ml is _____ grams, then _____ grams are needed for 100 ml. Weigh out the required amount and add it to the flask. Stir with a glass rod until the powder is dissolved.
   b. Attach a pipette bulb or propipette to the 5-ml pipette. Pipette 5 ml of the nutrient broth into each test tube, and cap each tube. Label the tubes "Nutrient broth." Place two in the To Be Autoclaved rack. Label the remaining tube "Not sterilized," and incubate it at room temperature until the next laboratory period.
   c. Add agar (1.5% w/v) to the remaining 85 ml of nutrient broth. What quantity of agar will you need to add? _____
   d. Bring the broth to a boil, and continue boiling carefully until all the agar is dissolved. *Be careful:* Do not let the solution boil over. Stir often to prevent burning and boiling over.
   e. Stopper the flask, label it "Nutrient agar," and place the flask and tube in the To Be Autoclaved basket.
   f. Listen to the instructor's demonstration about use of the autoclave.
   g. After autoclaving, allow the flask and tubes to cool to room temperature, or place them in a 45°C water bath. What effect does the agar have on the culture medium? _____
2. Pouring plates:
   Obtain a flask of melted nutrient agar from the 45°C water bath. If the agar has solidified, you will have to reheat it to liquefy it. To what temperature will it have to be heated? _____
   The sterile nutrient agar must be poured into Petri dishes *aseptically*—that is, without letting microbes into the nutrient medium. *Read the following procedure before beginning* so that you can work quickly and efficiently.
   a. Set four sterile, unopened Petri dishes in front of you with the cover (larger half) on top. Have a lighted laboratory burner within reach on your workbench.

Keep the burner away from your hair and in the center of the bench.

b. Holding the flask at an angle, remove the stopper with the fourth and fifth fingers of your other hand. Heat the mouth of the flask by passing it through the flame three times (**FIGURE 3.2a**). Why is it necessary to keep the flask at an angle through this procedure? _____

_____

c. Remove the cover from the first dish with the hand holding the stopper. Quickly and neatly pour melted nutrient agar into the dish until the bottom is just covered to a depth of approximately 5 mm (**FIGURE 3.2b**). Keep the flask at an angle, and replace the dish cover; move on to the next dish until all the agar is poured.

d. When all the agar is poured, gently swirl the agar in each dish to cover any empty spaces; do not allow the agar to touch the dish covers.

e. To decrease condensation, leave the Petri plate covers slightly ajar for about 15 minutes until the agar solidifies.

f. Place the empty flask in the discard area.

3. Culturing microbes from the environment:

a. Design your own experiment. The purpose is to sample your environment and your body. Use your imagination. Here are some suggestions:

Environmental Sample	
Lab or washroom or anyplace on campus	Inoculate a plate from an environmental surface, such as the floor or a workbench, by wetting a cotton swab in sterile water, swabbing the environmental surface, and then swabbing the surface of the agar. Why is the swab first moistened in sterile water? _____ After inoculation, discard the swab in the container of disinfectant.
Air	You might leave one nutrient agar plate open to the air for 30 to 60 minutes.
Body Sample	
Any part of your body	Inoculate a plate from any part of your body by wetting a cotton swab in sterile water, swabbing the surface, and then swabbing the surface of the agar. Why is the swab first moistened in sterile water? _____ After inoculation, discard the swab in the container of disinfectant.
Hair	Place a hair on the agar.
Fingers	Touch the agar with your fingers.

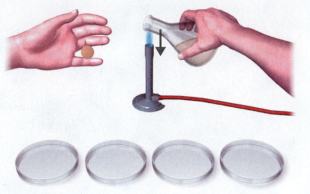

(a) Remove the stopper, and flame the mouth of the flask.

(b) Remove the cover from one dish, and pour nutrient agar into the dish bottom.

FIGURE 3.2 Petri plate pouring.

b. Inoculate two plates from the environment and two plates from your body. Inoculate one nutrient broth tube using a swab. Wet a swab in sterile water and swab the area to be sampled. After swabbing the agar surface, place the swab in the nutrient broth and leave it there during incubation. You may need to break off part of the wooden handle to fit the swab into the nutrient broth tube.

c. The plates and tube should be incubated at the approximate temperature of the environment sampled. Incubate bacteria from your body at or close to your body temperature. What is human body temperature? ____°C

d. Incubate all plates inverted so that water will condense in the lid instead of on the surface of the agar. Why is condensation on the agar undesirable? _____

_____

e. Incubate all inoculated media at the appropriate temperature until the next laboratory period.

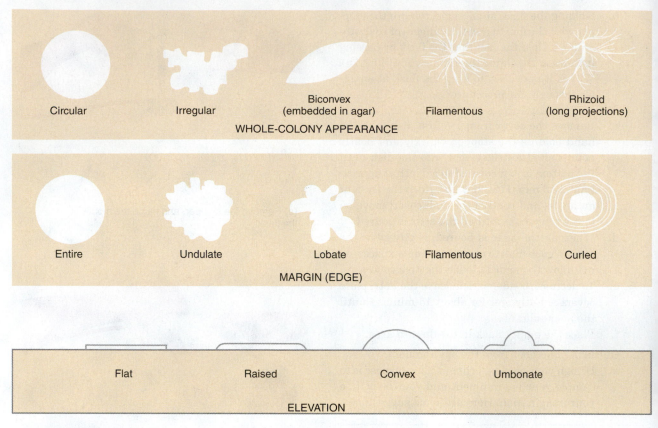

FIGURE 3.3 Colony descriptions. Colonies can be shiny, matte, or granular. Some examples appear in Figure 3.1.

PROCEDURE Second Period

1. Observe and describe the resulting growth on the plates. Note each different-appearing colony, and describe the colony pigment (color) and morphology using the characteristics given in **FIGURE 3.3**. Determine the approximate number of each type of colony. When many colonies are present, record TNTC (too numerous to count) as the number of colonies.

2. Describe the appearance of the nutrient broth labeled "Not sterile" and the broth you inoculated. Are they uniformly cloudy or turbid? _____

Look for clumps of microbial cells, called **flocculent.** Is there a membrane, or **pellicle,** across the surface of the broth? _____ See whether microbial cells have settled on the bottom of the tube, forming a **sediment** (**FIGURE 3.4**).

3. Discard the plates and tubes properly.

 Place contaminated plates in the biohazard bag for autoclaving.

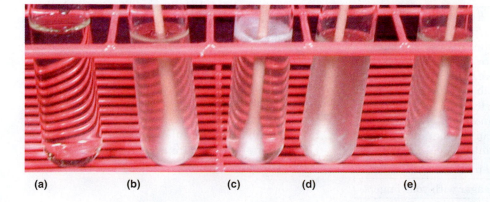

(a)   (b)   (c)   (d)   (e)

FIGURE 3.4 Bacterial growth in broth. The nutrient broth in tube **(a)** is clear, indicating no bacterial growth. Bacterial growth is determined by **(b)** turbidity, **(c)** pellicle on the surface, **(d)** flocculent suspended in the broth, and **(e)** a sediment at the bottom.

LABORATORY REPORT
Microbes in the Environment

PURPOSE _____

_____

_____

EXPECTED RESULTS

1. In the environments you sampled, which will have the greatest amount of growth on the nutrient agar plates?

    _____

2. Do you expect turbidity in the unsterilized nutrient broth that was incubated? _____

3. Will nutrient broth or nutrient agar provide more information on the variety of bacteria in an environment?

    _____

RESULTS

Fill in the following table with descriptions of the bacterial colonies on your plates and your classmates' plates. Use a separate line for each different-appearing colony.

	Colony Description					
	Estimated Diameter	Whole-Colony Appearance	Margin	Elevation	Pigment	Number of This Type
Area sampled:						
_____						
Incubated at						
_____°C for						
_____ days						

	Colony Description					
	Estimated Diameter	Whole-Colony Appearance	Margin	Elevation	Pigment	Number of This Type
Area sampled: _____ Incubated at _____°C for _____ days						

Area sampled: _____ Incubated at _____°C for _____ days						

Area sampled: _____ Incubated at _____°C for _____ days						

Nutrient broth. Area sampled: _____. Incubated at _____°C for _____ days

	Not-Sterilized Broth	Sterilized Broth Not Inoculated	Inoculated Broth
Growth present			
Turbidity present			
Flocculent present			
Sediment present			
Pellicle present			
Color			

CONCLUSIONS

1. Do the results you obtained agree with your expected results?

    1. _____

    2. _____

    3. _____

2. What is the minimum number of different bacteria present on one of your plates? _____

    How do you know? _____

3. Explain any growth in the sterilized and not-sterilized broths. _____

    _____

4. Which environment has the most total bacteria? _____ The least? _____ Provide a reason for

    the differences in total bacteria in these two places. _____

    _____

    _____

5. Which environment has the most different bacteria? _____ The least? _____ Provide a reason

    for the differences in bacteria in these two places. _____

    _____

    _____

QUESTIONS

1. What is the advantage of using Petri plates rather than test tubes in microbiology? _____

_____

2. What are bacteria using for nutrients in nutrient agar? _____

_____

What is the purpose of the agar? _____

_____

CRITICAL THINKING

1. Did all the organisms living in or on the environments sampled grow on your nutrient agar? Briefly explain.

2. How could you determine whether the turbidity in your nutrient broth tube was from a mixture of different microbes or from the growth of only one kind of microbe?

CLINICAL APPLICATION

The effectiveness of sanitation procedures in hospitals is determined by sampling surfaces for bacteria. Samples can be taken by rubbing a sterile, moistened cotton swab on a surface and inoculating a nutrient agar plate or with a Rodac environmental sampling plate. The 67-mm plates are designed so that an agar medium can be overfilled, producing a dome-shaped surface that can be pressed onto a surface for sampling microbes. As the sanitation control officer, you test each method to determine which you will use. What can you conclude from these results?

	Number of Plates with Colonies	Number of Plates Inoculated
Swab onto nutrient agar	5	50
Rodac plate containing nutrient agar	6	40

Transfer of Bacteria: Aseptic Technique

Study without thinking is worthless; thinking without study is DANGEROUS.
–CONFUCIUS

OBJECTIVES

After completing this exercise, you should be able to:

1. Provide the rationale for aseptic technique.
2. Differentiate among the following: broth culture, agar slant, and agar deep.
3. Aseptically transfer bacteria from one form of culture medium to another.

BACKGROUND

In the laboratory, bacteria must be cultured to facilitate identification and to examine their growth and metabolism. Bacteria are **inoculated,** or introduced, into various forms of culture media to keep them alive and to study their growth—without introducing unwanted microbes, or **contaminants,** into the media. **Aseptic technique** is used in microbiology to exclude contaminants.

All culture media are **sterilized,** or rendered free of all life, prior to use. Sterilization is usually accomplished using an autoclave. Containers of culture media, such as test tubes or Petri plates, should not be opened until you are ready to work with them, and even then you should not leave them open.

Culture Media

Broth cultures provide large numbers of bacteria in a small space and are easily transported. **Agar slants** are test tubes containing solid culture media that were left at an angle while the agar solidified. Agar slants, like Petri plates, provide a solid growth surface, but slants are easier to store and transport than Petri plates. Agar is allowed to solidify in the bottom of a test tube to make an **agar deep.** Deeps are often used to grow bacteria that require less oxygen than is present on the surface of the medium. Semisolid agar deeps containing 0.5–0.7% agar instead of the usual 1.5% agar can be used to determine whether a bacterium is motile. Motile bacteria will move away from the point of inoculation, giving the appearance of an inverted pine tree.

Inoculation

Aseptic transfer and inoculation are usually performed with a sterile, heat-resistant, noncorroding Nichrome wire attached to an insulated handle. When the end of the wire is bent into a loop, it is called an **inoculating loop;** when straight, it is an **inoculating needle** (FIGURE 4.1). For special purposes, cultures may also be transferred with sterile cotton swabs, pipettes, glass rods, or syringes. You will learn these techniques in later exercises.

Whether to use an inoculating loop or a needle depends on the form of the medium and the intent of the procedure; after completing this exercise, you will be able to decide which instrument to use.

MATERIALS

FIRST PERIOD

Tubes containing nutrient broth (3)
Tubes containing nutrient agar slants (3)
Tubes containing nutrient semisolid agar deeps (3)
Inoculating loop
Inoculating needle
Test-tube rack

SECOND PERIOD

Methylene blue

CULTURES

Lactococcus lactis broth
Bacillus subtilis broth
Escherichia coli slant

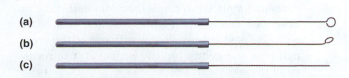

FIGURE 4.1 Inoculating loops and needle. (a) An inoculating loop. **(b)** A variation of the inoculating loop, in which the loop is bent at a 45° angle. **(c)** An inoculating needle.

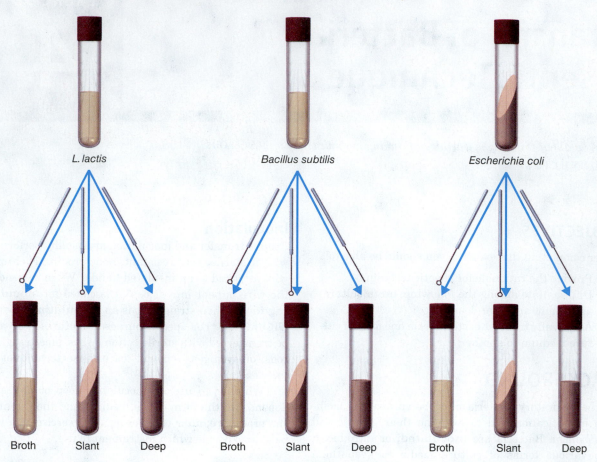

L. lactis *Bacillus subtilis* *Escherichia coli*

Broth Slant Deep Broth Slant Deep Broth Slant Deep

FIGURE 4.2 The procedure. Inoculate one tube of each medium with each of the cultures, using a loop or needle as indicated. Work with only one culture at a time to avoid contamination.

TECHNIQUES REQUIRED

Compound light microscopy (Exercise 1)
Wet mount (Exercise 2)

PROCEDURE First Period

1. Disinfect your work area (see Student Safety Contract step 4 on page 5). What is the disinfectant used in your lab?

   _____

2. Work with only one of the bacterial broth cultures at a time to prevent any mix-ups or cross-contamination (**FIGURE 4.2**). Label one tube of each medium with the name of the first broth culture, your name, the date, and your lab section. Inoculate each tube as described below (see Figure 4.2), and then work with the next culture. Begin with one of the broth cultures, and gently tap the bottom of it to resuspend the sediment.

3. To inoculate nutrient broth, hold the inoculating loop in your dominant hand and one of the broth cultures of bacteria in the other hand.

   a. Sterilize your inoculating loop by holding it in the hottest part of the flame (at the edge of

   the inner blue area, page 30) or the electric incinerator until it is red-hot (**FIGURE 4.3a**). The entire wire should get red. Why? _____

   b. Holding the loop like a pencil or paintbrush, curl the little finger of the same hand around the cap of the broth culture. Gently pull the cap off the tube while turning the culture tube (**FIGURE 4.3b**). Do not set down the cap. Why not? _____

   _____

   c. Holding the tube at an angle, pass the mouth of the tube through the flame three times (**FIGURE 4.3c**). What is the purpose of flaming the mouth of the tube? _____
   Always hold culture tubes and uninoculated tubes at about a 20° angle to minimize the amount of dust that could fall into them. Do not tip the tube too far, or the liquid will leak out around the loose-fitting cap. Do not let the top edge of the tube touch anything.

   d. Immerse the sterilized, cooled loop into the broth culture to obtain a loopful of culture (**FIGURE 4.3d**). Why must the loop be cooled first?

   _____

(a) Sterilize the loop by holding the wire in the flame until it is red-hot.

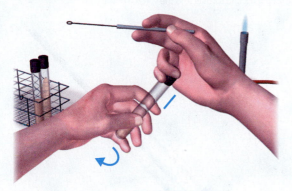

(b) While holding the sterile loop and the bacterial culture, remove the cap as shown.

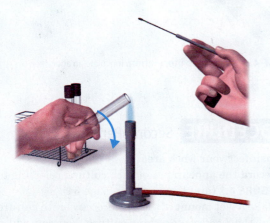

(c) Briefly pass the mouth of the tube through the flame three times before inserting the loop for an inoculum.

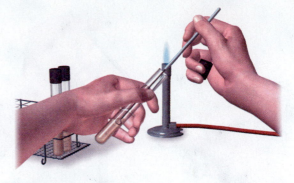

(d) Get a loopful of culture.

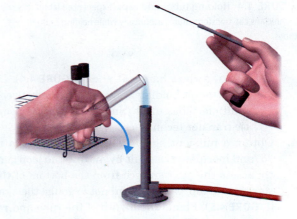

(e) Heat the mouth of the tube, and replace the cap.

FIGURE 4.3 Inoculating procedures.

Remove the loop, and, while holding the loop, flame the mouth of the tube and recap it by turning the tube into the cap (**FIGURE 4.3e**). Place the tube in your test-tube rack.

e. Remove the cap from a tube of sterile nutrient broth as previously described, and flame the mouth of the tube. Immerse the inoculating loop into the sterile broth, and then withdraw it from the tube. Flame the mouth of the tube and replace the cap. Return the tube to the test-tube rack.

f. Reflame the loop until it is red and let it cool. Some microbiologists prefer to hold several

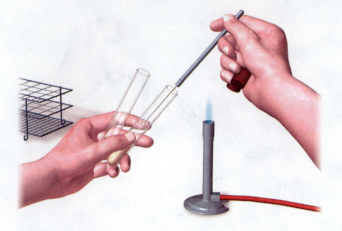

FIGURE 4.4 Holding tubes and caps. Experienced laboratory technicians can transfer cultures aseptically while holding multiple test tubes.

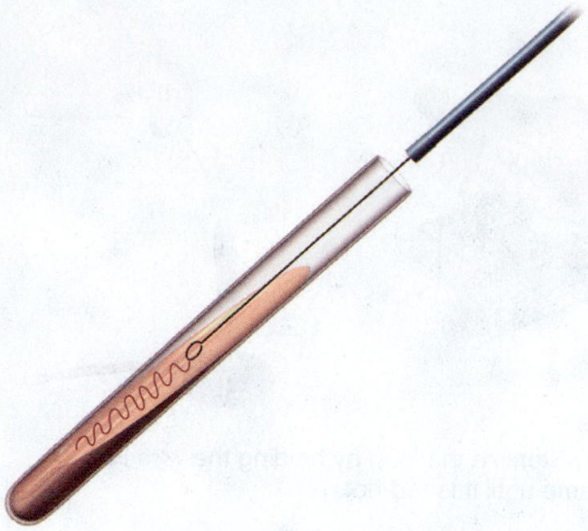

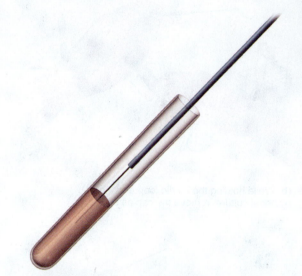

FIGURE 4.5 Agar slant inoculation. Inoculate a slant by streaking the loop back and forth across the surface of the agar.

tubes in their hands at once (**FIGURE 4.4**). *Do not* attempt holding and transferring between multiple tubes until you have mastered aseptic transfer techniques.

4. Obtain a nutrient agar slant. Repeat steps 3a to 3d, and inoculate the slant by moving the loop gently across the agar surface from the bottom of the slant to the top, being careful not to gouge the agar (**FIGURE 4.5**). Flame the mouth of the tube and replace the cap. Flame your loop and let it cool.

5. Obtain a nutrient semisolid agar deep, and, using your inoculating *needle,* repeat steps 3a to 3d. Inoculate the semisolid agar deep by plunging the needle straight down the middle of the deep and then pulling it out through the same stab, as shown in **FIGURE 4.6.** Flame the mouth of the tube and replace the cap. Flame your needle and let it cool.

6. Using the other broth culture, label one tube of each medium as described in step 2; inoculate a broth culture, agar slant, and semisolid agar deep, as described in steps 3, 4, and 5, using your inoculating loop and needle.

7. Label one tube of each medium with *E. coli* as described in step 2. To transfer *E. coli,* flame your loop and allow it to cool. Flame the mouth of the tube, and use your inoculating loop to carefully touch the culture on the agar. Do not gouge the agar. Flame the mouth of the tube and replace the cap. Inoculate a broth and a slant as described in steps 3 and 4. Inoculate a semisolid agar deep with an inoculating needle. Carefully touch the culture on the slant, and inoculate the deep as described in step 5.

8. Incubate all tubes at 35°C until the next period.

9. Disinfect your work area.

FIGURE 4.6 Agar deep inoculation. Inoculate an agar deep by stabbing into the agar with a needle.

PROCEDURE Second Period

1. Disinfect your work area.

2. Record the appearance of each culture, referring to **FIGURE 4.7** (and Figure 3.4 on page 34).

3. Make a wet mount of the *Lactococcus* broth culture and the *Lactococcus* slant culture. You can add a drop of methylene blue to the wet mount to stain the bacteria. Place a coverslip on the wet mounts, observe them with the oil immersion objective, and compare them.

Arborescent
(branched)

Beaded

Echinulate
(pointed)

Filiform
(even)

Rhizoid
(rootlike)

Spreading

FIGURE 4.7 Patterns of growth on agar slants.

LABORATORY REPORT

Transfer of Bacteria: Aseptic Technique

PURPOSE _____

_____

_____

EXPECTED RESULTS

1. Describe the growth pattern of a motile organism in a semisolid deep. _____

_____

2. How will the arrangement of the *Lactococcus lactis* in broth differ from the arrangement in the slant culture?

_____

RESULTS

Nutrient Broth

Describe the nutrient broth cultures.

Bacterium	Is It Turbid?	Is Flocculent, Pellicle, or Sediment Present?	Pigment
Lactococcus lactis			
Bacillus subtilis			
Escherichia coli			

Nutrient Agar Slant

Sketch the appearance of each culture. Note any pigmentation.

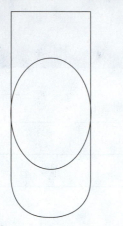 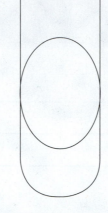 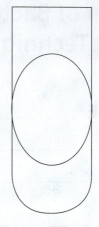

Bacteria: *Lactococcus lactis* *Bacillus subtilis* *Escherichia coli*

Pattern of growth: _____ _____ _____

Nutrient Semisolid Agar Deep

Show the location of bacterial growth, and note any pigment formation.

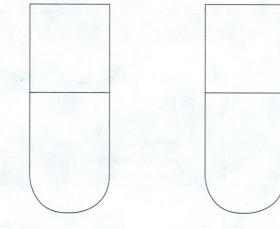 

Bacteria: *Lactococcus lactis* *Bacillus subtilis* *Escherichia coli*

Comparison of Broth and Slant Cultures under the Microscope

| | *Lactococcus lactis* | |
	Broth Culture	Slant Culture
Morphology		
Arrangement		

CONCLUSIONS

1. Do your results agree with your expected results? Briefly explain. _____

_____

2. Did growth occur at different levels in the agar deep? _____

_____

_____

3. Were any of the bacteria growing in the semisolid agar deeps motile? _____ Explain. _____

_____

_____

_____

4. When is a loop preferable for transferring bacteria? When is a needle preferable? Use examples from your

results to answer. _____

_____

_____

QUESTIONS

1. What other methods can be used to determine motility? _____

_____

2. What is the primary use of slants? _____

Of deeps? _____

Of broths? _____

3. What is the purpose of flaming the loop before use? After use? _____

_____

_____

4. Why must the loop be cool before you touch it to a culture? Should you set down the loop to let it cool? How do

you determine when the loop is cool? _____

_____

_____

5. Why is aseptic technique important? _____

_____

_____

CRITICAL THINKING

1. Why was the arrangement of *Lactococcus* in the broth culture different from the arrangement on the slant culture in the second period?

2. For a bacterium, what evolutionary advantage is associated with forming a pellicle in a liquid medium?

3. How can you tell that the media provided for this exercise were sterile?

CLINICAL APPLICATION

Your microbiology lab maintains reference bacterial cultures, which are regularly transferred to new nutrient agar slants. After incubation, the slants were observed, and the result indicated that an improper technique had been used. What led to this conclusion?

PART 3

Staining Methods

EXERCISES

In 1877, Robert Koch wrote,

> *How many incomplete and false observations might have remained unpublished instead of swelling the bacterial literature into a turbid stream, if investigators had checked their preparations with each other?**

To solve this problem, he introduced into microbiology the procedures of air-drying, chemical fixation, and staining with aniline dyes.

In addition to making a lasting preparation, staining bacteria enhances the contrast between bacteria and the surrounding material and permits observation of greater detail and resolution than do wet-mount procedures. Microorganisms are prepared for staining by smearing them onto a microscope slide. In these exercises, bacteria will be transferred from growth or culture media to microscope slides using an inoculating loop. Staining techniques may involve **simple stains,** in which only one reagent is used and all bacteria are usually stained similarly (Exercises 5 and 6), or **differential stains,** in which multiple reagents are used and bacteria react to the reagents differently (Exercises 7 and 8). Structural stains are used to identify specific parts of microorganisms (Exercise 9).

---

*Quoted in H. A. Lechevalier and M. Solotorovsky. *Three Centuries of Microbiology.* New York: Dover Publications, 1974, p. 79.

Most of the bacteria that are grown in laboratories are rods or cocci. Rods and cocci will be used throughout this part. Staining and microscopic examination are usually the first steps in identifying microorganisms. In Exercise 10, Morphologic Unknown, you will be asked to characterize selected bacteria by their microscopic appearance.

CASE STUDY: A Summer Rash!

Peter and Mary live in Minneapolis, Minnesota, with their two active sons. Peter is a microbiologist at General Mills and loves his job. Mary is a teacher at a neighborhood elementary school. It had been a hot summer, and the kids were ready for bed after playing outside all day. Five-year-old son Shawn had been scratching his skin, and when Peter looked closely, he saw a small, pimple-like rash over Shawn's chest and arms. He quickly examined his three-year-old son and, to his dismay, found a similar rash on Todd's abdomen. He immediately started to worry about viral diseases (such as measles) and bacterial diseases (impetigo, erysipelas, or even necrotizing fasciitis). Mary quickly reported that she had seen a similar rash in her third-grade class and suggested that they call their neighbors, the Fredricksons, whose children had been playing with Shawn and Todd. The Fredricksons examined their children and were surprised to discover that one of them had a light rash. All three children were active with no apparent pain and did not have an elevated temperature.

Mary, the voice of reason, suggested that they call the Urgent Care hotline before rushing off to an expensive visit to the Emergency Room. Sally, the nurse practitioner who answered their call, asked Mary several questions about the rash and the children's activities. It was the middle of July, and because of the hot weather, the children had been spending many afternoons playing in the backyard plastic pool. Peter had emptied the pool each morning and filled it with fresh water, which heated up nicely by the afternoon. Sally suspected that the three children had dermatitis from their repeated use of the pool, which never really dried out and never was disinfected. Sally did not recommend the use of antibiotics but advised the parents simply to watch the rash and to minimize the children's scratching. She indicated that the rash would go away in about a week and said they did not need to come in.

Peter, ever the microbiologist, took several swab cultures of the pool, selecting areas by the seams and slimy areas that appeared to be a biofilm. He took the swabs to work and cultured them on nutrient agar. After 24 hours, bacteria had grown on the plates and produced a bluish-green, water-soluble pigment. The culture was Gram stained, and a gram-negative, very small bacillus was visible. The children's rash lasted several days but then gradually went away without complications.

Questions

1. Why did the Gram stain results rule out impetigo, erysipelas, and necrotizing fasciitis?
2. Give a staining technique that would help determine whether the culprit was a *Bacillus*?
3. How did the bacteria enter the skin?

Preparation of Smears and Simple Staining

OBJECTIVES

After completing this exercise, you should be able to:

1. Prepare and fix a smear.
2. List the advantages of staining microorganisms.
3. Explain the basic mechanism of staining.
4. Perform a simple direct stain.

BACKGROUND

To stain bacteria, a thin film of bacterial cells, called a **smear,** must be placed on a slide.

Smear Preparation

A smear is made by spreading a small amount of a bacterial broth culture on a clean slide and allowing it to dry. If the bacteria are growing on solid media, remove a small amount of the bacterial culture, mix with a drop of water on the slide, and allow it to to air-dry. The smear must be **fixed** to kill the bacteria; coagulated proteins from the cells will cause cells to stick to the slide. Fixing denatures bacterial enzymes, preventing them from digesting cell parts, which causes the cell to break, a process called *autolysis.* Fixing also preserves microbes with minimal shrinkage or distortion when stained. The dry smear is passed through a Bunsen burner flame several times to **heat-fix** the bacteria. Alternatively, the dry smear can be placed on a 60°C slide warmer for 10 minutes or on the wire mesh of an electric incinerator (see the figure on page 30). Heat-fixing may not kill all the bacteria. Or, to chemically fix the bacteria without using heat, cover the smear with 95% methanol for 1 minute.

Simple Stain

Most stains used in microbiology are synthetic aniline (coal tar derivative) dyes derived from benzene. The dyes are usually salts, although a few are acids or bases, composed of charged colored ions. The ion that is colored is referred to as a **chromophore.** For example,

Methylene blue chloride $\rightleftharpoons$ Methylene blue$^+$ + Cl$^-$
(Chromophore)

If the chromophore is a positive ion like the methylene blue in the equation shown, the stain is considered a **basic stain;** if it is a negative ion, it is an **acidic stain.** Most bacteria are stained when a basic stain permeates the cell wall and adheres by weak ionic bonds to the bacterial cell, which is slightly negatively charged.

Staining procedures that use only one stain are called **simple stains.** A simple stain that stains the bacteria is a **direct stain,** and a simple stain that stains the background but leaves the bacteria unstained is a **negative stain** (Exercise 6). Simple stains can be used to determine cell morphology, size, and arrangement.

MATERIALS

Methylene blue
Wash bottle of distilled water
Slide
Inoculating loop
95% methanol (optional)

CULTURES

Staphylococcus epidermidis slant
Bacillus megaterium broth

TECHNIQUES REQUIRED

Compound light microscopy (Exercise 1)
Inoculating loop (Exercise 4)

PROCEDURE

Preparing Smears

1. Clean your slide well with abrasive soap or cleanser; rinse and dry. Even new slides may need to washed because of an oily film. Handle clean slides by the end or edge. Use a marker to make two dime-sized circles on the bottom of each slide so they will not wash off. Label each circle according to the bacterial culture used (**FIGURE 5.1a**).
2. Sterilize your inoculating loop by holding it in the hottest part of the flame (at the edge of the inner blue area) or the electric incinerator until it is red-hot (see the figure on page 30). Heat to redness. Why? _____

   Allow the loop to cool so that bacteria picked up with the loop won't be killed. Allow the loop to cool

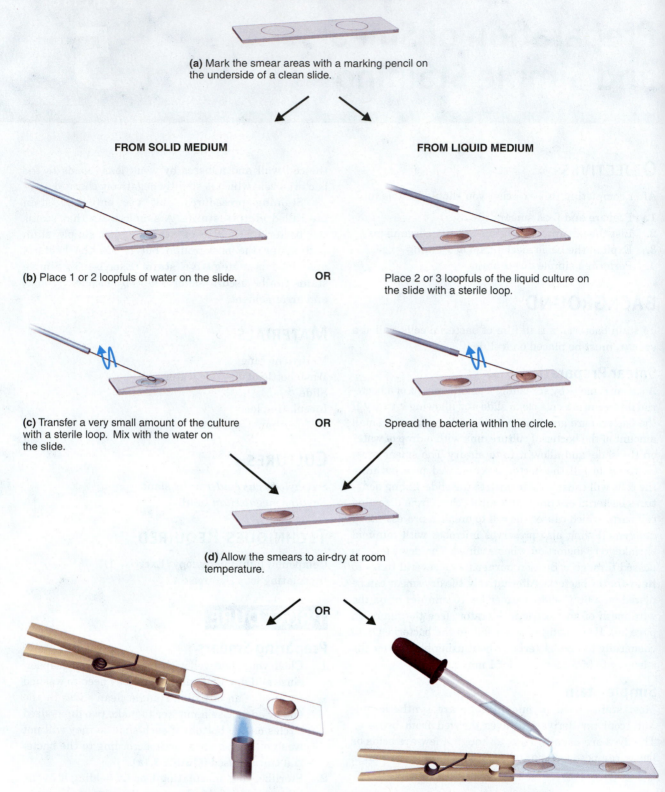

(a) Mark the smear areas with a marking pencil on the underside of a clean slide.

FROM SOLID MEDIUM

FROM LIQUID MEDIUM

(b) Place 1 or 2 loopfuls of water on the slide.

OR

Place 2 or 3 loopfuls of the liquid culture on the slide with a sterile loop.

(c) Transfer a very small amount of the culture with a sterile loop. Mix with the water on the slide.

OR

Spread the bacteria within the circle.

(d) Allow the smears to air-dry at room temperature.

OR

(e) Pass the slide through the flame of a burner two or three times.

(f) Cover the smears with 95% methanol for 1 minute, and then let the smears air-dry.

FIGURE 5.1 Preparing a bacterial smear.

without touching it or setting it down. Cooling takes about 30 seconds. You will determine the appropriate time with a little practice.

The loop must be cool before you insert it into a medium. A hot loop will spatter the medium and move bacteria into the air.

3. Prepare smears.
   a. For the bacterial culture on **solid media,** place 1 or 2 loopfuls of distilled water in the center of one circle, using the sterile inoculating loop (**FIGURE 5.1b**). Which bacterium is on a solid medium? _____
   b. Sterilize your loop.

Always sterilize your loop after using it and before setting it down.

c. Using the cooled loop, scrape a *small* amount of the culture off the slant—do not take the agar (**FIGURE 5.1c** and **FIGURE 5.2**). If you hear the sizzle of boiling water when you touch the agar with the loop, resterilize your loop, let it cool, and begin again. Why? _____

Try not to gouge the agar. Emulsify (to a milky suspension) the cells in the drop of water, and spread the suspension to fill a majority of the circle. The smear should look like diluted skim milk. Sterilize your loop again.

d. Make a smear of bacteria from the **broth culture** in the center of the other circle. Flick the tube of broth culture lightly with your finger to resuspend sedimented bacteria, and place 2 or 3 loopfuls of the culture in the circle (see Figure 5.1b). Sterilize your loop between each loopful. Spread the culture within the circle (see Figure 5.1c). Sterilize your loop.

e. Let the smears dry (**FIGURE 5.1d**). *Do not* blow on the slide; this will spread bacteria into the air. *Do not* flame or heat the wet slide. Heat will distort the cells' shapes and cause spattering in the air.

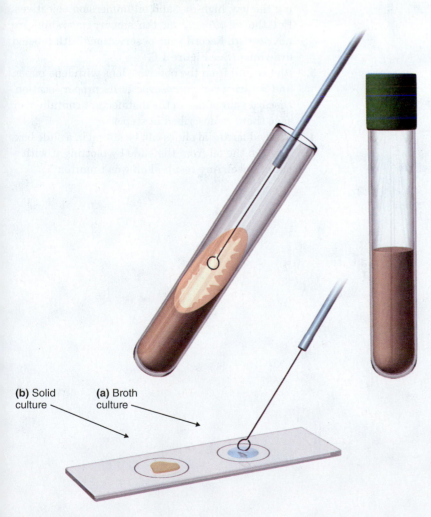

(b) Solid culture

(a) Broth culture

FIGURE 5.2 Transferring bacteria.
(a) Transfer 2 or 3 loopfuls of microbial suspension to a slide. (b) Gently scrape bacteria from the agar surface, and transfer the bacteria to a loopful of water on a slide. Be careful to avoid gouging the agar.

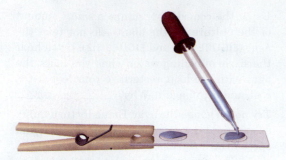

(a) Cover the smear with methylene blue and leave it for 30–60 seconds.

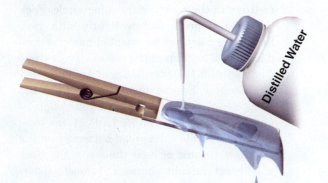

(b) Gently wash off the methylene blue with water by squirting the water so it runs through the smear.

(c) Blot it dry.

FIGURE 5.3 Simple staining.

f. Hold the slide with forceps, and fix the smears by one of the following methods:

(1) Pass the slide quickly through the blue flame with smear side up two or three times (**FIGURE 5.1e**), or place it on a 60° slide warmer for 10 minutes, or on an electric incinerator exterior.

(2) Cover the smear with 95% methanol for 1 minute (**FIGURE 5.1f**). Tip the slide to let the alcohol run off, and let the slide air-dry before staining. Do not fix until the smears are completely dry. Why? _____

_____

Staining Smears

1. Stain smears (**FIGURE 5.3**).
   a. Use a clothespin, forceps, or sliderack to hold the slide.
   b. Cover the smear with methylene blue, and leave it for 30 to 60 seconds (**FIGURE 5.3a**).
   c. Carefully wash the excess stain off with distilled water from a wash bottle. Let the water run through the smear (**FIGURE 5.3b**).
   d. Gently blot the smear with a paper towel or absorbent paper, and let it dry (**FIGURE 5.3c**).
2. Examine your stained smears microscopically using the low, high-dry, and oil immersion objectives. Put the oil *directly* on the smear; coverslips are not needed. Record your observations with labeled drawings. (See Figure 1.6.)
3. Blot the oil from the objective lens with lens paper, and return your microscope to its proper location. Discard your slides in the disinfectant container, or save them as described in step 4.
4. Stained bacterial slides can be stored in a slide box. Remove the oil from the slide by blotting it with a paper towel. Any residual oil won't matter.

LABORATORY REPORT
Preparation of Smears and Simple Staining

PURPOSE _____

_____

_____

EXPECTED RESULTS

1. Sketch a coccus:

2. Sketch a rod:

3. Which species used in this exercise is a rod? _____

RESULTS

Sketch a few bacteria viewed with the oil immersion objective lens.

| Bacterium: | *Staphylococcus epidermidis* | *Bacillus megaterium* |

Total
magnification: _____ ×   _____ ×

Morphology (shape): _____   _____

Arrangement of cells
relative to one another: _____   _____

CONCLUSIONS

1. Which bacterium is a rod? _____

2. Which bacterium is larger? _____

QUESTIONS

1. Of what value is a simple stain? _____

   _____

2. What is the purpose of heat-fixing the smear? _____

   _____

3. Another method of fixing smears is to use methanol instead of heat. How does alcohol chemically fix the

   bacteria? _____

   _____

4. In heat-fixing, what would happen if too much heat were applied? _____

   _____

CRITICAL THINKING

1. Methylene blue can be prepared as a basic stain or an acidic stain. How would the pH of the stain affect the staining of bacteria?

2. Can dyes other than methylene blue be used for direct staining? Briefly explain.

CLINICAL APPLICATIONS

1. Bacteria can be seen without staining. Why, then, was Koch's recommendation of fixing and staining important for the discovery of the bacterial causes of diseases?

2. Quality control staff in a sterilization unit of a hospital used a simple stain to determine whether bacteria were present in sterilized materials. A simple stain of sterile saline used for respiratory therapy revealed the presence of bacteria. Is the saline contaminated?

Negative Staining

OBJECTIVES

After completing this exercise, you should be able to:

1. Explain the application and mechanism of the negative stain technique.
2. Prepare a negative stain.

BACKGROUND

The **negative stain** technique does not stain the bacteria but instead stains the background. The bacteria will appear clear against a stained background. Colloidal stains such as India ink are used in the negative stain technique. The (1.0-μm) particles in the stain do not enter the bacteria and stain only the background. Negatively charged chromophores such as eosin can also be used:

$$\text{Sodium eosinate} \rightleftharpoons \text{Na}^+ + \underset{\text{(Chromophore)}}{\text{Eosin}^-}$$

The negatively charged chromophore is repelled by the cell. We will use a colloidal suspension of nigrosin. Nigrosin particles (0.2–1 μm) don't enter the cells, and at pH 7 the particles are negatively charged, causing them to remain a small distance away from the cells.

No heat-fixing or strong chemicals are used, so the bacteria are less distorted than in other staining procedures. The negative stain technique is useful in situations where other staining techniques don't clearly indicate cell morphology or size. Negative stains are especially useful for **coccobacilli,** which are short, oval bacilli that are difficult to stain.

MATERIALS

Nigrosin
Clean slides (6)
Distilled water
Sterile toothpicks

CULTURES

Bacillus subtilis
Staphylococcus epidermidis

TECHNIQUES REQUIRED

Compound light microscopy (Exercise 1)
Smear preparation (Exercise 5)

PROCEDURE

1. Make sure slides are clean and grease-free.
2. Refer to **FIGURE 6.1** as you read the following steps.
   a. Place a *small* drop or loopful of nigrosin at the end of the slide. For cultures on solid media, add a loopful of distilled water and emulsify a small amount of the culture in the nigrosin-water drop. For broth cultures, mix a loopful of the culture into the drop of nigrosin (**FIGURE 6.1a**). Do not spread the drop or let it dry.
   b. Using the end edge of another slide, spread out the drop (**FIGURE 6.1b** and **c**) to produce a smear varying from opaque black to gray. The angle of the spreading slide will determine the thickness of the smear.
   c. Let the smear air-dry (**FIGURE 6.1d**). *Do not heat-fix it.*
   d. Prepare a negative stain of the other culture.
3. Examine the stained slides microscopically, using the low, high-dry, and oil immersion objectives (**FIGURE 6.2** and **FIGURE 6.3**). As a general rule, if a few short rods are seen with small cocci, the morphology is rod-shaped. The apparent cocci are short rods viewed from the end or products of the cell division of small rods.
4. a. Place a small drop of nigrosin and a loopful of water at the end of a slide.
   b. Scrape the base of your teeth and gums with a sterile toothpick.
   c. Mix the nigrosin-water with the toothpick to get an emulsion of bacteria from your mouth.

 Discard the toothpick in disinfectant.

(a) Place a small drop of nigrosin near one end of a slide. Mix a loopful of broth culture in the drop. For organisms taken from a solid medium, mix a loopful of water in the nigrosin.

(b) Gently draw a second slide across the surface of the first until it contacts the drop. The drop will spread across the edge of the top slide.

(c) Push the top slide to the left along the entire surface of the bottom slide.

(d) Let the smear air-dry.

FIGURE 6.1 Preparing a negative stain.

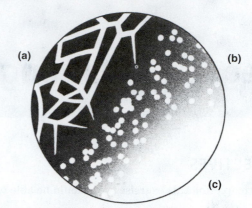

FIGURE 6.2 **A negative stain viewed under a microscope.**
(a) This part of the smear is too heavy: Cracks in the stain can be seen.
(b) Colorless cells are visible here. **(c)** Too little stain is in this area of the smear.

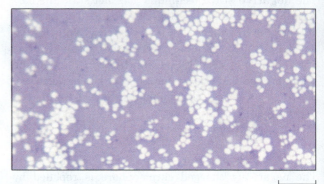

7 µm

FIGURE 6.3 **Negative stain of *Staphylococcus aureus*.**

    d. Follow steps 2b and 2c to complete the negative stain. Observe the stain, and describe your results.

Discard your slides in disinfectant.

    5. Wipe the oil off your microscope. Use lens paper to clean the lenses. Return the microscope to its proper place.

LABORATORY REPORT
Negative Staining

PURPOSE _____

_____

_____

EXPECTED RESULTS

1. Sketch a negatively stained rod:

2. Which organism used in this exercise is a rod? _____

RESULTS

Sketch a few bacteria (viewed with the oil immersion objective lens).

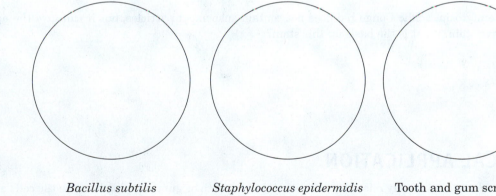

| Specimen: | *Bacillus subtilis* | *Staphylococcus epidermidis* | Tooth and gum scraping |

Total
magnification: ____× ____× ____×

Morphology and
arrangement: _____ _____ _____

CONCLUSIONS

1. Which cell is a rod? _____ How does its appearance differ from that of the rod you

stained with methylene blue (simple, direct stain)? _____

2. What range of bacterial morphologies did you observe in the tooth and gum scraping? _____

_____

QUESTIONS

1. Why is the size more accurate in a negative stain than in a direct stain? _____

_____

2. Why were you instructed to discard the toothpick in disinfectant? _____

_____

CRITICAL THINKING

1. What microscopic technique gives a field similar in appearance to that seen in the negative stain?

2. Carbolfuchsin can be used as a direct stain and as a negative stain. As a direct stain, the pH is _____.

3. Unlike nigrosin, acidic Congo red does not contain suspended particles, but it can give the appearance of a negative stain. What is the basis for this stain?

CLINICAL APPLICATION

Treponema denticola is easily missed in a direct stain of a tooth or gum scraping because the cells are distorted by the staining procedure. Why can this bacterium be seen with negative staining?

Gram Staining

OBJECTIVES

After completing this exercise, you should be able to:

1. Explain the rationale and procedure for the Gram stain.
2. Perform and interpret Gram stains.

BACKGROUND

The Gram stain is a useful stain for identifying and classifying bacteria. The **Gram stain** is a differential stain that enables you to classify bacteria as either gram-positive or gram-negative. The Gram-staining technique was discovered by Hans Christian Gram in 1884, when he attempted to stain cells and found that some lost their color when excess stain was washed off.

The staining technique consists of the following steps:

1. Apply **primary stain** (crystal violet). All bacteria are stained purple by this basic dye.
2. Apply **mordant** (Gram's iodine). The iodine combines with the crystal violet in the cell to form a crystal violet–iodine complex (CV–I).
3. Apply **decolorizing agent** (ethanol or ethanol–acetone). The primary stain is washed out of some bacteria (decolorized), whereas others are unaffected.
4. Apply **secondary stain,** or **counterstain** (safranin). This basic dye stains the decolorized bacteria red.

The most important determining factor in the procedure is that bacteria differ in their *rate* of decolorization. Those that decolorize easily are referred to as **gram-negative,** whereas those that decolorize slowly and retain the primary stain are called **gram-positive.**

Bacteria stain differently because of chemical and physical differences in their cell walls. Bacterial cell walls are complex lattice structures composed of layers of **peptidoglycans** and other compounds. Gram-positive cell walls contain multiple layers of peptidoglycans, whereas gram-negative bacteria contain a thin layer of peptidoglycan surrounded by an outer layer of lipoproteins, phospholipids, and lipopolysaccharides. The lipopolysaccharide functions as an endotoxin released when bacteria are dead.

Crystal violet is picked up by the cell. Iodine reacts with the dye in the cytoplasm to form a CV–I that is larger than the crystal violet that entered the cell. The alcohol decolorizing agent dehydrates the cell wall of the gram-positive bacteria. The CV–I cannot be washed out of gram-positive cells. In gram-negative cells, the decolorizing agent dissolves the outer lipopolysaccharide layer, and the CV–I washes out through the thin layer of peptidoglycan. The safranin stains the decolorized bacteria red.

The Gram stain is most consistent when done on young cultures of bacteria (less than 24 hours old). When bacteria die, their cell walls degrade and may not retain the primary stain, giving inaccurate results. Because Gram staining is usually the first step in identifying bacteria, you should memorize the procedure.

MATERIALS

Gram-staining reagents:
  Crystal violet
  Gram's iodine
  Ethanol
  Safranin
Wash bottle of distilled water
Slides (3)

CULTURES

Staphylococcus epidermidis
Escherichia coli
Bacillus subtilis

TECHNIQUES REQUIRED

Compound light microscopy (Exercise 1)
Smear preparation (Exercise 5)
Simple staining (Exercise 5)

PROCEDURE

1. Prepare and fix smears (see Figure 5.1 on page 52). Clean the slides well, and make a circle on the underside of each slide with a marker. Label each slide for one of the cultures. You can put two smears on one slide. Air-dry both smears and fix them.

(a) Starting with a fixed smear, cover the smear with crystal violet for 30 seconds.

(b) Gently wash off the crystal violet with water by squirting the water so it runs through the smear.

(c) Cover the smear with Gram's iodine for 10 seconds.

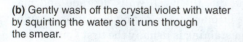

(d) Gently wash the smear with water.

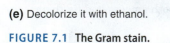

(e) Decolorize it with ethanol.

(f) Gently wash off the ethanol.

(g) Cover the smear with safranin for 30 seconds.

(h) Wash the smear with water.

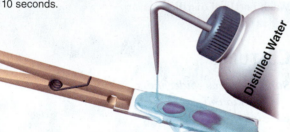

(i) Blot it dry.

FIGURE 7.1 The Gram stain.

2. Prepare a Gram stain of one smear. Use forceps, a clothespin, or slide rack to hold the slides.
   a. Cover the smear with crystal violet, and leave it for 30 seconds (**FIGURE 7.1a**).
   b. Wash the slide carefully with distilled water from a wash bottle. Do not squirt water directly onto the smear (**FIGURE 7.1b**).
   c. Cover the smear with Gram's iodine for 10 seconds (**FIGURE 7.1c**).
   d. Wash off the iodine by tilting the slide and squirting water above the smear so that the water runs over the smear (**FIGURE 7.1d**).
   e. Decolorize it with 95% ethanol (**FIGURE 7.1e**). Let the alcohol run through the smear until no large amounts of purple wash out (usually 10 to 20 seconds). The degree of decolorizing depends on the thickness of the smear. This is a critical step. *Do not overdecolorize*. However, experience is the only way you will be able to determine how long to decolorize. Very thick smears will give inaccurate results. Why? ____
   f. Immediately wash gently with distilled water (**FIGURE 7.1f**). Why? _____
   _____

   g. Add safranin for 30 seconds (**FIGURE 7.1g**).
   h. Wash the slide with distilled water, and blot it dry with a paper towel or absorbent paper (**FIGURE 7.1h** and **i**).
3. Repeat step 2 to stain your remaining slides.
4. Examine the stained slides microscopically, using the low, high-dry, and oil immersion objectives. Put the oil directly on the smear. Record your observations. (See **FIGURE 7.2**.) Do they agree with those given in your textbook?

   _____

   If not, try to determine why. Some common sources of Gram-staining errors are the following:
   a. The loop was too hot.
   b. Excessive heat was applied during heat-fixing.
   c. The decolorizing agent (ethanol) was left on the smear too long.
   d. The culture was too old.
   e. The smear was too thick.
5. Place the slides in the disinfectant.

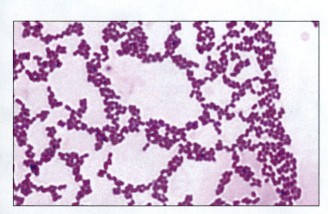

(a) Gram-positive cocci.   ⊢———⊣
                                                  10 μm

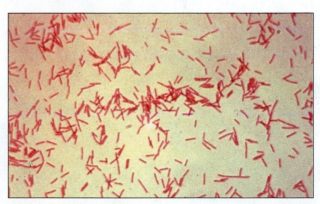

(b) Gram-negative bacilli.   ⊢———⊣
                                                  10 μm

FIGURE 7.2 Gram-stained bacteria

Name: _____   Date: _____   Lab Section: _____

LABORATORY REPORT
Gram Staining

PURPOSE _____

_____

_____

EXPECTED RESULTS

1. What color will a gram-negative cell stain? _____

2. What color will a gram-positive cell stain? _____

3. Which organism used in this exercise is a coccus? _____

RESULTS

Sketch a few bacteria (viewed with the oil immersion objective lens).

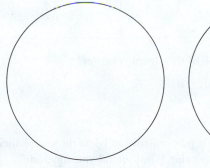 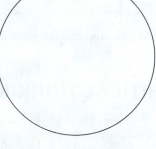

Bacterium:	*Staphylococcus epidermidis*	*Bacillus subtilis*	*Escherichia coli*
Total magnification:	____×	____×	____×
Morphology and arrangement:	_____	_____	_____
Color:	_____	_____	_____
Gram stain:	_____	_____	_____

CONCLUSIONS

1. Of these three bacteria, which organism is the largest? _____

2. Did your results agree with the information in your textbook? _____ If not, why not?

   _____

QUESTIONS

1. Why do gram-positive cells more than 24 hours old stain gram-negative? _____

   _____

2. Can iodine be added before the primary stain in a Gram stain? _____

3. List the steps of the Gram-staining procedure in order (omit washings), and fill in the color of gram-positive cells and gram-negative cells after each step.

Step	Chemical	Appearance	
		Gram-Positive Cells	**Gram-Negative Cells**
1			
2			
3			
4			

4. Which step can you omit without affecting determination of the Gram reaction?

CRITICAL THINKING

1. Suppose you performed a Gram stain on a sample from a pure culture of bacteria and observed a field of red and purple cocci. Adjacent cells were not always the same color. What would you conclude?

2. Suppose you are viewing a Gram-stained field of red rods and purple cocci through the microscope. What do you conclude?

3. Human cells can be stained with crystal violet and safranin, so why can't human cells be Gram stained?

CLINICAL APPLICATION

Considering that it isn't possible to identify bacteria from a Gram stain, why might a physician perform a Gram stain on a sample before prescribing an antibiotic?

Acid-Fast Staining

OBJECTIVES

After completing this exercise, you should be able to:

1. Apply the acid-fast procedure.
2. Explain what is occurring during the acid-fast staining procedure.
3. Perform and interpret an acid-fast stain.

BACKGROUND

The **acid-fast stain** is a differential stain. In 1882, Paul Ehrlich discovered that *Mycobacterium tuberculosis* (the causative agent of tuberculosis) retained the primary stain even after washing with an acid-alcohol mixture. (We hope you can appreciate the phenomenal strides that were made in microbiology in the 1880s. Most of the staining and culturing techniques used today originated during that period.) Most bacteria are decolorized by acid-alcohol; only the families Mycobacteriaceae, Nocardiaceae, Gordoniaceae, Dietziaceae, and Tsukamurellaceae (*Bergey's Manual**) are acid-fast. The acid-fast technique has great value as a diagnostic procedure because both *Mycobacterium* and *Nocardia* contain **pathogenic** (disease-causing) species. The acid-fast staining technique can be used to diagnose tuberculosis and leprosy; the latter is caused by *Mycobacterium leprae.*

The cell walls of acid-fast organisms contain a waxlike lipid called **mycolic acid,** which renders the cell wall impermeable to most stains. Mycolic acid enhances the inflammation response initiating the formation of a tubercle, a walled-off lesion. Mycolic acid in the cell wall is thought to be a factor in the resistance of *Mycobacterium* to harsh, dry environments. The cell wall is so impermeable, in fact, that before mycobacteria are cultured, a clinical specimen is usually treated with strong sodium hydroxide to remove debris and contaminating bacteria. The mycobacteria are not killed by this procedure.

Today, the techniques developed by Franz Ziehl and Friedrich Neelsen and by Joseph J. Kinyoun

are the most widely used acid-fast stains. In the **Ziehl–Neelsen procedure,** the smear is flooded with carbolfuchsin (a dark red dye containing 5% phenol), which has a high affinity for the mycolic acid in the bacterial cell. The smear is heated to facilitate penetration of the carbolfuchsin stain into the bacteria. The stained smears are washed with an acid-alcohol mixture that easily decolorizes most bacteria except the acid-fast microbes. Methylene blue is then used as a counterstain to enable you to observe the non–acid-fast organisms. In the **Kinyoun modification,** called a *cold stain,* the concentrations of phenol and carbolfuchsin are increased so that heating isn't necessary.

The mechanism of the acid-fast stain is probably the result of the relative solubility of carbolfuchsin and the impermeability of the cell wall. Fuchsin is more soluble in phenol (carbolic acid) than in water, and phenol solubilizes more easily in lipids than in acid-alcohol. Therefore, carbolfuchsin has a higher affinity for lipids than for acid-alcohol and will remain with the cell wall when washed with acid-alcohol. We will use the Kinyoun modification cold-stain technique in this exercise.

MATERIALS

Acid-fast staining reagents:
    Kinyoun's carbolfuchsin
    Acid-alcohol
    Methylene blue
Wash bottle of distilled water
Slide

CULTURES

Mycobacterium smegmatis
Escherichia coli

DEMONSTRATION SLIDE

Acid-fast sputum slide

TECHNIQUES REQUIRED

Compound light microscopy (Exercise 1)
Smear preparation (Exercise 5)
Simple staining (Exercise 5)

Bergey's Manual of Systematic Bacteriology, 2nd ed., 5 vols. (2004), is the standard reference for classification of prokaryotic organisms. *Bergey's Manual of Determinative Bacteriology,* 9th ed. (1994), is the standard reference for identification of culturable bacteria and archaea.

(a) Starting with a fixed smear, cover the smear with carbolfuchsin for 5 minutes.

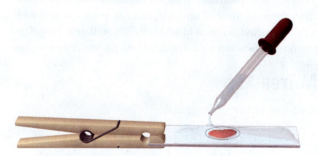

(b) Gently wash off the carbolfuchsin with water by squirting the water so it runs through the smear.

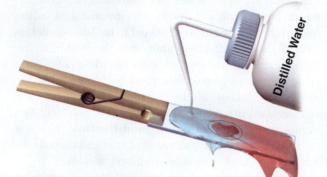

(c) Cover the smear with acid-alcohol for 1 minute.

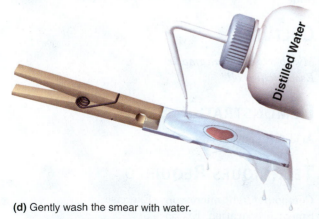

(d) Gently wash the smear with water.

(e) Cover the smear with methylene blue for 1 minute.

(f) Wash the smear with water.

(g) Blot it dry.

FIGURE 8.1 **The acid-fast stain.**

PROCEDURE

1. Prepare and fix a smear of each culture (see Figure 5.1 on page 52).
2. Cover the smears with carbolfuchsin, and let the smears stand for 5 minutes (**FIGURE 8.1a**).
3. Gently wash the slide with distilled water from a wash bottle. Do not squirt water directly onto the smear (**FIGURE 8.1b**).
4. Without drying it, wash the smear with decolorizer (acid-alcohol) for 1 minute or until no more red color runs off when the slide is tipped (**FIGURE 8.1c**).
5. Wash the smear carefully with distilled water (**FIGURE 8.1d**).
6. Counterstain the smear for about 1 minute with methylene blue (**FIGURE 8.1e**).
7. Wash the smear with distilled water, and blot it dry (**FIGURE 8.1f** and **g**).
8. Examine the acid-fast stained slide microscopically, and record your observations. Discard your slide in the disinfectant.
9. Observe the demonstration slide (**FIGURE 8.2**).

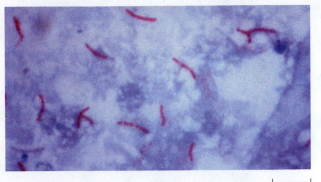

7 μm

FIGURE 8.2 Acid-fast stain of a sputum specimen.
Mycobacterium tuberculosis stained red; their cell walls are acid-fast.

LABORATORY REPORT
Acid-Fast Staining

PURPOSE _____

_____

_____

EXPECTED RESULTS

1. What color will an acid-fast organism stain in the acid-fast stain? _____

2. What color will a non–acid-fast cell stain in the acid-fast stain? _____

RESULTS

Sketch a few bacteria (viewed with the oil immersion objective lens).

Bacterium:	*Mycobacterium smegmatis*	*Escherichia coli*	Demonstration Slide

Bacterium: *Mycobacterium smegmatis*   *Escherichia coli*   Demonstration Slide

Total magnification: ____×   ____×   ____×

Morphology: _____   _____   _____

Color: _____   _____   _____

Specimen: _____   _____   _____

CONCLUSIONS

1. Which bacterium was acid-fast? _____

2. Did the acid-fast stained sputum slide indicate a possible positive test for tuberculosis? _____

QUESTIONS

1. What are the large blue-stained areas on the sputum slide? _____

2. What is the decolorizing agent in the Gram stain? _____

   In the acid-fast stain? _____

3. What diseases are diagnosed using the acid-fast procedure? _____

4. What is phenol (carbolic acid), and what is its *usual* application? _____

   _____

5. Carbolfuchsin used in the original acid-fast stain consisted of 0.3 g basic fuchsin in 10 ml 95% ethanol, 5 ml phenol, and 95 ml water. The recipe for the Kinyoun modification is 4 g basic fuchsin in 20 ml ethanol, 8 g phenol, and 100 ml water. The original acid-fast stain required heating the smear to force the carbolfuchsin into the wall. Why can heat be eliminated in the Kinyoun modification that you used? _____

   _____

CRITICAL THINKING

1. How might the acid-fast characteristic of *Mycobacterium* enhance the organism's ability to cause disease?

2. In 1882, after experimenting with staining *Mycobacterium,* Paul Ehrlich wrote that only alkaline disinfectants would be effective against *Mycobacterium.* How did he reach this conclusion without testing the disinfectants?

CLINICAL APPLICATIONS

1. Clinical specimens suspected of containing *Mycobacterium* are digested with sodium hydroxide (NaOH) for 30 minutes prior to staining. Why is this technique used? Why isn't this technique used for staining other bacteria?

2. The acid-fast stain is used to detect *Cryptosporidium* protozoa oocysts in fecal samples. Which of the following would you expect to be a major component of the oocyst wall: carbohydrates, lipids, or proteins? _____

   What disease is caused by *Cryptosporidium*? _____

Structural Stains (Endospore, Capsule, and Flagella)

OBJECTIVES

After completing this exercise, you should be able to:

1. Prepare and interpret endospore, capsule, and flagella stains.
2. Recognize the different types of flagellar arrangements.
3. Identify functions of endospores, capsules, and flagella.

BACKGROUND

Structural stains can be used to identify and study the structure of bacteria. Currently, most of the fine structural details are examined using an electron microscope, but historically, staining techniques have given much insight into bacterial fine structure. We will examine a few structural stains that are still useful today. These stains are used to observe endospores, capsules, and flagella.

Endospores

Endospores are formed by several genera in the orders Bacillales and Clostridiales.* *Bacillus* and *Clostridium* are the most familiar genera. Endospores are called "resting bodies" because they do not metabolize and are resistant to heating, various chemicals, and many harsh environmental conditions. Endospores are not for reproduction; they are formed when water or essential nutrients are not available. Once an endospore forms in a cell, the cell will disintegrate (**FIGURE 9.1a**). Endospores can remain dormant for long periods of time. However, an endospore may return to its vegetative or growing state.

Taxonomically, it is helpful to know whether a bacterium is an endospore-former and also the position of the endospores in the bacterium (**FIGURE 9.1b** through **e**). Endospores are impermeable to most stains, so heat is usually applied to drive the stain into the endospore. Once stained, the endospores do not readily decolorize. We will use the **Schaeffer–Fulton** endospore stain.

Capsules

Many bacteria secrete chemicals that adhere to their surfaces, forming a viscous coat. This structure is called a **capsule** when it is round or oval in shape, and a **slime layer** when it is irregularly shaped and loosely bound to the bacterium. The ability to form capsules is genetically determined, but the size of the capsule is influenced by the medium on which the bacterium is growing. Most capsules are composed of polysaccharides, which are water soluble and uncharged. Because of the capsule's nonionic nature, simple stains will not adhere to it. Most capsule-staining techniques stain

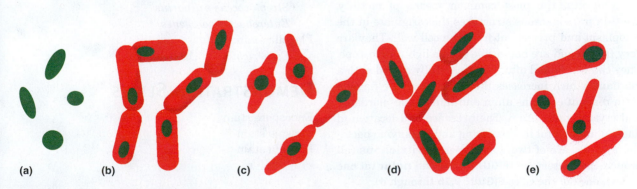

FIGURE 9.1 Some examples of bacterial endospores. (a) Free endospores after the cell has disintegrated.
(b) Subterminal endospores (*Bacillus macerans*). **(c)** Central, swollen endospores (*Clostridium perfringens*).
(d) Central endospores (*Bacillus polymyxa*). **(e)** Terminal, swollen endospores (*Clostridium tetani*).

*Bergey's Manual of Systematic Bacteriology, 2nd ed. (2004).

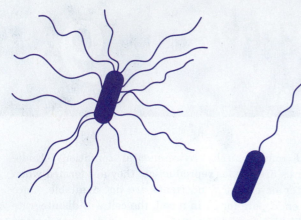

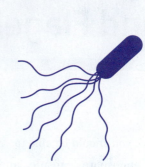

(a) Peritrichous flagella

(b) Monotrichous flagellum

(c) Lophotrichous flagella

(d) Amphitrichous flagella

FIGURE 9.2 **Flagella arrangements.** Polar arrangements are shown in **(b)**, **(c)**, and **(d)**.

the bacteria and the background, leaving the capsules unstained—essentially, a "negative" capsule stain.

Capsules play an important role in the **virulence** (disease-causing ability) of some bacteria. For example, when bacteria such as *Streptococcus pneumoniae* have a capsule, the body's white blood cells cannot phagocytize the bacteria efficiently, and disease occurs. When *S. pneumoniae* lack a capsule, they are easily engulfed and are not virulent.

Flagella

Many bacteria are **motile,** which means they have the ability to move from one position to another in a directed manner. Most motile bacteria possess flagella, but other forms of motility occur. Myxobacteria exhibit gliding motion, and spirochetes undulate using axial filaments. Flagella can enhance a bacterium's virulence by allowing the bacterium to avoid phagocytosis by white blood cells.

Flagella, the most common means of motility, are thin proteinaceous structures that originate in the cytoplasm and project out from the cell wall. They are very fragile and are not visible with a light microscope. They can be stained after being carefully coated with a mordant, which increases their diameter. The flagella stain reagent contains alum and tannic acid mordants and crystal violet stain. The presence and location of flagella are helpful in identifying and classifying bacteria. Flagella are of two main types: **peritrichous** (all around the bacterium) (**FIGURE 9.2a**) and **polar** (at one or both ends of the cell) (**FIGURE 9.2b** through **d**).

Motility may be determined by observing hanging drop or wet-mount preparations of unstained bacteria (Exercise 2), flagella stains, or inoculation of soft (or semisolid) agar deeps (Exercise 4). If time does not permit doing flagella stains, observe the demonstration slides.

MATERIALS

Slides
Coverslips
Paper towels
Wash bottle of distilled water
Safety glasses
Endospore stain reagents: malachite green and safranin
Capsule stain reagents: Congo red, acid-alcohol, and acid fuchsin
Flagella stain reagent

CULTURES (AS NEEDED)

Endospore stain:
  Bacillus megaterium (24-hour)
  Bacillus subtilis (24-hour)
  Bacillus subtilis (72-hour)
Capsule stain:
  Streptococcus salivarius
  Enterobacter aerogenes
Flagella stain:
  Escherichia coli (18-hour)

DEMONSTRATION SLIDES

Endospore stain
Capsule stain
Flagella stain

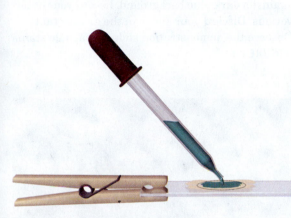

(a) Place a piece of absorbent paper over the smear.

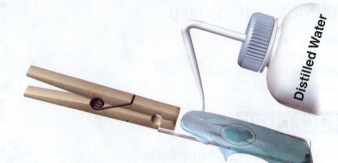

(d) Wash the smear with water.

(b) Cover the paper with malachite green.

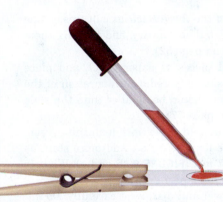

(e) Cover the smear with safranin for 30 seconds.

(c) Steam the slide for 5 minutes.

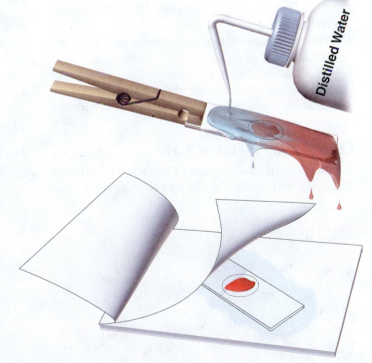

(f) Wash the smear with water and blot it dry.

FIGURE 9.3 **The endospore stain.**

TECHNIQUES REQUIRED

Compound light microscopy (Exercise 1)
Smear preparation (Exercise 5)
Simple staining (Exercise 5)
Negative staining (Exercise 6)

PROCEDURE

Endospore Stain (Figure 9.3)

Be careful. The malachite green has a messy habit of ending up everywhere. Most likely, you will end up with green fingers no matter how careful you are. Wear safety glasses!

1. Make smears of the three *Bacillus* cultures on one or two slides, let them air-dry, and heat-fix them (see Figure 5.1 on page 52).
2. Tear out small pieces of paper towel, and place them on each smear to reduce evaporation of the stain. The paper should be smaller than the slide (**FIGURE 9.3a** on page 77).
3. Cover the smears and paper with malachite green; steam the slide for 5 minutes. Add more stain as needed. *Keep it wet* (**FIGURE 9.3b** and **c**). What is the purpose of the paper?
4. Remove the towel and discard it carefully. *Do not put it in the sink.* Wash the stained smears well with distilled water (**FIGURE 9.3d**).
5. Counterstain with safranin for 30 seconds (**FIGURE 9.3e**).
6. Wash the smear with distilled water, and blot it dry (**FIGURE 9.3f**).
7. Examine the slide microscopically, and record your observations. Discard your slides in the disinfectant.
8. Observe the demonstration slides of the bacterial endospores (**FIGURE 9.4**).

Capsule Stain (Figure 9.5)

1. Draw two circles on a slide. Place a loopful of Congo red in each circle (**FIGURE 9.5a** on page 79).

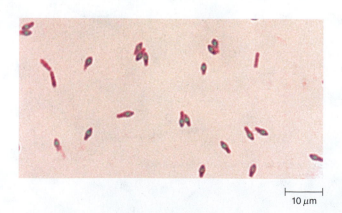

FIGURE 9.4 Endospore stain of *Bacillus* sp.

2. Prepare a thick smear of *S. salivarius* in the Congo red in one circle. Prepare a thick smear of *E. aerogenes* in the other circle. Let the smears air-dry (**FIGURE 9.5b** and **c**). What color are the smears? ___
3. Fix the smears with acid-alcohol for 15 seconds (**FIGURE 9.5d**). What color are the smears? _____
4. Wash the smears with distilled water, and cover them with acid fuchsin for 1 minute (**FIGURE 9.5e** and **f**).
5. Wash the smears with distilled water, and blot them dry (**FIGURE 9.5g**).
6. Examine the slide microscopically. The bacteria will stain red, and the capsules will be colorless against a dark blue background. Record your observations. Discard your slides in the disinfectant.
7. Observe the demonstration slides of capsule stains (**FIGURE 9.6**).

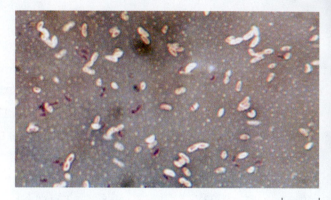

FIGURE 9.6 Capsule stain of *Streptococcus mutans*.

Flagella Stain (Figure 9.7)

1. Flagella stains require special precautions to avoid damaging the flagella. Scrupulously clean slides are essential, and the culture must be handled carefully to prevent flagella from coming off the cells.
2. Draw a circle on the underside of each slide approximately 1 cm from the end of the slide (**FIGURE 9.7a**). Place a drop of distilled water in the circle (**FIGURE 9.7b**).
3. With a sterile inoculating loop, gently touch a colony of the culture to be examined, and then lightly touch the drop of water without touching the slide (**FIGURE 9.7c**). Do not mix.
4. Tilt the slide so the drop will flow to the opposite end of the slide. Allow the slide to air-dry (**FIGURE 9.7d**). Do not heat-fix.
5. Cover the smear with flagella stain for 4 minutes (**FIGURE 9.7e**).

(a) Place one loopful of Congo red in each circle.

(b) Prepare a thick smear of bacteria in the Congo red.

(c) Allow the smears to air-dry.

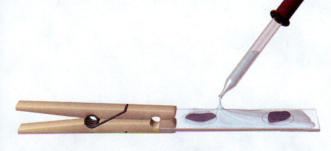

(d) Cover each smear with acid-alcohol for 15 seconds.

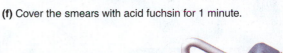

(e) Wash the smears with water.

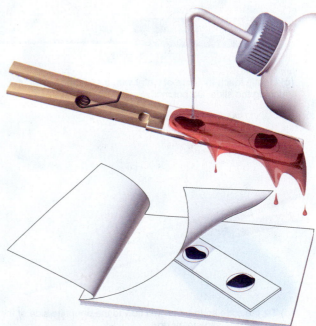

(f) Cover the smears with acid fuchsin for 1 minute.

(g) Wash the smears with water and blot them dry.

FIGURE 9.5 The capsule stain.

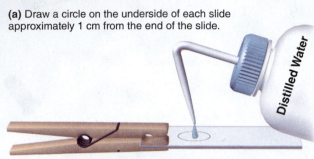

(a) Draw a circle on the underside of each slide approximately 1 cm from the end of the slide.

(b) Place a drop of distilled water in the circle.

(c) Lightly touch the drop of water with the bacteria without touching the slide. Do not mix.

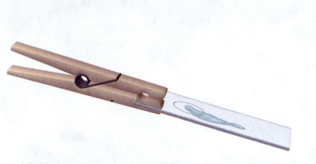

(d) Tilt the slide so the drop will flow to the opposite side of the slide. Allow the slide to air dry.

FIGURE 9.7 The flagella stain.

(e) Cover the dried smear with flagella stain for 4 minutes.

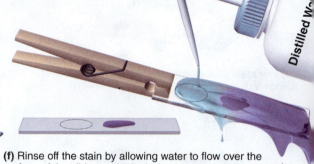

(f) Rinse off the stain by allowing water to flow over the surface of the slide and let it air dry.

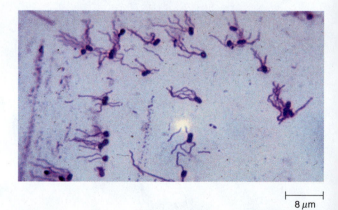

8 µm

FIGURE 9.8 Flagella stain of *Proteus vulgaris*. The flagella are peritrichous.

6. Carefully rinse the stain by allowing water to flow over the surface of the slide. Do not tilt the slide while rinsing (**FIGURE 9.7f**).

7. Once the stain is washed, gently tilt the slide so excess water will run off. Allow the slide to air-dry.

8. Examine the slide microscopically. Look for fields with several isolated cells rather than clumps of bacteria. What color should the flagella be? _____ The cells? _____ Discard your slides in the disinfectant.

9. Observe the demonstration slides illustrating various flagellar arrangements (**FIGURE 9.8**; see also Figure 9.2 on page 76).

LABORATORY REPORT
Structural Stains
(Endospore, Capsule, and Flagella)

PURPOSE _____

_____

_____

EXPECTED RESULTS

1. After staining, endospores will be _____ in a(n) _____ cell.

2. After staining, capsules will be _____ around a(n) _____ cell.

RESULTS

Endospores

Sketch your results, and label the color in each diagram. Label the vegetative cells and endospores.

Bacterium:	*Bacillus megaterium*	*Bacillus subtilis* (24-hour)	*Bacillus subtilis* (72-hour)

Bacterium: *Bacillus*   *Bacillus subtilis*   *Bacillus subtilis*
                 megaterium   (24-hour)   (72-hour)

Total
magnification: ____×   ____×   ____×

Endospore
position: _____   _____   _____

Demonstration Slides

Bacterium: _____   _____   _____

Total magnification: ___×   ___×   ___×

Endospore
position: _____   _____   _____

Capsules

Sketch and label the capsules and bacterial cells.   Demonstration Slides

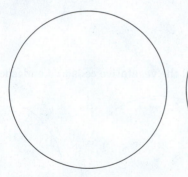  

Bacterium: *Streptococcus salivarius*   *Enterobacter aerogenes*   _____

Total
magnification: ___×   ___×   ___×

Capsules
present: _____   _____   _____

Flagella

Sketch and label the flagella and bacteria.

Demonstration Slides

Bacterium: *Escherichia coli*   _____   _____

Total
magnification: ____×   ____×   ____×

Flagella
position: _____   _____   _____

CONCLUSIONS

1. How did the 24-hour and the 72-hour *Bacillus* cultures differ in appearance? How do you account for this

    difference? _____

    _____

    _____

2. Of what morphology are most bacteria that possess flagella? _____

3. What prevents the cell from appearing green in the finished endospore stain? _____

    _____

    _____

QUESTIONS

1. What are the Gram reactions of *Clostridium* and *Bacillus*? _____

2. How might a capsule contribute to pathogenicity? _____

    How might flagella contribute to pathogenicity? _____

3. Of what advantage to *Clostridium* is an endospore? _____

    _____

    _____

4. The flagella stain uses crystal violet, as does the Gram stain. Why do all cells stain purple in the flagella stain but not in the Gram stain? _____

_____

_____

CRITICAL THINKING

1. How would an endospore stain of *Mycobacterium* appear?

2. What type of culture medium would increase the size of a bacterial capsule?

3. Describe the microscopic appearance of encapsulated *Streptococcus* if stained with safranin and nigrosin.

4. In the Dorner endospore stain, a smear covered with carbolfuchsin is steamed, then decolorized with acid-alcohol and counterstained with nigrosin. Describe the microscopic appearance after this procedure.

CLINICAL APPLICATION

Endospores appear as colorless areas in a simple stain and in a Gram stain. Why is it necessary to perform an endospore stain to identify *Clostridium difficile* in healthcare settings?

Morphologic Unknown

Dishonesty is **KNOWING BUT IGNORING** *the fact that the data are contradictory.*
Stupidity is not recognizing the contradictions. – **ANONYMOUS**

OBJECTIVE

After completing this exercise, you should be able to:

Identify the morphology and staining characteristics of an unknown organism.

BACKGROUND

Differential staining is usually the first step in identifying bacteria. Morphology and structural characteristics obtained from microscopic examination are also useful for identification. *Bergey's Manual** is the most widely used reference for bacterial identification. In *Bergey's Manual,* bacteria are identified by differential staining, morphology, and several other characteristics. Additional testing is needed to identify bacterial species. You will learn these techniques later in this course.

You will be given an unknown culture of bacteria. Determine its morphologic and structural characteristics. The culture contains one species (rod or coccus) and is less than 24 hours old.

Quality control (QC) is an essential component of the microbiology laboratory. The accuracy of information from stains and other tests depends on a variety of factors, including the specimen, reagents, procedures, and the person doing the test. Generally during staining, known positive and negative bacteria are stained at the same time as unknown bacteria, as a check on reagents and procedures. In this exercise, you will use your results from Exercises 7 through 9 as your QC.

MATERIAL

Staining reagents

CULTURE

24-hour unknown slant culture of bacteria # _____

TECHNIQUES REQUIRED

Compound light microscopy (Exercise 1)
Hanging-drop and wet-mount procedures (Exercise 2)
Smear preparation (Exercise 5)
Simple staining (Exercise 5)
Negative staining (Exercise 6)
Gram staining (Exercise 7)
Acid-fast staining (Exercise 8)
Endospore, capsule, and flagella staining (Exercise 9)

PROCEDURE

1. Record the number of your unknown in your Laboratory Report.
2. Determine the morphology, Gram reaction, and arrangement of your unknown. Perform a Gram stain and, if needed, an endospore stain, acid-fast stain, and capsule stain. When are the latter needed? ___

   _____
3. Motility may be determined by a flagella stain, wet mount, or hanging drop.
4. Tabulate your results in the Laboratory Report.

---

*Bergey's Manual of Systematic Bacteriology, 2nd ed., 5 vols. (2004), is the standard reference for classification of prokaryotic organisms. *Bergey's Manual of Determinative Bacteriology,* 9th ed. (1994), is the standard reference for identification of culturable bacteria and archaea.

| LABORATORY REPORT
Morphologic Unknown | EXERCISE
10 |

PURPOSE _____
_____
_____

EXPECTED RESULTS

1. All bacteria have endospores. Agree/disagree

2. Some gram-negative bacteria are acid-fast. Agree/disagree

RESULTS

Write "not necessary" by any category that does not apply.

Unknown # _____

Gram Stain

Sketch your unknown and controls.   Controls from Exercise 7

Bacterium:	*Staphylococcus epidermidis*	*Escherichia coli*	Unknown
Total magnification:	___×	___×	___×
Gram reaction:	_____	_____	_____
Morphology:	_____	_____	_____
Predominant arrangement:	_____	_____	_____

Acid-Fast Stain

Sketch your unknown and controls.   Controls from Exercise 8

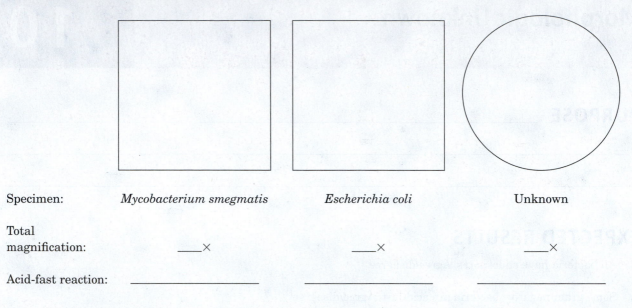

Specimen:	*Mycobacterium smegmatis*	*Escherichia coli*	Unknown

Total
magnification: ____×   ____×   ____×

Acid-fast reaction: _____   _____   _____

Endospore Stain

Sketch your unknown and controls.   Controls from Exercise 9

Specimen:	24-hr *Bacillus*	72-hr *Bacillus*	Unknown

Total
magnification: ____×   ____×   ____×

Endospores present: _____   _____   _____

Capsule Stain

Sketch your unknown and control.   Control from Exercise 9

Specimen: *Streptococcus salivarius*   Unknown

Total
magnification: ___×   ___×

Capsules present: _____   _____

Motility

Is your unknown motile? _____

How did you determine this? _____

CONCLUSIONS

1. What are the Gram reaction, morphology, and arrangement of your unknown culture? _____

    _____

2. Is your unknown acid-fast? _____

3. Does your unknown bacterium have endospores? _____ Capsules? _____ Flagella? _____

4. Observation of your slant after another 24 hours may be necessary if you did not see endospores initially. Why?

    _____

QUESTIONS

1. Which two stains done in this experiment are differential stains? _____

    _____

2. Do all bacteria make endospores? _____

3. What is the Gram reaction of acid-fast bacteria? _____

CRITICAL THINKING

1. Using *Bergey's Manual* and your text, place the following genera in this flowchart:

   - *Bacillus*
   - *Escherichia*
   - *Neisseria*
   - *Staphylococcus*
   - *Corynebacterium*
   - *Mycobacterium*
   - *Sporosarcina*

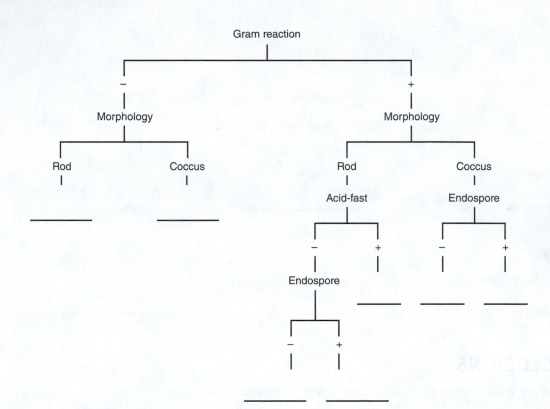

2. Which genera listed in the previous question should you test for capsules? _____

   For motility? _____

3. Using your textbook and this lab manual, fill in the Morphology column in the following table. Then use the information to construct a flowchart for these bacteria. Draw your flowchart in the box below the table.

	Morphology	Gram Reaction	Motile	Capsule	Arrangement	Endospore
Clostridium		+	+	−	Pairs, chains	+
Enterobacter		−	+	−	Singles, pairs	−
Klebsiella		−	−	+	Singles, pairs	−
Lactobacillus		+	Rarely	−	Chains	−
Staphylococcus		+	−	−	Pairs, clusters	−
Streptococcus		+	−	Some species	Pairs, chains	−

4. Provide an example from the previous question to show that microscopic examination alone cannot be used to identify bacteria.

CLINICAL APPLICATION

Assume you have performed a Gram stain on a sample of pus from a patient's urethra. You see red, nucleated cells (>10 μm) and purple rods (2.5 μm). What can you conclude?

PART 4

Cultivation of Bacteria

EXERCISES

Through their microscopes, early researchers saw a mixture of rod-, coccal-, and spiral-shaped bacteria. Hence, most biologists agreed with Linnaeus that bacteria belonged in the class "Chaos." In the 1870s, however, a few bacteriologists recognized that bacteria are not all the same and that, as Anton deBary wrote in 1885, these shapes "form the characteristic features for the recognition and differentiation of species."* This resulted in the pure culture approach to studying the characteristics of a single cell.

A **pure culture** consists of only one species of microorganism. Pure cultures are needed to characterize and identify microbes. The earliest methods of obtaining pure cultures involved inoculating small amounts of a sample taken from a diseased animal into a flask of blood. The microbes were transferred from the flask to another flask until only one microbe was seen, a procedure that was tedious and often ineffective. Scientists in Koch's laboratory added agar to solidify natural materials such as blood and gelatin. The next development came in 1895, when a company called Difco (Differential Ferments Company) began producing enzymes to help digestion. Difco used enzymes to digest proteins. The first solid culture medium was composed of these digests, which included peptone, and agar. By 1916, it was known that the

> *significant nutritive factor in bacteriological culture media was the peptone.... The role of peptone in culture media is twofold. It furnishes nitrogenous food in the form of amino acids and also [pH] buffer substance in the form of phosphate salts.†*

---

*Quoted in M. Penn and M. Dworkin. "Robert Koch and Two Visions of Microbiology." *Bacteriological Reviews* 40(2): 276–283, June 1976.
†I. J. Klinger. "The Effect of Hydrogen Ion Concentration." *Journal of Bacteriology* 2: 351–353, 1917.

This discovery led to development of nutrient agar. Further experimentation led to the addition of sugars and dyes to isolate and differentiate specific bacteria growing on the media.

Bacteria can be separated on solid media to isolate them in pure cultures. You will learn to isolate bacteria and to obtain pure cultures in the exercises in Part 4.

CASE STUDY: Critical Care in a Nursery

At a local hospital, a significant increase in births resulted in a crowded nursery. An outbreak of skin infections resembling impetigo occurred in five of the infants. One of the pediatric hospitalists was concerned and ordered lab tests on the infants; the skin lesions were swabbed and cultured on mannitol salt agar (MSA). Bacteria grew on the medium, and many of the cultures changed it to a yellow color around the colonies. Colonies from the MSA plate were selected with an inoculating loop and were used to make a streak plate on a nutrient agar plate. Golden yellow colonies grew on the plate, and the microbiology lab used additional laboratory procedures to identify the isolates. An infection-control specialist was consulted, who recommended that all personnel in the nursery have swabs taken of their nasal passages and skin. A complete infection control survey of the equipment and nursery area was also performed. All physicians and nurses were required to attend an in-house infection-control training session where the value of proper handwashing technique was described. The babies were treated with antibiotics, and all recovered with no evidence of resistance to methicillin.

Questions

1. What bacterium was isolated?
2. Mannitol salt agar is a selective and differential medium. Explain.
3. What is a healthcare-associated infection?
4. What is methicillin-resistance Staphylococcus aureus (MRSA), and why is it of such concern to clinicians?

Isolation of Bacteria by Dilution Techniques

OBJECTIVES

After completing this exercise, you should be able to:

1. Isolate bacteria by using the streak plate and pour plate techniques.
2. Prepare and maintain a pure culture.

BACKGROUND

In nature, most microbes are found growing in environments that contain many different organisms. Unfortunately, mixed cultures are of little use in studying microorganisms because of the difficulty they present in determining which organism is responsible for any observed activity. To study growth characteristics, pathogenicity, metabolism, antibiotic susceptibility, or other characteristic requires a pure culture—one that contains a single kind of microbe. Because bacteria are too small to separate directly without sophisticated micromanipulation equipment, indirect methods of separation must be used.

In the 1870s, Joseph Lister attempted to obtain pure cultures by performing serial dilutions until each of his containers theoretically contained one bacterium. However, success was very limited, and **contamination,** the presence of unwanted microorganisms, was common. In 1880, Robert Koch prepared solid media, after which microbiologists could separate bacteria by dilution and trap them on the solid media. An isolated bacterium grows into a visible colony that consists of one kind of bacterium.

Isolating Bacteria

Currently three dilution methods are commonly used for isolating bacteria: the streak plate, the spread plate, and the pour plate. In the **streak plate technique,** a loop is used to streak the mixed sample many times over the surface of a solid culture medium in a Petri plate. Theoretically, the process of streaking the loop repeatedly over the agar surface causes the bacteria to fall off the loop one by one and ultimately to be distributed over the agar surface, where each cell or group

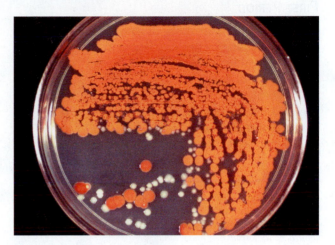

FIGURE 11.1 Bacteria growing into visible colonies on solid media. The appearance of colonies can be used to distinguish different bacteria.

of the same bacteria as a pair or chain develops into a colony. The streak plate is the most common isolation technique in use today (**FIGURE 11.1**).

Counting Bacteria

The spread plate and pour plate are quantitative techniques to determine the number of bacteria in a sample. The number of bacteria in a sample can be determined by counting the number of colony-forming units. The sample must be diluted, because even 1 milliliter of milk or water could yield 20,000 colonies—too many to count. A series of dilutions is made and cultured because the number of bacteria in the sample is not known. (The procedure for these serial dilutions is set forth in Appendix B.) In the **spread plate technique,** a small amount of a previously diluted specimen is spread over the surface of a solid medium using a spreading rod.

In the **pour plate technique,** a small amount of diluted sample is mixed with melted agar and poured into empty, sterile Petri dishes. After incubation, bacterial growth is visible as colonies in and on the agar of a pour plate. To determine the number of colony-forming units in the original sample, a plate with between 25 and 250 colonies is selected. Using fewer than 25 colonies is inaccurate because a single contaminant causes at least a 4% error. A plate with more than 250 colonies is difficult to count. The number of bacteria in the original sample is calculated using the following equation:

$$\text{Colony-forming units per ml} = \frac{\text{Number of colonies}}{\text{Dilution* × Amount plated}}$$

MATERIALS

FIRST PERIOD

Petri plate containing nutrient agar
Tubes containing melted nutrient agar (3)
99-ml water dilution blanks (3)
Sterile Petri dishes (3)
250-ml beaker
Sterile 1-ml pipettes (4)
Propipette or pipette bulb
Safety goggles

In this exercise, 1 ml of sample is put into each plate. Dilution refers to the dilution of the sample (Appendix B). For example, if 37 colonies were present on the 1:100 plate, the calculation would be as follows:

Colony-forming units per ml =

$$\frac{37}{1:100 \times 1} = 37 \times 100 = 3700 = 3.70 \times 10^3$$

SECOND PERIOD

Petri plate containing nutrient agar

CULTURE

Mixed broth culture of bacteria

TECHNIQUES REQUIRED

Compound light microscopy (Exercise 1)
Aseptic technique (Exercise 4)
Pipetting (Appendix A)
Serial dilution techniques (Appendix B)

PROCEDURE First Period

Streak Plate

1. Label the bottom of one nutrient agar plate with your name and lab section, the date, and the source of the inoculum.
2. Flame the inoculating loop to redness, allow it to cool, and aseptically obtain a loopful of the broth culture.
3. The streaking procedure may be done with the Petri plate on the workbench (**FIGURE 11.2a**) or held in your hand (**FIGURE 11.2b**).
   a. To streak a plate (**FIGURE 11.3**), lift one edge of the Petri plate cover, and streak the first sector by making as many streaks as possible without overlapping previous streaks. Do not gouge the agar while streaking the plate. Hold the loop

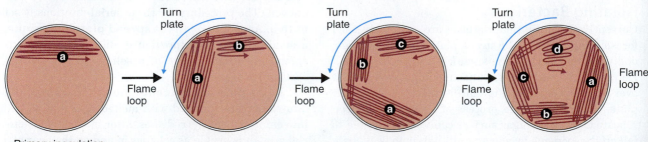

FIGURE 11.2 Inoculation of a solid medium in a Petri plate. Lift one edge of the cover while the plate (a) rests on the table or (b) is held.

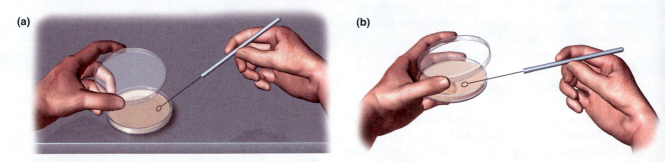
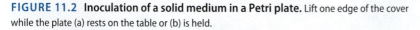

FIGURE 11.3 Streak plate technique for pure culture isolation of bacteria. The direction of streaking is indicated by the arrows. Between each section, sterilize the loop and reinoculate with a fraction of the bacteria by going back across the previous section.

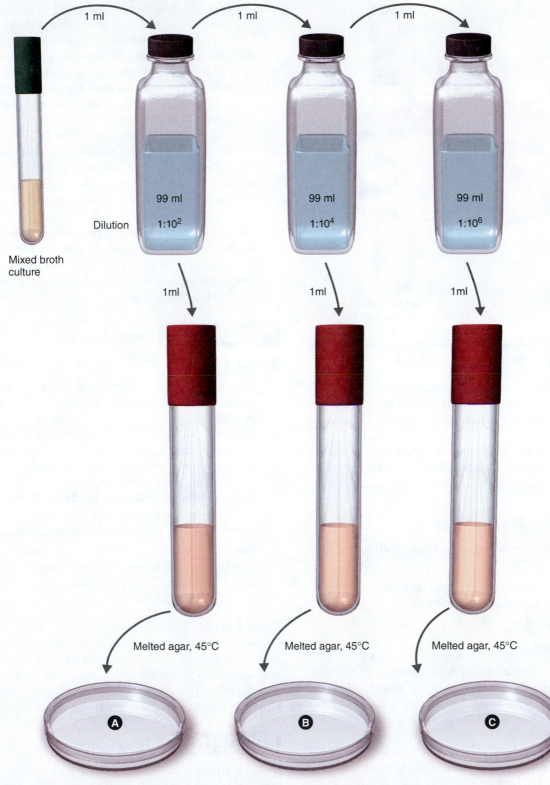

FIGURE 11.4 Pour plate technique. Bacteria are diluted through a series of dilution blanks. One ml of the dilution blank is transferred to the melted nutrient agar. The agar and bacteria are poured into sterile Petri dishes. The bacteria will form colonies where they are trapped in or on the agar.

as you would hold a pencil or paintbrush, and gently touch the surface of the agar.

b. Flame your loop and let it cool. Turn the plate so the next sector is on top. Streak through one area of the first sector, and then streak a few times away from the first sector.

c. Flame your loop, turn the plate again, and streak through one area of the second sector. Then streak the third sector.

d. Flame your loop, streak through one area of the third sector, and then streak the remaining area of the agar surface, being careful not

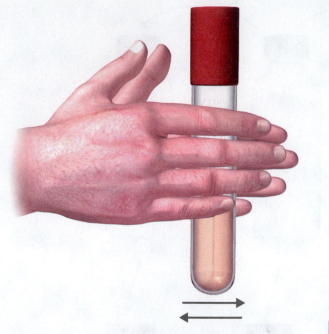

FIGURE 11.5 Mixing in agar media. Mix the inoculum in a tube of melted agar by rolling the tube between your hands.

to make additional contact with any streaks in the previous sections. Flame your loop before setting it down. Why? _____

_____

4. Incubate the plate in an inverted position at 35°C until discrete, isolated colonies develop (usually 24–48 hours). Why inverted? _____

_____

Pour Plate (Figure 11.4)

1. Label the bottoms of three empty, sterile Petri dish-

 Wear safety goggles when pipetting and handling liquid cultures.

es with your name and lab section and the date. Label one dish "A," another "B," and the third one "C." Place the labeled dishes on your workbench right-side up. Label the dilution blanks "1," "2," and "3" (**FIGURE 11.4** on page 97).

2. Fill a beaker with hot (45–50°C) water (about 3–6 cm deep), and place three tubes of melted nutrient agar in the beaker.

3. Select a mixed broth culture.

4. Serial dilutions and plate pouring:
   a. Remove a pipette, attach a bulb, and aseptically transfer 1 ml of the broth to dilution bottle number 1. Mix well.
   b. Using a different pipette, transfer 1 ml from dilution number 1 to dilution bottle number 2. Mix well. With the same pipette, transfer 1 ml

from dilution bottle number 1 to a tube of melted agar. Mix well, as shown in **FIGURE 11.5**. Aseptically pour the agar into plate A.

   c. Using a third pipette, transfer 1 ml from bottle number 2 to dilution bottle number 3. Mix well. Transfer 1 ml from bottle number 2 to a tube of melted agar, mix well, and pour the agar into Petri plate B.
   d. Using a fourth pipette, transfer 1 ml from bottle number 3 to the remaining tube of melted agar, mix well, and pour into plate C.
   e. Discard all pipettes in the disinfectant container, and place the dilution bottles in the To Be Autoclaved area.

5. Discard the tubes properly. Let the agar harden in the plates, and then incubate them at 35°C in an inverted position until growth is seen. *Suggestion:* When incubating multiple plates, use a rubber band to hold them together.

PROCEDURE Second Period

Streak Plate

1. Record the results of your streak plate. Use proper terms to describe the colonies (refer to Figure 3.3 on page 34).

2. Prepare a subculture of one colony. Sterilize your loop by flaming it. Let it cool. To subculture, touch the center of a small, isolated colony located on a streak line, and then aseptically streak a sterile nutrient agar plate (see Figure 11.3). How can you tell whether you touched only one colony and whether you have a pure culture?_____

3. Incubate the plate at 35°C until you observe good growth. Describe the growth (see Figure 3.3):

_____

_____

Pour Plate

Count the number of colonies on the pour plates. Remember that more than 250 are too numerous to count and fewer than 25 are too few to count. Use the formula on page 96 to determine the number of colony-forming units per milliliter in the original culture.

PROCEDURE Third Period

Streak Plate

Record the appearance of your subculture.

LABORATORY REPORT

Isolation of Bacteria by Dilution Techniques

PURPOSE _____

_____

_____

EXPECTED RESULTS

1. What will a streak plate with two species of bacteria look like? _____

_____

2. Where will the colonies be located in the pour plate? _____

_____ Where will they be the largest? _____

RESULTS

Streak Plate

Sketch the appearance of the streak plates.

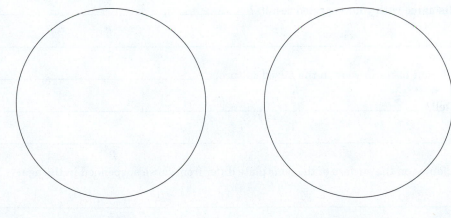

Mixed culture   Subculture

Fill in the following table using colonies from the most isolated streak areas.

| Culture | Colony Description (Describe each different-appearing colony.) | | | | |
	Estimated Diameter	Whole-Colony Appearance	Margin	Elevation	Pigment
Mixed culture					

Pour Plate

Dilution	Number of Colonies
10^2	
10^4	
10^6	

Calculate the number of colony-forming units per milliliter in the mixed culture. Which plate will you use for your calculations? _____ Show your calculations.

_____ Colony-forming units per ml

CONCLUSIONS

1. Do your results agree with your expected results? _____

_____

2. How many different bacteria were in the mixed culture? _____

How can you tell? _____

_____

3. How do the colonies on the surface of the pour plate differ from those suspended in the agar? _____

_____

_____

4. How could your streak plate technique be improved? _____

_____

_____

5. Did you get a pure culture in your subculture? _____ How do you know? _____

_____

QUESTIONS

1. What is a contaminant? _____

2. How would you determine whether a colony was a contaminant on a streak plate? _____

_____

On a pour plate? _____

_____

3. What would happen if the plates were incubated a week longer? _____

A month? _____

CRITICAL THINKING

1. Can some bacteria grow on the streak plate and not be seen if the pour plate technique is used? Explain.

2. What is a disadvantage of the streak plate technique? Of the pour plate technique?

3. Will the isolated colonies always be in the fourth sector on the streak plate?

CLINICAL APPLICATION

The following data were obtained from pour plates used to test the effectiveness of a food preservative. Two samples of cottage cheese were inoculated with bacteria; the preservative was added to one sample. After incubation, samples of the cottage cheese (control) and samples treated with the preservative (experimental) were diluted and plated on nutrient agar. Calculate the number of bacteria. Was the preservative effective?

	Dilution	Amount Plated (ml)	Number of Colonies	CFU/ml
Control	1:400	1	160	
Experimental	1:200	0.1	32	

Special Media for Isolating Bacteria

OBJECTIVES

After completing this exercise, you should be able to:

1. Differentiate selective from differential media.
2. Provide an application for enrichment and selective media.

BACKGROUND

One of the major limitations of dilution techniques used to isolate bacteria is that organisms present in limited amounts may be diluted out on plates filled with dominant bacteria. For example, if the culture to be isolated has 1 million of bacterium A and only 1 of bacterium B, bacterium B will probably be limited to the first sector in a streak plate. To help isolate organisms found in the minority, various enrichment and selective culturing methods are available that either enhance the growth of some organisms or inhibit the growth of other organisms. **Selective media** contain chemicals that prevent the growth of unwanted bacteria without inhibiting the growth of the desired organism. **Enrichment media** contain chemicals that enhance the growth of desired bacteria. Other bacteria will grow, but the growth of the desired bacteria will be increased.

Another category of media useful in identifying bacteria is **differential media.** These media contain various nutrients that allow the investigator to distinguish one bacterium from another by how they metabolize or change the media with a waste product.

Because multiple methods and multiple media exist, you must be able to match the correct procedure to the desired microbe. For example, if bacterium B is salt-tolerant, a high concentration (>5%) of salt could be added to the culture medium. Physical conditions can also be used to select for a bacterium. If bacterium B is heat-resistant, the specimen could be heated before isolation. Dyes such as phenol red, eosin, or methylene blue are sometimes included in differential media. Products of bacterial metabolism can react with these dyes to produce a color change in the medium or colonies (**FIGURE 12.1** and **FIGURE 12.2**). You will study bacterial metabolism in the exercises in Part 5. The dyes (eosin and methylene blue) in eosin methylene blue (EMB) agar are also selective. These dyes inhibit the growth of some bacteria. Three culture media will be compared in this exercise (**TABLE 12.1**).

MATERIALS

FIRST PERIOD

Petri plates containing nutrient agar (2)
Petri plates containing mannitol salt agar (2)
Petri plates containing EMB agar (2)

SECOND PERIOD

Gram-staining reagents
`BSL-1` Demonstration plate (sealed)
with *Staphylococcus aureus* on one half and *Staphylococcus epidermidis* on the other half

TABLE 12.1	MAJOR CHEMICAL COMPONENTS OF MEDIA USED IN THIS EXERCISE		
Chemical Components	Nutrient Agar	Mannitol Salt Agar	EMB Agar
Peptone	0.5%	1.0%	1.0%
NaCl	0.8%	7.5%	
Agar	1.5%	1.5%	1.5%
Mannitol		1.0%	
Lactose, sucrose			0.5% each
Eosin			0.04%
Methylene blue			0.0065%
Phenol red		0.025%	

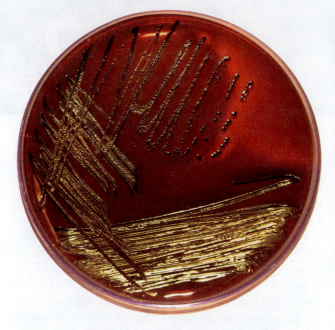

FIGURE 12.1 **EMB agar.** Colonies of *Escherichia* and *Citrobacter* develop a metallic green sheen on EMB agar.

FIGURE 12.2 *Staphylococcus aureus* on mannitol salt agar. Bacteria that produce acid from mannitol will cause the phenol red to turn yellow.

CULTURES

Escherichia coli
Alcaligenes faecalis
Staphylococcus aureus `BSL-2`
Staphylococcus epidermidis
Unknown mixed culture

TECHNIQUES REQUIRED

Compound light microscopy (Exercise 1)
Smear preparation (Exercise 5)
Gram staining (Exercise 7)
Aseptic technique (Exercise 4)
Inoculating loop technique (Exercise 4)
Streak plate procedure (Exercise 11)

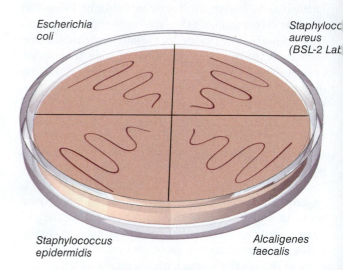

Escherichia coli

Staphyloco aureus (BSL-2 Lab

Staphylococcus epidermidis

Alcaligenes faecalis

FIGURE 12.3 **Inoculating the plates.** Divide a Petri plate into four sections by drawing lines on the bottom of the plate. Inoculate each section by streaking it with an inoculating loop.

PROCEDURE First Period

1. Using a marker, divide one nutrient agar plate into four sections by labeling the bottom. (BSL-1 labs will need to mark three sections on the plate.) Repeat to mark one mannitol salt plate and one EMB plate. Label one quadrant on each plate for each culture.
2. Streak each culture on the agar, as shown in **FIGURE 12.3**.
3. Label the remaining nutrient agar, mannitol salt, and EMB plates with the number of your unknown. Record the number of the unknown in the Laboratory Report. The unknown contains two different bacteria.
4. Streak your "unknown" onto the appropriate plates. (Use the streaking technique shown in Figure 11.3 on page 96.)

5. Incubate the plates in an inverted position at 35°C.

PROCEDURE Second Period

1. Record the results after 24 to 48 hours.
2. Perform a Gram stain on one colony of each different organism. Why don't you have to perform a Gram stain on each organism from every medium? _____

_____

_____

LABORATORY REPORT
Special Media for Isolating Bacteria

PURPOSE _____
_____
_____

HYPOTHESES

1. Bacteria growing on EMB agar will be gram- _____.

2. Which media will be selective? _____ Differential? _____

RESULTS

Organism	Nutrient Agar		Mannitol Salt Agar		EMB Agar		Gram Stain Results
	Growth: +/−	Appearance	Growth: +/−	Appearance	Growth: +/−	Appearance	
E. coli							
S. epidermidis							
A. faecalis							
S. aureus BSL-2							
Unknown Number ____							

CONCLUSIONS

1. Do your results support your hypotheses? _____

2. What two organisms are in your mixed culture? _____

   Could you identify them from the Gram stain? _____

   How did you identify them? _____

3. How did the results observed on the mannitol salt agar and EMB agar correlate to the Gram reaction of the bacteria? _____

4. Which medium is selective? _____

QUESTIONS

1. What is the purpose of peptone in the media? _____

_____

2. What is the purpose of agar in the media? _____

_____

CRITICAL THINKING

1. What ingredient makes mannitol salt selective? _____

2. Fill in the blanks in this diagram to make a key to these bacteria:
   - *Escherichia coli*
   - *Alcaligenes faecalis*
   - *Staphylococcus aureus*
   - *Staphylococcus epidermidis*

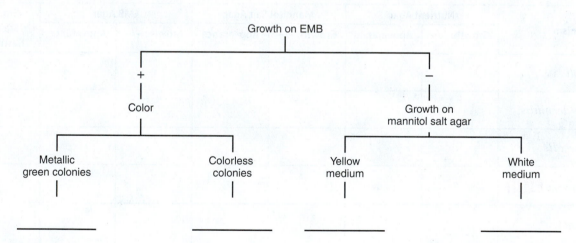

Circle or highlight the gram-positive bacteria in one color; use a different color to circle or highlight the gram-negative bacteria.

CLINICAL APPLICATION

Clinical samples are routinely inoculated onto differential and selective media. Determine whether the media listed below are selective or differential.

a. Nutrient agar that contains vancomycin is selective/differential for _____.

b. EMB agar that contains sorbitol is used to select/differentiate *E. coli* O157. Explain: _____

_____

c. Nutrient agar with tellurite is selective/differential for _____.

PART 5

Microbial Metabolism

EXERCISES

This statement from the first edition of *Bergey's Manual* summarizes the need for studying microbial metabolism:

> *The earliest writers classified the bacteria solely on their morphological characters. A more detailed classification was not possible because the biologic characters of so few of the bacteria had been determined. With the accumulation of knowledge of the biologic characters of many bacteria it was realized that it is just as incorrect to group all rods under a single genus as to group all quadruped animals under one genus.*[*]

"Biologic characters" refers to information about the metabolism of bacteria.

The chemical reactions that occur within all living organisms are referred to as **metabolism.** Metabolic processes involve **enzymes,** which are proteins that catalyze biological reactions. The majority of enzymes function inside a cell—that is, they are **endoenzymes.** Many bacteria make some enzymes, called **exoenzymes,** that are released from the cell to catalyze reactions outside the cell (see the illustration on the next page). Because many bacteria share the same colony and cell morphology, additional factors, such as metabolism, are used to identify them. Moreover, studying the metabolic activities of microbes helps us understand their role in ecology.

---

[*]Society of American Bacteriologists (D. H. Bergey, Chairperson). *Bergey's Manual of Determinative Bacteriology.* Baltimore: Williams & Wilkins, 1923, p. 1.

Exercises 13 through 17 in Part 5 introduce concepts in microbial metabolism and laboratory tests used to detect various metabolic activities. In Exercise 18, you will identify unknown bacteria on the basis of metabolic characteristics.

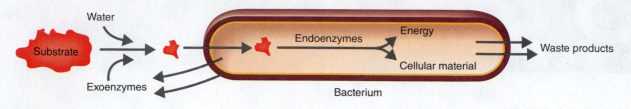

Exoenzymes breaking down large molecules outside a cell. Smaller molecules released by this reaction are taken into the cell and further degraded by endoenzymes.

CASE STUDY: Intestinal Gas

The human large intestine (colon) contains up to 10^{12} bacteria/g (dry weight) of contents, representing 500 species. The metabolism of these bacteria affects host intestinal function and health. Your diet provides nutrients not only for you but also for these bacteria. An important metabolic function of intestinal bacteria is metabolism of nondigestible carbohydrates such as starch and cellulose. Bacteria produce short-chain fatty acids including butyric acid and propionic acid, which provide energy for human colonic cells. The odors of feces can be attributed to indole and H_2S produced by bacteria from dietary proteins. Intestinal bacteria may produce highly carcinogenic nitrosamines from amino acid residues and nitrite.

Use the following choices to answer the questions.
  a. Anaerobic respiration
  b. Deamination of amino acids
  c. Decarboxylation of amino acids
  d. Desulfurization of amino acids
  e. Fermentation

Questions

  1. Butyric acid and propionic acid are produced from starch by what process?
  2. Nitrite is produced from nitrate by what process?
  3. Amino acid residues are produced by removing CO_2 by what process?
  4. How do bacteria produce indole from tryptophan?
  5. How do bacteria produce H_2S from cysteine and methionine?

Carbohydrate Catabolism

The men who try to do something and fail are INFINITELY better than those who try nothing and succeed. – LLOYD JONES

OBJECTIVES

After completing this exercise, you should be able to:

1. Define the following terms: *carbohydrate, catabolism,* and *hydrolytic enzymes.*
2. Differentiate oxidative from fermentative catabolism.
3. Perform and interpret starch hydrolysis and OF tests.

BACKGROUND

Chemical reactions that release energy from the decomposition of complex organic molecules are referred to as **catabolism.** Most bacteria catabolize carbohydrates for carbon and energy. **Carbohydrates** are organic molecules that contain carbon, hydrogen, and oxygen in the ratio $(CH_2O)_n$. Carbohydrates can be classified based on size: monosaccharides, oligosaccharides, and polysaccharides. **Monosaccharides** are simple sugars containing from three to seven carbon atoms. **Oligosaccharides** are composed of two to about 20 monosaccharide molecules; *disaccharides* are the most common oligosaccharides. **Polysaccharides** consist of 20 or more monosaccharide molecules.

Starch Hydrolysis

Exoenzymes are mainly **hydrolytic enzymes** that leave the cell and break down, by the addition of water, large substrates into smaller components, which can then be transported into the cell. Amylase hydrolyzes the polysaccharide starch into smaller carbohydrates.

Glucose, a monosaccharide, can be released by hydrolysis (**FIGURE 13.1**). In the laboratory, the presence of an exoenzyme is determined by looking for a change in the substrate outside a bacterial colony.

OF Medium

Glucose can enter a cell and be catabolized; some bacteria, using endoenzymes, catabolize glucose oxidatively, producing carbon dioxide and water. **Oxidative catabolism** requires the presence of molecular oxygen (O_2). Most bacteria, however, can ferment glucose without using oxygen. **Fermentative catabolism** does not require oxygen but may occur in its presence. The metabolic end-products of fermentation are small organic molecules, usually organic acids. Some bacteria produce gases from the fermentation of carbohydrates.

Whether an organism is oxidative or fermentative can be determined by using Rudolph Hugh and Einar Leifson's OF basal media with the desired carbohydrate added. **OF medium** is a nutrient semisolid agar deep containing a *high* concentration of carbohydrate and a *low* concentration of peptone (**FIGURE 13.2**). The peptone will support the growth of bacteria that don't use the carbohydrate. Two tubes are used: one open to the air and one sealed to keep air out. OF medium contains the indicator bromthymol blue, which turns yellow in the presence of acids, indicating catabolism of the carbohydrate. Alkaline conditions, caused by the use of peptone and not of carbohydrate, are indicated by a dark blue color due to ammonia production. If the carbohydrate is metabolized in both tubes, fermentation has occurred. An organism that can use the carbohydrate

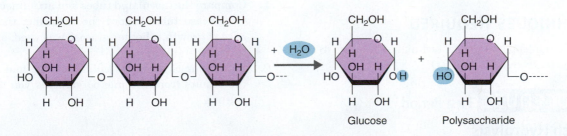

FIGURE 13.1 Starch hydrolysis. A molecule of water is used when starch is hydrolyzed.

109

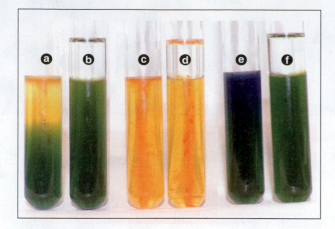

FIGURE 13.2 **Reactions in OF-glucose medium.** The species in tubes **(a)** and **(b)** is an oxidizer. The organism in tubes **(c)** and **(d)** is a fermenter. The culture in tubes **(e)** and **(f)** does not use glucose.

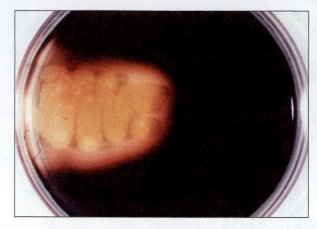

FIGURE 13.3 **Starch hydrolysis test.** After incubation, add iodine to the plate to detect the presence of starch. The clearing around the bacteria indicates that the starch was hydrolyzed.

only under aerobic conditions will produce acid in the open tube only. Acids are produced as intermediates in respiration, and the indicator will turn yellow in the top of the open tube. This organism is called "oxidative" in an OF test.

Carbohydrate catabolism will be demonstrated in this exercise. These differential tests are important in identifying bacteria.

MATERIALS

FIRST PERIOD

Petri plate containing nutrient starch agar
OF-glucose deeps (2)
Mineral oil

SECOND PERIOD

Gram's iodine
OF-glucose deep

CULTURES (AS ASSIGNED)

Bacillus subtilis
Escherichia coli
Alcaligenes faecalis
Micrococcus luteus
Pseudomonas aeruginosa **BSL-2**

TECHNIQUES REQUIRED

Inoculating loop and needle technique (Exercise 4)
Aseptic technique (Exercise 4)

PROCEDURE First Period

Starch Hydrolysis

1. With a marker, divide the starch agar into three sectors by labeling the bottom of the plate.

2. Using a loop, streak a single line of *Bacillus, Escherichia,* and *Pseudomonas* **BSL-2** in the appropriate sector.

3. Incubate the plate, inverted, at 35°C for 24 hours. After growth occurs, the plate may be refrigerated at 5°C until the next lab period.

OF-Glucose

1. Using an inoculating needle, inoculate two tubes of OF-glucose media with the assigned bacterial culture (*Escherichia, Micrococcus, Alcaligenes,* or *Pseudomonas* **BSL-2**).

2. Pour about 5 mm of mineral oil over the medium in one of the tubes. Replace the cap.

3. Incubate both tubes at 35°C until the next lab period.

PROCEDURE Second Period

Starch Hydrolysis

Record any bacterial growth, and then flood the plate with Gram's iodine (**FIGURE 13.3**). Areas of starch hydrolysis will appear clear, while unchanged starch will form a dark blue complex with the iodine. Record your results.

OF-Glucose

1. Compare the inoculated tubes and an uninoculated OF-glucose tube. Record the following: the presence of growth, whether glucose was used, and the type of metabolism.

2. From your classmates' tubes, observe and record the results from the microorganisms you did not culture.

LABORATORY REPORT
Carbohydrate Catabolism

PURPOSE _____

_____

_____

HYPOTHESES

1. All bacteria will use glucose. Agree/disagree

2. All bacteria will hydrolyze starch. Agree/disagree

RESULTS

Record your results in the following tables.

Starch Hydrolysis

Organism	Growth	Color of Medium around Colonies after Addition of Iodine	Starch Hydrolysis: (+) = yes (−) = no
Bacillus subtilis			
Escherichia coli			
Pseudomonas aeruginosa BSL-2			

OF-Glucose

Color of uninoculated medium: _____

Organism	Growth		Color		Fermenter (F), Oxidative (O), Neither (−)
	Aerobic Tube	Anaerobic Tube	Aerobic Tube	Anaerobic Tube	
Micrococcus luteus					
Alcaligenes faecalis					
Escherichia coli					
Pseudomonas aeruginosa BSL-2					

CONCLUSIONS

1. Do you accept your hypotheses? _____

   _____

2. Which organism(s) gave a positive test for starch hydrolysis? _____

   _____

3. Which organism(s) used peptone and not glucose? _____

   _____

QUESTIONS

1. What would be found in the clear area that would not be found in the blue area of a starch agar plate after the

   addition of iodine? _____

2. How can you tell that amylase is an exoenzyme and not an endoenzyme? _____

   _____

3. How can you tell from OF-glucose medium whether an organism uses glucose aerobically? _____

   _____

   Ferments glucose? _____

   Doesn't use glucose? _____

4. If an organism grows in the OF-glucose medium that is exposed to air, is the organism oxidative or fermentative?

   Explain. _____

   _____

5. Why is it important to first determine whether growth occurred in a differential medium, such as starch agar,

   before examining the plate for starch hydrolysis? _____

   _____

CRITICAL THINKING

1. Aerobic organisms degrade glucose, producing carbon dioxide and water. What acid turns the indicator yellow?

2. How can organisms that don't use starch grow on a starch agar plate?

3. If iodine were not available, how would you determine whether starch hydrolysis had occurred?

4. Locate the H^+ and OH^- ions from the water molecule that was split (hydrolyzed) in this reaction, showing
   esculin hydrolysis.

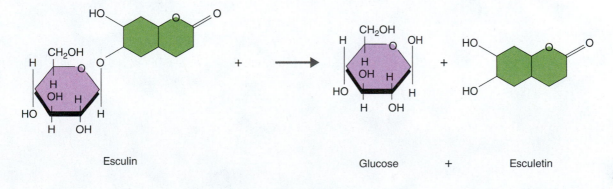

Esculin   Glucose   +   Esculetin

CLINICAL APPLICATION

You have isolated a bacterium from a patient's infected wound. The bacterium is oxidative and does not hydrolyze
starch. Using your results from this exercise, identify the bacterium.

Fermentation

OBJECTIVES

After completing this exercise, you should be able to:

1. Define *fermentation*.
2. Perform and interpret carbohydrate fermentation tests.
3. Perform and interpret the MR and V–P tests.
4. Perform and interpret the citrate test.

BACKGROUND

Once a bacterium has been determined to be fermentative by the OF test, further tests can determine which carbohydrates, in addition to glucose, are fermented; in some instances, the end-products can also be determined. Many carbohydrates—including monosaccharides such as glucose, disaccharides such as sucrose, and polysaccharides such as starch—can be fermented. Many bacteria produce organic acids (for example, lactic acid) and hydrogen and carbon dioxide gases from carbohydrate fermentation (**FIGURE 14.1**).

Fermentation Tubes

A **fermentation tube** is used to detect acid and gas production from carbohydrates. The fermentation medium contains peptone, an acid–base indicator (phenol red), an inverted tube to trap gas, and 0.5–1.0% of the desired carbohydrate. In **FIGURE 14.2**, the phenol red indicator is red (neutral) in an uninoculated fermentation tube; fermentation that results in acid production will turn the indicator yellow (pH of 6.8 or below). Some bacteria produce acid and gas. When gas is produced during fermentation, some will be trapped in the inverted, or Durham, tube. Fermentation occurs with or without oxygen present; however, during prolonged incubation periods (greater than 24 hours), many bacteria will begin growing oxidatively on the peptone after exhausting the carbohydrate supplied, causing neutralization of the indicator and turning it red because of ammonia production.

MRVP Broth

Fermentation processes can produce a variety of end-products, depending on the substrate, the incubation, and the organism. In some instances, large amounts of acid may be produced, and in others a majority of neutral products may result (**FIGURE 14.3a**). The **MRVP broth** is used to distinguish organisms that produce large amounts of acid from glucose and organisms that produce the neutral product *acetoin* (**FIGURE 14.4**). MRVP medium is a glucose-supplemented nutrient broth used for the **methyl red (MR) test** and the **Voges–Proskauer (V–P) test.** If an organism produces a large amount of organic acid from glucose, the medium will remain red when methyl red is added in a positive MR test, indicating that the pH is below 4.4. Methyl red is orange-red between pH 4.4 and 6.0. If neutral products are produced, methyl red will turn yellow, indicating a pH above 6.0. The production of acetoin is detected by the addition of potassium hydroxide and α-naphthol. If acetoin is present, the upper part of the medium will turn red; a negative V–P test will turn the medium light brown. The chemical process is shown in **FIGURE 14.3b**.

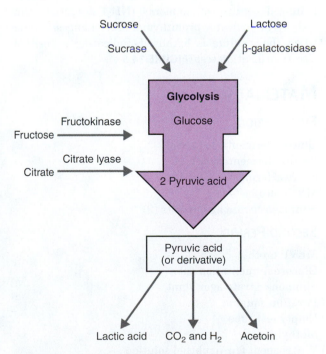

FIGURE 14.1 Fermentation. Bacteria are often identified by their enzymes. These enzymes can be detected by observing a bacterium's ability to grow on specific compounds. For example, *E. coli* and *Salmonella* are distinguished because *E. coli* can ferment lactose, and typical *Salmonella* cannot. Fermentation can occur whether O_2 is present or not present.

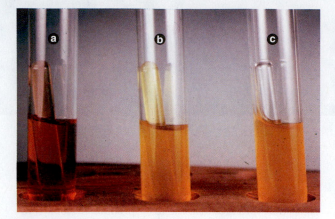

FIGURE 14.2 Carbohydrate fermentation tube. (a) The phenol red indicator is red in a neutral or alkaline solution. **(b)** Phenol red turns yellow in the presence of acids. **(c)** Gases are trapped in the inverted tube while the indicator shows the production of acid.

Citrate Agar

The ability of some bacteria to ferment citrate can be useful for identifying bacteria. When citric acid or sodium citrate is in solution, it loses a proton or Na^+ to form a citrate ion. Bacteria with the enzyme *citrate lyase* can break down citrate to form pyruvate, which can be reduced in fermentation. Simmons citrate agar (**TABLE 14.1**) is used to determine citrate use. When bacteria use citrate and ammonium, the medium is alkalized because of ammonia (NH_3) produced from NH_4^+. The indicator bromthymol blue changes to blue when the medium is alkalized, indicating a positive citrate utilization test (**FIGURE 14.5**).

MATERIALS

FIRST PERIOD

Glucose fermentation tube
Lactose fermentation tube
Sucrose fermentation tube
MRVP broths (4)
Simmons citrate agar slants (2)

SECOND PERIOD

MRVP broth
Glucose fermentation tube
Simmons citrate agar slant
Parafilm squares
Empty test tube
Methyl red
V–P reagent I, α-naphthol solution
V–P reagent II, potassium hydroxide (40%)

CULTURES (AS ASSIGNED)

Escherichia coli
Enterobacter aerogenes

TABLE 14.1	SIMMONS CITRATE AGAR
Ingredient	**Amount**
Sodium citrate	0.2%
Sodium chloride	0.5%
Monoammonium phosphate	0.1%
Dipotassium phosphate	0.1%
Magnesium sulfate	0.02%
Agar	1.5%
Bromthymol blue	0.0008%

Alcaligenes faecalis
Proteus vulgaris **BSL-2**

TECHNIQUES REQUIRED

Inoculating loop technique (Exercise 4)
Aseptic technique (Exercise 4)

PROCEDURE First Period

Fermentation Tubes

1. Use a loop to inoculate the fermentation tubes with the assigned bacterial culture.
2. Incubate the tubes at 35°C. Examine them at 24 and 48 hours for growth, acid, and gas production. Compare them to an uninoculated fermentation tube. Why is it important to record the presence of growth? _____

MRVP Tests

1. Using a loop, inoculate two MRVP tubes with *Escherichia* and two with *Enterobacter*.
2. Incubate the tubes at 35°C for 48 hours or longer. Why is time of incubation important (see Figure 14.4)? _____

Citrate Test

1. Using a loop, inoculate one citrate slant with *Escherichia coli* and the other slant with *Enterobacter aerogenes*.
2. Incubate the tubes at 35°C until the next lab period.

PROCEDURE Second Period

Fermentation Tubes

1. Record your results for the bacterial cultures, and obtain results for the other species from your classmates (see Figure 14.2).
2. Observe and record results for the BSL-2 level culture.

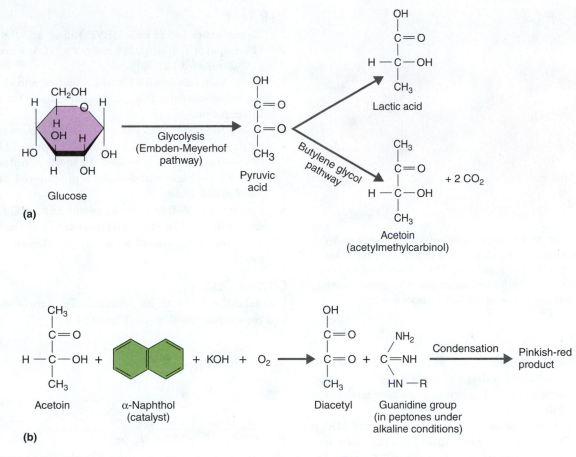

(a)

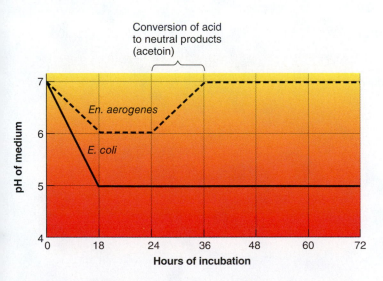

(b)

FIGURE 14.3 MRVP test. (a) Organic acids, such as lactic acid, or neutral products, such as acetoin, may result from fermentation. **(b)** Potassium hydroxide (KOH) and α-naphthol are used to detect acetoin.

FIGURE 14.4 Methyl red test. Some bacteria produce large amounts of acid from pyruvic acid (pH < 5.5). Other bacteria make neutral products such as acetoin (pH > 6.0).

FIGURE 14.5 Citrate test. Use of citric acid as the sole carbon source in Simmons citrate agar causes the indicator to turn blue **(b).** Tube **(a)** is citrate-negative.

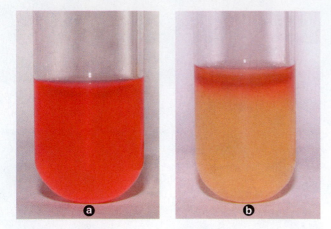

FIGURE 14.6 **MRVP broth.** The MR and V–P tests must be done in separate tubes of MRVP broth inoculated with a bacterial culture and incubated 24 to 48 hours. **(a)** Red, after addition of methyl red, indicates a positive methyl red test. **(b)** After the addition of V–P reagents I and II, a positive V–P test develops a red color when exposed to oxygen.

MR Test

1. To one tube of each MRVP set (*Escherichia* and *Enterobacter*), add 5 drops of methyl red. Record the resulting color. Red indicates a positive methyl red test (**FIGURE 14.6a**).
2. Pour half of the contents of an uninoculated MRVP broth into an empty test tube. Save the other half for the V–P control (V–P test step 4). Perform the MR test on this for a control. Record your results.

V–P Test

1. To the other set of two MRVP tubes, add 0.6 ml (12 drops) of V–P reagent I and 0.2 ml (2 or 3 drops) of V–P reagent II.
2. Cover each tube with a Parafilm square, and shake the tubes carefully. Discard the Parafilm in the disinfectant jar.
3. Leave the caps off to expose the media to oxygen so that the acetoin will oxidize (see Figure 14.3b). Allow the tubes to stand for 15 to 30 minutes. A positive V–P test will develop a pinkish-red color. See **FIGURE 14.6b**.
4. Perform the V–P test on the remaining half of the uninoculated MRVP broth (from step 2 of the MR test, described above) as a control. Record your results.

Citrate Test

Compare the tubes to an uninoculated citrate slant, and record your results (see Figure 14.5).

LABORATORY REPORT
Fermentation

PURPOSE _____

_____

_____

HYPOTHESES

1. All bacteria will ferment the three sugars. Agree/disagree

2. A bacterium that doesn't ferment glucose will not ferment lactose. Agree/disagree

RESULTS

Fermentation Tubes

Color of uninoculated medium: _____

Organism		Carbohydrate											
		Glucose				Lactose				Sucrose			
		Growth	Color	Acid	Gas	Growth	Color	Acid	Gas	Growth	Color	Acid	Gas
Escherichia coli	24 hr												
	48 hr												
Enterobacter aerogenes	24 hr												
	48 hr												
Alcaligenes faecalis	24 hr												
	48 hr												
Proteus vulgaris BSL-2	24 hr												
	48 hr												

MRVP Tests

Organism	Growth	MR		V–P	
		Color	+ or −	Color	+ or −
Escherichia coli					
Enterobacter aerogenes					
Controls					

Citrate Test

Organism	Growth	Color	+ or −
Escherichia coli			
Enterobacter aerogenes			
Controls			

CONCLUSIONS

1. Do you accept your hypotheses? _____

   _____

2. Do all bacteria ferment lactose? _____

3. *E. coli* and *E. aerogenes* are small gram-negative rods. How can you differentiate them? _____

   _____

QUESTIONS

1. Why are fermentation tubes evaluated at 24 and 48 hours? _____

    _____

    What would happen if an organism used up all the carbohydrate in a fermentation tube? _____

    _____

    What would the organism use for energy? _____

    _____

    What color would the indicator be then? _____

    _____

2. If an organism metabolizes glucose aerobically, what result will occur in the fermentation tubes? _____

    _____

    _____

3. Which of these media is selective? _____ Why is it selective? _____

    _____

CRITICAL THINKING

1. Could an organism be a fermenter and also be both MR and V–P negative? Explain.

2. How could you determine which of the following carbohydrates a bacterium fermented: mannitol, sorbitol, adonitol, or arabinose?

3. If a bacterium cannot ferment glucose, why not test its ability to ferment other carbohydrates?

4. If a bacterium cannot use glucose, what result will occur in a fermentation tube?

5. If a bacterium cannot use citrate, what result will occur in citrate agar?

CLINICAL APPLICATION

You put an opened bottle of orange drink in the refrigerator last week. Now the drink has bubbles and an off-taste. Explain these changes.

Protein Catabolism, Part 1

OBJECTIVES

After completing this exercise, you should be able to:

1. Determine gelatin hydrolysis.
2. Test for the presence of urease.
3. Determine protein hydrolysis in litmus milk.

BACKGROUND

Proteins are large organic molecules. Cellular enzymes and many structures, for example, are proteins. The subunits that make up a protein are called **amino acids** (**FIGURE 15.1**). Amino acids consist of carbon, hydrogen, oxygen, nitrogen, and, sometimes, sulfur. Amino acids bond together by **peptide bonds** (**FIGURE 15.2**), forming a small chain (a **peptide**) or a larger molecule (a **polypeptide**).

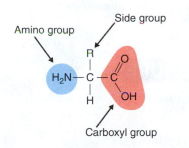

FIGURE 15.1 General structural formula for an amino acid. The letter *R* stands for any of a number of groups of atoms. Different amino acids have different R groups.

Bacteria can hydrolyze the peptides or polypeptides to release amino acids (see Figure 15.2). They use the amino acids as carbon and energy sources when carbohydrates are not available. However, amino acids are used primarily in anabolic reactions.

Large protein molecules, such as gelatin, are hydrolyzed by exoenzymes, and the smaller products of hydrolysis are transported into the cell. Hydrolysis of gelatin can be demonstrated by growing bacteria in nutrient gelatin. **Nutrient gelatin** dissolves in warm water (50°C), solidifies (**gels**) when cooled below 25°C, and liquefies (**sols**) when heated to about 25°C. When the exoenzyme *gelatinase* hydrolyzes gelatin, it liquefies and does not solidify even when cooled below 20°C (**FIGURE 15.3**).

Bacteria can hydrolyze the protein in milk called **casein.** Casein hydrolysis can be detected in litmus milk. Litmus milk consists of skim milk and the indicator litmus. The medium is opaque due to casein in colloidal suspension and the litmus is blue. After **peptonization** (hydrolysis of the milk proteins), the medium becomes clear due to hydrolysis of casein to soluble amino acids and peptide fragments. Litmus milk is also used to detect **lactose fermentation,** because litmus turns pink in the presence of acid. Excessive amounts of acid will cause **coagulation** (curd formation) of the milk. **Catabolism of amino acids** will result in an alkaline (purple) reaction. Additionally, some bacteria can **reduce litmus** (Exercise 17), causing the litmus indicator to turn white in the bottom of the tube.

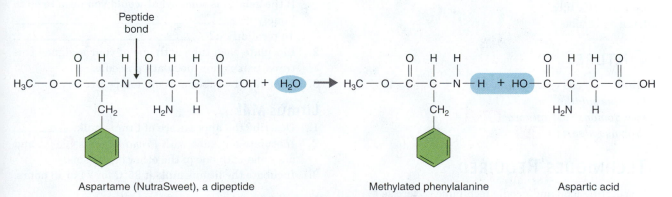

FIGURE 15.2 Hydrolysis of peptides. A molecule of water is used when the peptide is hydrolyzed.

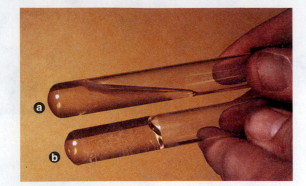

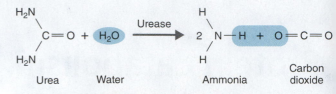

FIGURE 15.4 **Urea hydrolysis.**

FIGURE 15.3 **Gelatin hydrolysis.** After hydrolysis (**a**), gelatin remains liquid. Tube (**b**) is unhydrolyzed gelatin.

Urea is a waste product of protein digestion in most vertebrates and is excreted in the urine. Presence of the enzyme **urease,** which liberates ammonia from urea (**FIGURE 15.4**), is a useful diagnostic test for identifying bacteria. **Urea agar** contains peptone, glucose, urea, and phenol red. The pH of the prepared medium is 6.8 (phenol red turns yellow). During incubation, bacteria possessing urease produce ammonia, which raises the pH of the medium, turning the indicator fuchsia (hot pink) at pH 8.4 (**FIGURE 15.5**).

We will investigate bacterial action on nutrient gelatin, urea agar, and litmus milk in this exercise. (We will investigate amino acid metabolism in Exercise 16, Protein Catabolism, Part 2.)

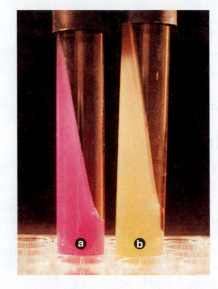

FIGURE 15.5 **Urease production. (a)** Hydrolysis of urea produces ammonia, which turns the indicator fuchsia. Tube (**b**) is the negative control.

MATERIALS

FIRST PERIOD

Tubes containing nutrient gelatin (2)
Tubes containing litmus milk (2)
Tubes containing urea agar (2)

SECOND PERIOD

Nutrient gelatin tube
Litmus milk tube
Urea agar tube

CULTURES

Staphylococcus epidermidis
Bacillus subtilis
Pseudomonas aeruginosa `BSL-2`
Proteus vulgaris `BSL-2`

TECHNIQUES REQUIRED

Inoculating loop and needle technique (Exercise 4)
Aseptic technique (Exercise 4)

PROCEDURE First Period

Label one tube of each medium with your assigned culture.

Gelatin Hydrolysis

1. Examine the nutrient gelatin: Is it solid or liquid?
   _____
   What is the temperature of the laboratory? _____
   If the gelatin is solid, what would you need to do to liquefy it? _____
   To resolidify it? _____
2. Inoculate each tube with one of your assigned cultures, using your inoculating needle.
3. Incubate the tubes at room temperature.

Litmus Milk

1. Describe the appearance of litmus milk. _____
2. Inoculate one tube with *Pseudomomas* `BSL-2` and one tube with one of the other organisms.
3. Incubate the litmus milk at 35°C for 24 to 48 hours.

Urease Test

1. Inoculate each urea agar slant with one of your assigned cultures, using your inoculating loop.
2. Incubate the tubes for 24 to 48 hours at 35°C.

PROCEDURE Second Period

Gelatin Hydrolysis

1. Record your observations at 2 to 4 days and again at 4 to 7 days. Do not agitate the tube when the gelatin is liquid. Why? _____

   _____

   If the gelatin has liquefied, place the tube in a beaker of crushed ice for a few minutes. Is the gelatin still liquefied? _____

2. Record your results. Indicate liquefaction or hydrolysis by (+) (see Figure 15.3).

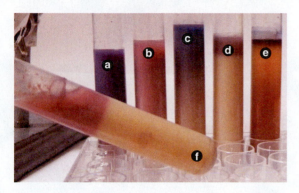

FIGURE 15.6 Litmus milk. (a) Uninoculated control; **(b)** Acid;
(c) Alkaline; **(d)** Litmus reduced; **(e)** Peptomized; **(f)** Coagulated

Litmus Milk

Record the results. Litmus is pink under acidic conditions, purple in alkaline conditions, and white when reduced. Record any pH change, peptonization, coagulation, and litmus reduction. See **FIGURE 15.6**.

Urease Test

Record the results of your urea agar: (+) for the presence of urease (phenol red turned to fuchsia) and (−) for no urease (see Figure 15.5). What color is phenol red at pH 6.8 or below? _____
At pH 8.4 or above? _____

LABORATORY REPORT
Protein Catabolism, Part 1

PURPOSE _____

_____

_____

HYPOTHESES

1. After hydrolysis, the pH of urea agar is _____.

2. After hydrolysis, gelatin will _____ at 22°C.

RESULTS

Fill in these tables.

Control	Appearance of Uninoculated Medium
Gelatin	
Litmus milk	
Urea agar	

Test	Results							
	Pseudomonas aeruginosa BSL-2		Proteus vulgaris BSL-2		Staphylococcus epidermidis		Bacillus subtilis	
	2–4 days	4–7 days	2–4 days	4–7 days	2–4 days	4–7 days	2–4 days	4–7 days
Gelatin hydrolysis Growth								
Hydrolysis								
Litmus milk pH								
Other changes								
Urease test Growth								
Color								
Hydrolysis								

CONCLUSIONS

1. Do you accept your hypotheses? _____

2. Do all bacteria produce urease? _____ Gelatinase? _____ How do you know? _____

_____

QUESTIONS

1. Nutrient gelatin can be incubated at 35°C. What would you have to do to determine hydrolysis after incubation

   at 35°C? _____

2. Phenol red is the indicator in the urease test and in fermentation tubes (Exercise 14). Complete the table.

	Color of Uninoculated Medium or Negative Results	+ Results	
		Color	pH
Fermentation tubes			
Urease test			

CRITICAL THINKING

1. Why is agar used as a solidifying agent in culture media instead of gelatin?

2. What would you find in the liquid of hydrolyzed gelatin?

3. A person changing a baby's wet diaper smells ammonia. Why?

CLINICAL APPLICATION

Helicobacter pylori bacteria grow in the human stomach. These bacteria produce a large amount of urease. Of what value is this urease to *Helicobacter*?

Protein Catabolism, Part 2

OBJECTIVES

After completing this exercise, you should be able to:

1. Define *deamination* and *decarboxylation*.
2. Explain the derivation of H_2S in decomposition.
3. Perform and interpret an indole test.
4. Interpret results from MIO medium.

BACKGROUND

Once amino acids are taken into a bacterial cell, various metabolic processes can occur using endoenzymes. Before an amino acid can be used as a carbon and energy source, the amino group must be removed. The removal of an amino group is called **deamination.** The amino group is converted to ammonia, which can be excreted from the cell. Deamination results in the formation of an organic acid. Deamination of the amino acid phenylalanine can be detected by forming a colored ferric ion complex with the resulting acid (**FIGURE 16.1**). Deamination can also be ascertained by testing for the presence of ammonia using *Nessler's reagent,* which turns deep yellow in the presence of ammonia.

Various amino acids may be decarboxylated, yielding products that can be used for synthesis of other cellular components. **Decarboxylation** is the removal of carbon dioxide from an amino acid. The presence of a specific decarboxylase enzyme results in the breakdown of the amino acid with the formation of the corresponding amine, liberation of carbon dioxide, and a shift in pH to alkaline. Media for decarboxylase reactions consist of glucose, nutrient broth, a pH indicator, and the desired amino acid. In **FIGURE 16.2**, bromcresol purple is used as a pH indicator, and a positive decar-

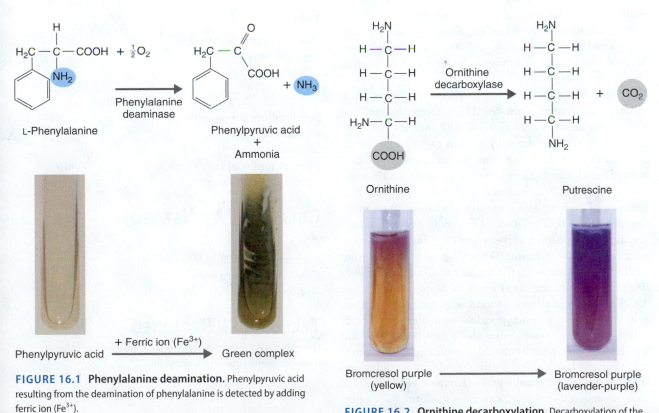

FIGURE 16.1 **Phenylalanine deamination.** Phenylpyruvic acid resulting from the deamination of phenylalanine is detected by adding ferric ion (Fe^{3+}).

L-Phenylalanine

Phenylpyruvic acid + Ammonia

Phenylpyruvic acid + Ferric ion (Fe^{3+}) → Green complex

FIGURE 16.2 **Ornithine decarboxylation.** Decarboxylation of the amino acid ornithine causes a change in the bromcresol purple indicator.

Ornithine → Ornithine decarboxylase → Putrescine + CO_2

Bromcresol purple (yellow) → Bromcresol purple (lavender-purple)

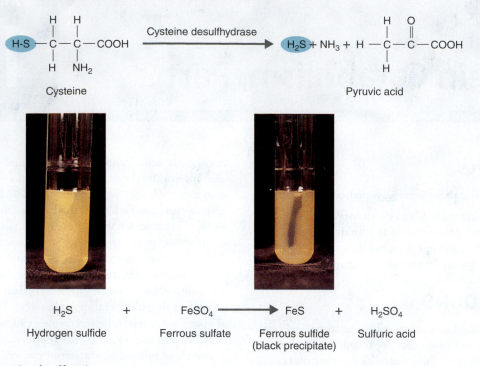

$$H_2S \quad + \quad FeSO_4 \longrightarrow FeS \quad + \quad H_2SO_4$$

Hydrogen sulfide   Ferrous sulfate   Ferrous sulfide   Sulfuric acid
(black precipitate)

FIGURE 16.3 Cysteine desulfuration. The release of H_2S from the amino acid cysteine is indicated by the formation of ferrous sulfide.

boxylase test yielding excess amines is indicated by purple. (Bromcresol purple is yellow in acidic conditions.) The names given to some of the amines, such as putrescine, indicate how foul smelling they are. Cadaverine was the name given to the foul-smelling diamine derived from decarboxylation of lysine in decomposing bodies on a battlefield.

Some bacteria have *desulfhydrase* enzymes that remove **hydrogen sulfide (H_2S)** from the sulfur-containing amino acids: cysteine and methionine. The production of hydrogen sulfide from cysteine is shown in **FIGURE 16.3**. H_2S can also be produced from the reduction of inorganic compounds, such as thiosulfate ($S_2O_3^{2-}$). H_2S is commonly called rotten-egg gas because of the copious amounts liberated when eggs are decomposed. To detect H_2S production, a heavy-metal salt containing ferrous ion (Fe^{2+}) is added to a nutrient culture medium. When H_2S is produced, the sulfide (S^{2-}) reacts with the metal salt to produce a visible black precipitate (see Figure 16.3).

The ability of some bacteria to deaminate the amino acid tryptophan, producing indole or a blue compound called indigo, is a useful diagnostic tool (**FIGURE 16.4**). The **indole test** is performed by inoculating a bacterium into tryptone broth and detecting indole via the addition of dimethylaminobenzaldehyde (**Kovacs reagent**). In this exercise, a differential screening medium, MIO agar, will be used. **MIO** is a single culture medium in which *m*otility, *i*ndole production, and *o*rnithine decarboxylase activity can be determined (**FIGURE 16.5**).

MATERIALS

FIRST PERIOD

Phenylalanine slants (2)
Peptone iron deeps (2)
MIO deeps (2)

SECOND PERIOD

Ferric chloride reagent
Kovacs reagent
MIO deep
Phenylalanine slant
Peptone iron deep

CULTURES (AS NEEDED)

Escherichia coli
Proteus vulgaris `BSL-2`
Enterobacter aerogenes
Bacillus subtilis

TECHNIQUES REQUIRED

Inoculating loop and needle technique (Exercise 4)
Aseptic technique (Exercise 4)

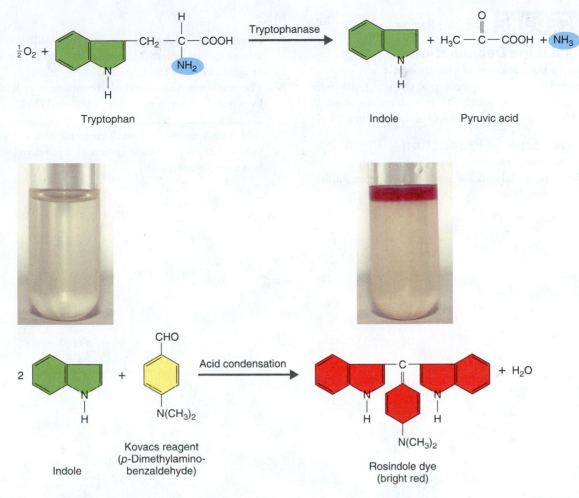

FIGURE 16.4 Indole production. Many bacteria produce indole from the amino acid tryptophan. Indole is indicated by the appearance of a red color after the addition of Kovacs reagent.

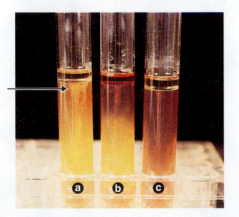

FIGURE 16.5 MIO medium. The culture in tube **(a)** is motile (at arrow). In tube **(b),** the red color after the addition of Kovacs reagent indicates indole production, and the removal of CO_2 from ornithine turns the indicator purple in tube **(c).**

PROCEDURE First Period

Phenylalanine Deamination

1. Streak one phenylalanine slant heavily with *Escherichia coli*. Streak one phenylalanine slant heavily with *Proteus vulgaris* `BSL-2`.
2. Incubate the tubes for 1 to 2 days at 35°C.

Hydrogen Sulfide Production

1. Using your inoculating needle, stab one peptone iron deep with *Escherichia coli*. Streak one phenylalanine slant heavily with *Proteus vulgaris* `BSL-2`.
2. Incubate the tubes at 35°C for up to 7 days.

MIO

1. With your inoculating needle, stab one MIO deep with *Enterobacter aerogenes* and the other with *Proteus vulgaris* `BSL-2` or *Bacillus subtilis*.
2. Incubate the tubes for 24 hours at 35°C.

PROCEDURE Second Period

Phenylalanine Deamination

Observe for the presence of growth. Add 4 or 5 drops of ferric chloride reagent to the top of the slant, allowing the reagent to run through the growth on the slant. A positive test gives a dark green color (see Figure 16.1).

Hydrogen Sulfide Production

Observe for the presence of growth. Blackening in the butt of the tube indicates a positive hydrogen sulfide test (see Figure 16.3).

MIO

1. Observe for the presence of growth. Motility is visible as diffusion out from the stab inoculation line or by clouding of the medium.

2. The ornithine decarboxylation reaction is indicated by a purple color; a negative is yellow. Why? _____
_____

3. Add 4 or 5 drops of Kovacs reagent. Mix the tube gently. A cherry red color indicates a positive indole test (see Figure 16.5b).

LABORATORY REPORT
Protein Catabolism, Part 2

PURPOSE _____

_____

_____

HYPOTHESES

1. Phenylalanine deaminase is indicated by a(n) _____ color in _____ .

2. Cysteine desulfhydrase is indicated by a(n) _____ color in _____ .

3. Ornithine decarboxylase is indicated by a(n) _____ color in _____ .

RESULTS

Fill in the following table.

Organism	Test						MIO					
	Phenylalanine Deaminase			Hydrogen Sulfide			Motility		Indole		Ornithine Decarboxylase	
	Original color: _____			Original color: _____			Original color: _____					
	Growth	Color	Reaction (+ or −)	Growth	Color	Reaction (+ or −)	Growth	Motility (+ or −)	Color	Reaction (+ or −)	Color	Reaction (+ or −)
Escherichia coli							Not tested		Not tested		Not tested	
Proteus vulgaris BSL-2												
Enterobacter aerogenes				Not tested								
Bacillus subtilis	Not tested			Not tested								

CONCLUSIONS

1. Do you accept your hypotheses? _____

_____

2. Biochemical tests are often used to identify bacteria. Complete the following identification scheme using your data.

What test from this exercise will separate *P. vulgaris* from the other bacteria?

What test from this exercise will separate these two species?

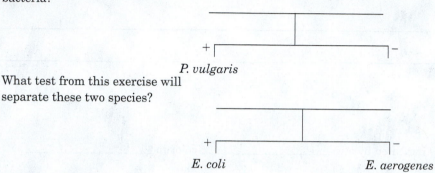

P. vulgaris

E. coli   *E. aerogenes*

QUESTIONS

1. When canned foods spoil, what causes the blackening of the cans? _____

_____

_____

_____

2. Show how lysine could be decarboxylated to give the end-products indicated.

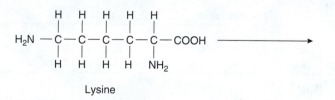

Lysine   Cadaverine   _____

3. Show how alanine could be deaminated to give the end-products indicated.

$$H_2N - \underset{\underset{H}{|}}{\overset{\overset{CH_3}{|}}{C}} - COOH \longrightarrow$$ +

Alanine   Pyruvic acid   _____

4. Of what value is deamination to a microbe? _____

_____

CRITICAL THINKING

1. Why look for black precipitate (FeS) in the butt instead of on the surface of an H_2S test?

2. Decarboxylation of an amino acid results in the evolution of carbon dioxide. Would a gas trap, such as that used in a fermentation test, be an accurate measure of decarboxylation?

3. Semisolid media such as MIO are used to determine motility. Another such medium is SIM (sulfide, indole, motility) medium. When would you use SIM instead of MIO?

CLINICAL APPLICATION

In blue diaper syndrome, a baby's urine turns blue in the diapers following the administration of oral tryptophan. The baby's serum and urine levels of tryptophan are very low. What is causing the blue pigment indigo to appear?

Respiration

OBJECTIVES

After completing this exercise, you should be able to:

1. Define the term *reduction*.
2. Compare and contrast *aerobic respiration, anaerobic respiration,* and *fermentation.*
3. Perform nitrate reduction, catalase, and oxidase tests.

BACKGROUND

Living organisms obtain energy by removing electrons, or **oxidizing** substrates. During glycolysis, glucose is oxidized to pyruvic acid. Oxidation requires an electron acceptor to combine with the liberated electrons. The electron acceptor is **reduced** when it accepts electrons. An oxidation reaction releases electrons, so it must always be paired with a reduction reaction to accept the electrons. In photosynthesis, chlorophyll loses electrons (oxidation), and the electrons are ultimately used to reduce carbon dioxide.

In fermentation, glucose is oxidized, and organic molecules found inside the cell, such as pyruvic acid, serve as electron acceptors. The electron acceptor is reduced and forms other compounds, such as lactic acid. The electron transport chain is used in oxidative metabolism, or **respiration.** Inorganic molecules from outside the cell usually serve as electron acceptors. Molecular oxygen (O_2) is the final electron acceptor in **aerobic respiration** (FIGURE 17.1).

Respiration

In aerobic bacteria, cytochromes in the plasma membrane carry electrons to O_2. Four general classes of bacterial cytochromes have been identified, and the **oxidase test** is used to determine the presence of one of these, cytochrome *c.* The oxidase test is useful in identifying bacteria because some bacteria do not have cytochrome *c.*

During respiration, hydrogen atoms may combine with oxygen, forming hydrogen peroxide (H_2O_2), which is lethal to the cell. Most aerobic organisms produce the enzyme **catalase,** which breaks down hydrogen peroxide to water and oxygen, as shown here:

$$2H_2O_2 \xrightarrow{\text{Catalase}} 2H_2O \quad + \quad O_2\uparrow$$
Hydrogen peroxide · · · · · · Water · · · · · · · · · · Oxygen

In the process of **anaerobic respiration,** inorganic compounds other than O_2 act as final electron acceptors. (See examples in Figure 17.1; also, a few bacteria can use organic electron acceptors in anaerobic respiration.) During anaerobic respiration, some bacteria reduce nitrates to nitrites; others further reduce nitrite to nitrous oxide, and some bacteria reduce nitrous oxide to nitrogen gas.

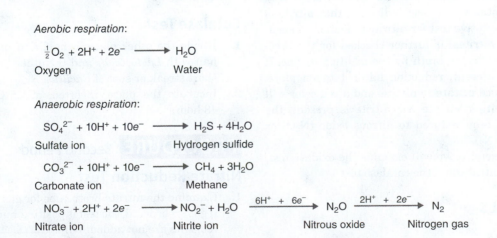

Aerobic respiration:

$$\tfrac{1}{2}O_2 + 2H^+ + 2e^- \longrightarrow H_2O$$
Oxygen · · · · · · · · · · · · · · · · · · Water

Anaerobic respiration:

$$SO_4^{2-} + 10H^+ + 10e^- \longrightarrow H_2S + 4H_2O$$
Sulfate ion · · · · · · · · · · · · · · · · Hydrogen sulfide

$$CO_3^{2-} + 10H^+ + 10e^- \longrightarrow CH_4 + 3H_2O$$
Carbonate ion · · · · · · · · · · · · · · Methane

$$NO_3^- + 2H^+ + 2e^- \longrightarrow NO_2^- + H_2O \xrightarrow{6H^+ + 6e^-} N_2O \xrightarrow{2H^+ + 2e^-} N_2$$
Nitrate ion · · · · · · · · · · · · · · Nitrite ion · · · · · · · · · · · · Nitrous oxide · · · · · · · Nitrogen gas

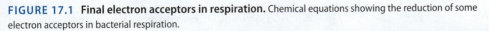

FIGURE 17.1 Final electron acceptors in respiration. Chemical equations showing the reduction of some electron acceptors in bacterial respiration.

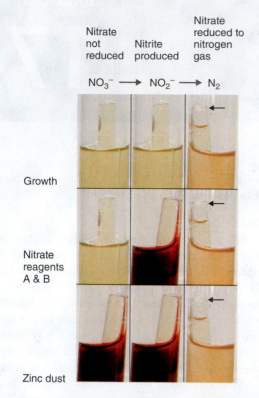

Nitrate not reduced → Nitrite produced → Nitrate reduced to nitrogen gas

$$NO_3^- \longrightarrow NO_2^- \longrightarrow N_2$$

Growth

Nitrate reagents A & B

Zinc dust

FIGURE 17.2 Nitrate test. Nitrate reagents A and B are added to nitrate broth after incubation to determine nitrate reduction. If the broth turns red after the reagents are added, nitrate ion was reduced to nitrite ion. Zinc dust is added if no color change occurs. If the broth turns red, nitrate ions are present. If nitrates were reduced to nitrous oxide or nitrogen gas, no color change occurs. If nitrates were reduced to nitrogen gas, the gas should be visible in the Durham (inverted) tube (at arrows).

Nitrate Reduction

Nitrate broth (nutrient broth plus 0.1% potassium nitrate) is used to determine a bacterium's ability to reduce nitrates (**FIGURE 17.2**). Nitrites are detected by adding dimethyl-α-naphthylamine and sulfanilic acid to nitrate broth. A red color, indicating that nitrite is present, is a positive test for nitrate reduction. A negative test (no nitrites) is further checked for the presence of nitrate in the broth by the addition of zinc. If nitrates are present, reduction has not taken place. Zinc will reduce nitrate to nitrite, and a red color will appear. If neither nitrate nor nitrite is present, the nitrogen has been reduced to nitrous oxide (N_2O) or nitrogen gas (N_2).

In this exercise, we will examine the oxidase test, reduction of nitrate, and the catalase test.

MATERIALS

FIRST PERIOD

Tubes containing nitrate broth (3)
Petri plates containing trypticase soy agar (2)

SECOND PERIOD

Tube containing nitrate broth
Nitrate reagent A (dimethyl-α-naphthylamine)
Nitrate reagent B (sulfanilic acid)
Hydrogen peroxide, 3%
Sterile toothpick
Oxidase reagent, oxidase strip, or oxidase DrySlide
Zinc dust

CULTURES

Bacillus subtilis
Escherichia coli
Pseudomonas aeruginosa `BSL-2`
Lactococcus lactis
Alcaligenes faecalis

TECHNIQUES REQUIRED

Inoculating loop technique (Exercise 4)
Aseptic technique (Exercise 4)
Plate streaking (Exercise 11)

PROCEDURE First Period

Nitrate Reduction Test

1. Label three tubes of nitrate broth, and inoculate one with *Lactococcus*, one with *Bacillus*, and one with *Pseudomonas* `BSL-2`.
2. Incubate the tubes at 35°C for 2 days.

Oxidase Test

1. Divide one plate in half. Label one half *"Escherichia"* and the other *"Alcaligenes."* Inoculate each organism on the appropriate half.
2. Incubate the plate, inverted, at 35°C for 24 to 48 hours.

Catalase Test

1. Divide the other plate in half, and inoculate one half with *Lactococcus* and the other half with one short streak or spot of *Bacillus*.
2. Incubate the plate, inverted, at 35°C for 24 to 48 hours.

PROCEDURE Second Period

Nitrate Reduction Test

1. Examine the nitrate broth tube for gas production. If gas is observed, the test is positive, and additional reagents are not added. Why? _____

2. Add 5 drops of nitrate reagent A and 5 drops of nitrate reagent B to each tube and to an uninoculated nitrate broth tube. Shake the tubes gently.

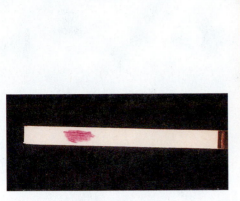

(a) Oxidase test strip.

(b) Oxidase reagent put on colonies.

FIGURE 17.3 **Oxidase test.** Cytochrome oxidase–positive bacteria turn oxidase reagent pink to purple to black.

3. A red color within 30 seconds is a positive test. If the broth turns red, what compound is present?

_____

4. If it does not turn red, add a small pinch of zinc dust; if it turns red after 20 to 30 seconds, the test is negative. If not, it is positive for nitrate reduction. Why? _____

_____

5. Record your results (see Figure 17.2).

Oxidase Test

1. To test for cytochrome oxidase, do *one* of the following:
   a. Place an oxidase test strip in a Petri dish, and moisten an area of the strip with water. Smear three or four well-isolated colonies onto the moistened area with a loop. Oxidase-positive bacteria will turn blue or purple within 10 to 20 seconds (**FIGURE 17.3a**).
   b. Drop oxidase reagent on the colonies, and observe them for a color change to pink within 1 minute and then a change from blue to black. Oxidase-negative colonies will not change color (**FIGURE 17.3b**).
   c. Using a loop, smear a little of one colony onto the oxidase DrySlide. Oxidase-positive bacteria will turn a dark-purple color within 2 minutes.

Catalase Test

1. To test for catalase, do *one* of the following:
   a. Drop hydrogen peroxide on the colonies, and observe them for bubbles.

FIGURE 17.4 **Catalase test.** Bubbles after the addition of hydrogen peroxide indicate a positive catalase test.

   b. With a sterile toothpick, touch the center of a colony and transfer it to a clean glass slide. Discard the toothpick in the To Be Autoclaved area. Drop hydrogen peroxide on the organisms, and observe them for bubbles (catalase-positive). See **FIGURE 17.4**. What gas is in the bubbles? _____

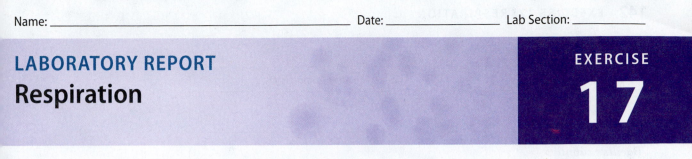

LABORATORY REPORT
Respiration

EXERCISE
17

PURPOSE _____

_____

_____

HYPOTHESES

1. The presence of catalase is indicated by _____ after addition of _____.

2. Some bacteria can reduce nitrate ions to _____, _____, or _____.

RESULTS

Nitrate Reduction Test

Organism	Color after Adding Nitrate Reagents A and B	Color after Adding Zinc	Gas
Uninoculated tube			
Lactococcus lactis			
Bacillus subtilis			
Pseudomonas aeruginosa BSL-2			

Oxidase Test

Organism	Color after Adding Reagent
Escherichia coli	
Alcaligenes faecalis	

Catalase Test

Organism	Appearance after Adding H_2O_2
Lactococcus lactis	
Bacillus subtilis	

CONCLUSIONS

1. Do you accept your hypotheses? _____

2. Which bacteria reduced nitrate? _____

3. Which organism was oxidase-positive? _____

4. Which species was catalase-positive? _____

QUESTIONS

1. Define the term *reduction*. _____

    _____

    _____

2. Why does hydrogen peroxide bubble when it is poured on a skin cut? _____

    _____

    _____

3. Differentiate aerobic respiration from anaerobic respiration. _____

    _____

    _____

4. Differentiate fermentation from anaerobic respiration. _____

    _____

    _____

CRITICAL THINKING

1. Would nitrate reduction occur more often in the presence or in the absence of molecular oxygen? Explain.

2. Is nitrate reduction beneficial to farmers?

CLINICAL APPLICATIONS

1. People with Takahara's disease don't produce catalase. Although often asymptomatic, this condition can lead to recurrent oral infections by peroxide-generating bacteria. How would you diagnose a catalase deficiency?

2. One manifestation of cytochrome oxidase deficiency in humans is severe muscle weakness. Briefly explain why.

3. *Clostridium* bacteria are killed in the presence of molecular oxygen. Would you expect *Clostridium* to produce catalase? Briefly explain.

Unknown Identification and *Bergey's Manual*

The strategy of DISCOVERY *lies in determining the sequence of choice of problems to solve. Now it is in fact very much more difficult to see a problem than to find a solution to it. The former requires* IMAGINATION, *the latter only ingenuity.* – JOHN BERNAL

OBJECTIVES

After completing this exercise, you should be able to:

1. Explain how bacteria are characterized and classified.
2. Use *Bergey's Manual*.
3. Identify an unknown bacterium.

BACKGROUND

In microbiology, a system of classification must be available to allow the microbiologist to categorize and classify organisms. Communication among scientists would be very limited if no universal system of classification existed. Classification of microorganisms is based on similarities in their ribosomal RNA (rRNA). Bacteria and archaea with similar rRNA sequences are grouped together into taxa. The taxa used for prokaryotes are domain, phylum, class, order, family, genus, and species. Bacteria are *identified* for practical purposes—for example, to determine an appropriate treatment for an infection. They are not necessarily identified by the same techniques by which they are *classified*. Most identification procedures are easily performed in a laboratory and use as few procedures or tests as possible.

The most important reference for bacterial taxonomy is **Bergey's Manual.**[*] *Bergey's Manual of Systematic Bacteriology* provides classification based on rRNA. *Bergey's Manual of Determinative Bacteriology* is a valuable reference in the lab because it provides identification (determinative) schemes based on such criteria as morphology, differential staining, oxygen requirements, and biochemical testing. Different tests are required, depending on the bacterium being identified. A test that is critical for distinguishing one bacterial group from another may be irrelevant for identifying other bacteria. The criteria listed in *Bergey's Manual* and used to develop the keys in Appendix H are relatively constant, although atypical bacteria will be found through repeated laboratory culture. This variability, however, only heightens the fun of classifying bacteria.

You will be given an unknown heterotrophic bacterium to characterize and identify. By using careful deduction and systematically compiling and analyzing data, you should be able to identify the bacterium.

Keys to species of bacteria are provided in Appendix H (see Figures H.1 and H.2). These keys are an example of a dichotomous classification system—that is, a population is repeatedly divided into two parts until a description identifies a single member. Such a key is sometimes called an "artificial key" because there is no single correct way to write one. You may want to check your conclusion with the species description given in *Bergey's Manual*.

To begin your identification, ascertain the purity of the culture you have been given, and prepare stock and working cultures. Avoid contaminating your unknown. Note growth characteristics and Gram-stain appearance for clues about how to proceed. After culturing and staining the unknown, you will be able to eliminate many bacterial groups. Final determination of your unknown will depend on your carefully selecting the relevant biochemical tests and weighing the value of one test over another in case of contradictions. Enjoy!

MATERIALS

Petri plates containing trypticase soy agar (2)
Trypticase soy agar slant (2)
All stains, reagents, and media previously used

CULTURES

Unknown bacterium # _____

TECHNIQUES REQUIRED

Compound light microscopy (Exercise 1)
Hanging-drop procedure (Exercise 2)
Wet-mount technique (Exercise 2)

---

[*]*Bergey's Manual of Systematic Bacteriology,* 2nd ed. (2004), is the reference for classification. *Bergey's Manual of Determinative Bacteriology,* 9th ed. (1994), is used for laboratory identification of culturable bacteria and archaea.

Inoculating loop and needle technique (Exercise 4)
Aseptic technique (Exercise 4)
Negative staining (Exercise 6)
Gram staining (Exercise 7)
Acid-fast staining (Exercise 8)
Endospore, capsule, and flagella staining (Exercise 9)
Plate streaking (Exercise 11)
OF test (Exercise 13)
Starch hydrolysis (Exercise 13)
MRVP tests (Exercise 14)
Fermentation tests (Exercise 14)
Citrate test (Exercise 14)
Protein catabolism (Exercises 15 and 16)
Catalase test (Exercise 17)
Nitrate reduction test (Exercise 17)
Oxidase test (Exercise 17)

PROCEDURE First Period

1. Streak your unknown onto the trypticase soy agar plates for isolation. Incubate one plate at 35°C and the other at room temperature for 24 to 48 hours. Note the growth characteristics (see Figure 3.3 on page 34) and the temperature at which each one grows best.
2. To help you begin, try to identify the bacteria in **FIGURE 18.1** using Appendix H (see Figures H.1 and H.2). The results of one test, such as the Gram stain, help you plan the next test. (The unknown in Figure 18.1 is a gram-positive coccus. According to Figure H.2, a catalase test should be done next.)

PROCEDURE Second Period

Aseptically inoculate two trypticase soy agar slants from a colony on your streak plate. Incubate them for 24 hours. Describe the resulting growth (see Figure 4.7 on page 43). Keep both slant cultures in the refrigerator. Designate one as your stock culture, the other as your working culture. Subculture your stock culture onto another slant when your working culture is contaminated or not viable. Keep the working culture in the refrigerator when it is not in use.

PROCEDURE Next Periods

1. Use your working culture for all identification procedures. When a new slant is made and its purity demonstrated, discard the old working culture. What should you do if you think your culture is contaminated? _____

   _____
2. Read the keys in Appendix H (see Figures H.1 and H.2) to develop ideas on how to proceed. Perhaps determining staining characteristics might be a good place to start. What shape is it? _____ What can be eliminated? _____

   Not all bacteria are included in Appendix H. You should check *Bergey's Manual of Determinative Bacteriology.*
3. After determining its staining and morphologic characteristics, determine which biochemical tests you will need. Do not be wasteful. Inoculate *only* what is needed. It is not necessary to repeat a test—do it once accurately. Do not perform unnecessary tests.
4. If you come across a new test—one not previously done in this course—determine whether it is essential. Can you circumvent it? _____ If not, consult your instructor.
5. Record your results in the Laboratory Report, and identify your unknown.

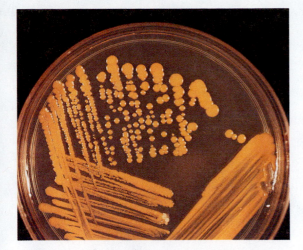

(a) The unknown organism is isolated in pure culture.

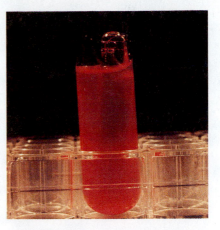

(d) Results of the previous tests lead to the next series of tests. This photograph shows the results of a glucose fermentation tube inoculated with the unknown bacterium.

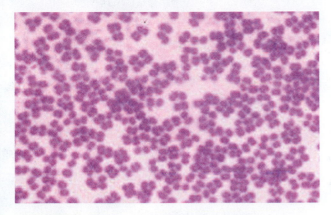

(b) A Gram stain is made from the pure culture.

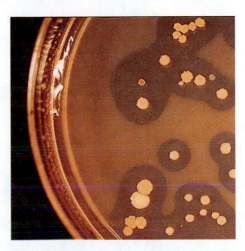

(e) Lipid hydrolysis test. The clearing around the colonies indicates hydrolysis of the lipid in the medium.

(c) Based on the result of the Gram stain, a catalase test is indicated.

FIGURE 18.1   Identification of an unknown. This is an example of how to identify an unknown bacterium. Use these test results and the criteria in *Bergey's Manual* to identify the bacterial species shown in the photographs.

LABORATORY REPORT
Unknown Identification
and *Bergey's Manual*

PURPOSE _____

_____

_____

RESULTS

Unknown # _____

Morphological, Staining, and Cultural Characteristics	
The cell	Colonies on trypticase soy agar
Staining characteristics	Diameter _____
Gram _____ Age _____	Appearance _____
Other _____ Age _____	Color _____
Shape _____	Elevation _____
Size _____	Margin _____
Arrangement _____	Consistency _____
Endospores (position) _____	Agar slant Age _____
_____	Amount of growth _____
Motility _____	Pattern _____
Determined by _____	Color _____

Make a table listing the tests you performed and their results. Remember to do only the necessary tests and to avoid repeating a test.

CONCLUSIONS

1. What organism is shown in Figure 18.1? _____

2. What organism was in unknown # _____? _____

QUESTIONS

1. On a separate sheet of paper, write your rationale for arriving at your conclusion.

2. Why is it necessary to complete the identification of a bacterium on the basis of its physiology rather than its

    morphology? _____

    _____

    _____

    _____

PART
6

Microbial Growth

EXERCISES

In previous exercises, we studied bacterial growth by observing colonies or turbidity, making visual determinations on an all-or-none basis with the naked eye. In Exercise 20, we will use instrumentation to detect the slight changes in bacterial numbers that occur over time—changes that are not visible to the naked eye—to measure lag and log phases of growth (see the graph on the next page).

Bacteria have nutritional, physical, chemical, and environmental requirements that must be met for growth to occur. Knowledge of the conditions necessary for microbial growth can facilitate culturing microorganisms and controlling unwanted organisms.

In 1861, Jean Baptiste Dumas presented a paper to the Academy of Sciences on behalf of Pasteur. In this paper, he stated:

> *The existence of infusoria having the characteristics of ferments is already a fact which seems to deserve attention, but a characteristic which is even more interesting is that these infusoria-animalcules live and multiply indefinitely in absence of the smallest amount of air or free oxygen. . . . This is, I believe, the first known example of animal ferments and also of animals living without free oxygen gas.**

When we speak of *air* as a growth requirement, we are referring to the oxygen in air. Furthermore, when we refer to *oxygen,* we usually mean the molecular oxygen (O_2) that acts as an electron acceptor in aerobic respiration. The significance of the presence or absence of oxygen is investigated in Exercise 19.

Most bacteria in aqueous environments such as tap water, oceans, and body fluids are not growing in colonies as they do on laboratory media. Nor are bacteria usually swimming around as they do on a microscope slide. Most bacteria grow in mixed assemblages of species called biofilms (Exercise 21) and are usually attached to a solid surface.

---

*In H. A. Lechevalier and M. Solotorovsky. *Three Centuries of Microbiology.* New York: Dover Publications, 1974, p. 45.

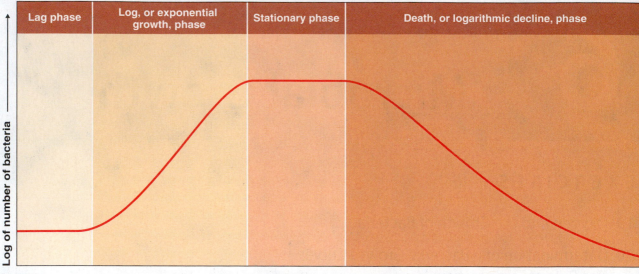

Bacterial growth curve, showing the four typical phases of growth.

CASE STUDY: Health Care Associated Pneumonia

Gary, a nurse, is concerned because 5 of the 12 patients he has cared for in the past month developed *Mycobacterium chelonae* pneumonia during their hospital stay. Working with the infection control officer, Gary looks for commonalities. All the patients had different physicians and diets, and all were assigned to three other nurses in addition to Gary. All of the patients were given ice chips and water. Gary sets up an experiment to determine the source of the bacteria. A total of 139 samples were collected from five different sources: the public drinking-water supply, ice, Gary's respiratory tract, hospital walls, and hospital floors.

    M. chelonae was isolated from three of the five samples analyzed. Twelve colonies grew from one of the public drinking-water samples, and 200 colonies grew from the ice machine at the nurses' station. *M. chelonae* was not isolated from commercially bagged ice.

    Gary is intrigued. Why are the bacteria in hospital ice but not commercial ice if *M. chelonae* is present in the drinking water? Gary sets up a growth curve experiment by inoculating *M. chelonae* into nutrient broth and incubating the broths at various temperatures.

	Starting Absorbance	Absorbance after 48 hr of Incubation
Temperature of incoming water (20°C)	0.002	0.007
Temperature of ice machine water reservoir (25°)	0.001	1.019
Ice temperature (0°C)	0.015	0.001
Human body temperature (37°C)	0.019	0.723

Question

Explain how you know the ice was the source of infection.

Oxygen and the Growth of Bacteria

What is education but a **PROCESS** *by which a person* **BEGINS** *to learn how to learn?* – **PETER USTINOV**

OBJECTIVES

After completing this exercise, you should be able to:

1. Identify the incubation conditions for each of the following types of organisms: obligate aerobes, obligate anaerobes, aerotolerant anaerobes, microaerophiles, and facultative anaerobes.
2. Describe three methods of culturing anaerobes.
3. Cultivate anaerobic bacteria.

BACKGROUND

The presence or absence of molecular oxygen (O_2) can be very important to the growth of bacteria (**FIGURE 19.1**). Some bacteria, called **obligate aerobic bacteria,** require oxygen. *Alcaligenes* is an example. Others, called **anaerobic bacteria,** do not use oxygen. One reason why **obligate anaerobes** cannot tolerate the presence of oxygen is that they lack catalase, and the resultant accumulation of hydrogen peroxide is lethal. **Aerotolerant anaerobes** cannot use oxygen but tolerate it fairly well, although their growth may be enhanced by microaerophilic conditions. Most of these bacteria use fermentative metabolism.

Some bacteria, the **microaerophiles,** grow best in an atmosphere with increased carbon dioxide (5% to 10%) and lower concentrations of oxygen. Microaerophiles will grow in a solid nutrient medium at a depth to which small amounts of oxygen have diffused into the medium (**FIGURE 19.1e**). To culture microaerophiles on Petri plates and nonreducing media, a **CO_2 jar** is used. Inoculated plates and tubes are placed in a large jar with a lighted candle or CO_2-generating packet. (See Figure 47.1 on page 374.)

The majority of bacteria are capable of living with or without oxygen; these bacteria are called **facultative anaerobes** (**FIGURE 19.1b**). *Escherichia coli* is a facultative anaerobe.

Six genera of bacteria lacking catalase are *Streptococcus, Enterococcus, Lactococcus, Leuconostoc, Lactobacillus,* and *Clostridium.* Species of *Clostridium* are obligate anaerobes, but members of the other five

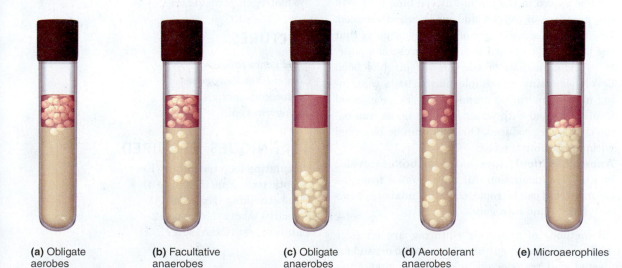

(a) Obligate aerobes

(b) Facultative anaerobes

(c) Obligate anaerobes

(d) Aerotolerant anaerobes

(e) Microaerophiles

FIGURE 19.1 Oxygen and bacterial growth. Shown is the effect of oxygen concentration on the growth of various types of bacteria in a tube of solid medium. Oxygen diffuses only a limited distance from the atmosphere into the solid medium. The resazurin dye in the medium turns pink when oxygen is present.

genera are aerotolerant anaerobes. The five genera of aerotolerant anaerobes lack the cytochrome system to produce hydrogen peroxide and therefore do not need catalase. Determining the presence or absence of catalase can help in identifying bacteria. When a few drops of 3% hydrogen peroxide are added to a microbial colony and catalase is present, molecular oxygen is released as bubbles.

$$2H_2O_2 \longrightarrow 2H_2O + O_2\uparrow$$

In the laboratory, we can culture anaerobes either by excluding free oxygen from the environment or by using reducing media. Many anaerobic culture methods involve both processes. Some of the anaerobic culturing methods are as follows:

1. **Reducing media** contain reagents that chemically combine with free oxygen, reducing the concentration of oxygen. In thioglycolate broth, **sodium thioglycolate** ($HSCH_2COONa$) will combine with oxygen. A small amount of agar is added to increase the viscosity, which reduces the diffusion of air into the medium. Usually dye is added to indicate where oxygen is present in the medium. Resazurin, which is pink in the presence of excess oxygen and colorless when reduced, or methylene blue (see method 2, below) are commonly used indicators.

2. Conventional, nonreducing media can be incubated in an anaerobic environment. Oxygen is excluded from a **Brewer anaerobic jar** by adding a GasPak (Figure 19.2). Carbon dioxide and hydrogen are given off when the GasPak sachet (of inorganic carbonate, activated carbon, ascorbic acid, and water) is exposed to air. A methylene blue indicator strip is placed in the jar; methylene blue is blue in the presence of oxygen and white when reduced. One of the disadvantages of the Brewer jar is that the jar must be opened to observe or use one plate.

   A modification of the Brewer jar has been developed using disposable plastic bags containing a CO_2-generating ampule. Iron (Fe^0, e.g., steel wool) activated with water removes O_2 as iron oxide (Fe_2O_3) is produced. One plate can be observed without opening the bag.

3. **Anaerobic incubators** and **glove boxes** can also be used for incubation. Air is evacuated from the chamber and can be replaced with a mixture of carbon dioxide and nitrogen.

All methods of anaerobic culturing are effective only if the specimen or culture of anaerobic organisms is collected and transferred in a manner that minimizes exposure to oxygen. In this exercise, we will try two methods of anaerobic culturing and will perform the catalase test.

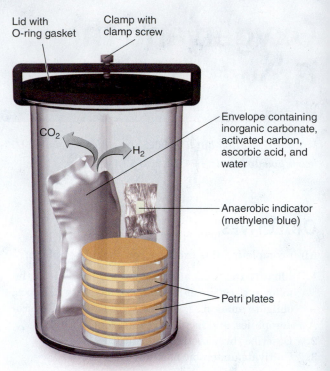

FIGURE 19.2 **Brewer anaerobic jar.** Carbon dioxide is generated when the chemical packet is exposed to air.

MATERIALS

FIRST PERIOD

Petri plates containing nutrient agar (2)
Tubes containing thioglycolate broth + indicator (4)
Brewer anaerobic jars (2 per lab section)

SECOND PERIOD

3% hydrogen peroxide (H_2O_2)

CULTURES

Alcaligenes faecalis
Clostridium sporogenes
Lactococcus lactis
Escherichia coli

TECHNIQUES REQUIRED

Inoculating loop technique (Exercise 4)
Aseptic technique (Exercise 4)
Plate streaking (Exercise 11)
Selective media (Exercise 12)
Catalase test (Exercise 17)

PROCEDURE First Period

1. Do not shake the thioglycolate. Why? _____
   What should you do to salvage it if you do shake it?
   _____
   _____

2. Label four tubes of thioglycolate broth, and aseptically inoculate one with a loopful of *Alcaligenes,*
   one with *Clostridium,* one with *Lactococcus,* and
   one with *Escherichia.*

3. Incubate the tubes at 35°C until the next lab period.

4. With a marker, divide two nutrient agar plates into
   four sectors on the bottom of the plates. Label one
   plate "Aerobic" and the other "Anaerobic."

5. Streak a single line of *Alcaligenes, Clostridium,
   Lactococcus,* and *Escherichia* in the appropriate
   sector (**FIGURE 19.3**).

6. Incubate the "Aerobic" plates, inverted, at 35°C until
   the next lab period. Place the "Anaerobic" plates, inverted, in the Brewer jar.

7. Your instructor will demonstrate how to use your
   anaerobic-incubation system.

8. Incubate the jar at 35°C until the next lab period.

PROCEDURE Second Period

1. Record the appearance of growth in each tube. Record the growth on each plate.

2. Perform the catalase test by adding a few drops of
   3% H_2O_2 to the different colonies on the nutrient

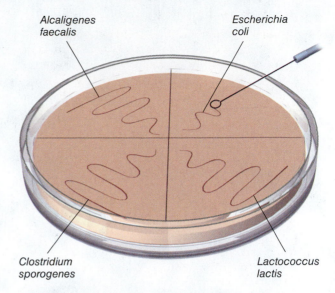

FIGURE 19.3 **Petri plate inoculation.** Divide a Petri plate into quadrants by drawing lines on the bottom of the plate. Inoculate each quadrant by streaking it with an inoculating loop.

agar. A positive catalase test produces a bubbling
white froth. A dissecting microscope (Appendix E)
can be used if more magnification is required to
detect bubbling. The catalase test may also be done
by transferring bacteria to a slide and adding the
H_2O_2 to it. Bacterial growth on blood agar must
be tested this way. Why? _____
_____

LABORATORY REPORT
Oxygen and the Growth of Bacteria

PURPOSE _____

_____

_____

EXPECTED RESULTS

Before the next lab period, indicate the amount of growth you *expect* to see on each plate and the catalase reaction.

Bacteria	Aerobic Growth	Anaerobic Growth	Catalase Reaction
Alcaligenes			
Clostridium			
Lactococcus			
Escherichia			

RESULTS

Thioglycolate

Sketch the location of bacterial growth, and note where the indicator has turned blue or pink.

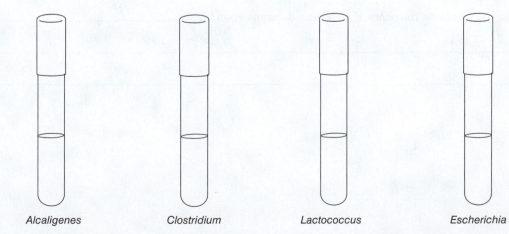

| *Alcaligenes* | *Clostridium* | *Lactococcus* | *Escherichia* |

Nutrient Agar Plates

Bacteria	Aerobic Growth	Anaerobic Growth	Catalase Reaction
Alcaligenes			
Clostridium			
Lactococcus			
Escherichia			

Does the anaerobic plate have an odor? _____

CONCLUSION

Did your results differ from your expected results? _____

Briefly explain why or why not. _____

_____

QUESTIONS

1. Can aerobic bacteria grow in the absence of O_2? _____ What would you need to do to determine whether bacteria growing on a Petri plate from the Brewer jar are anaerobes? _____

   _____

2. What does the appearance of a blue or pink color in a thioglycolate tube mean? _____

   _____

3. Why will obligate anaerobes grow in thioglycolate? _____

   _____

4. To what do you attribute the odors of anaerobic decomposition? _____

   _____

CRITICAL THINKING

1. *Bergey's Manual* describes *Streptococcus* and *Escherichia* as facultative anaerobes. How do the oxygen requirements of these organisms differ? Which one could correctly be called an aerotolerant anaerobe?

2. How can the aerobe *Pseudomonas aeruginosa* grow in the absence of oxygen?

3. The catalase test is often used clinically to distinguish two genera of gram-positive cocci, _____

   and _____. It is also used to distinguish two genera of gram-positive rods, _____

   and _____.

CLINICAL APPLICATION

The following genera of anaerobes are commonly found in fecal samples. Using *Bergey's Manual* and your text, place these genera in the flowchart to differentiate them.

- *Bacteroides*
- *Clostridium*
- *Enterococcus*
- *Fusobacterium*
- *Lactobacillus*
- *Veillonella*

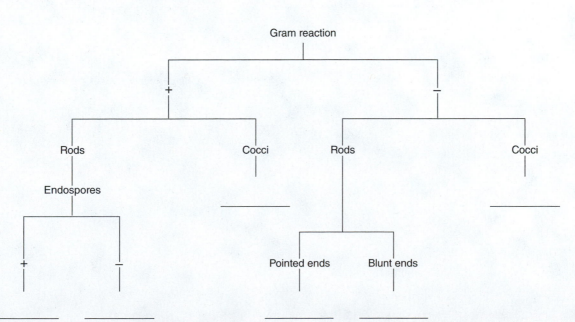

Determination of a Bacterial Growth Curve: The Role of Temperature

OBJECTIVES

After completing this exercise, you should be able to:

1. Measure bacterial growth turbidimetrically.
2. Interpret growth data plotted on a graph.
3. Determine the effect of temperature on bacterial growth.

BACKGROUND

The effect of temperature on bacteria can be determined by measuring the population growth rate. The phases of growth of a bacterial population are shown in the graph on page 152. During the log, or exponential growth, phase, the cells are growing at the fastest rate possible under the conditions provided. **Generation** or **doubling time** is the time it takes for one cell to divide into two cells—or the time required for the population of cells to double. The shorter the generation time, the faster the growth. The growth of a bacterial population can be determined by inoculating a growth medium with a few cells and counting the cells over time. However, bacteria in suspension in a broth scatter light and cause the transparent broth to appear turbid. This turbidity can be measured with a spectrophotometer (Appendix C) to determine bacterial growth. Although turbidity is not a direct measure of bacterial numbers, increasing turbidity does indicate growth.

Changes in the logarithmic absorbance scale on the spectrophotometer correspond to changes in the number of cells. Consequently, a growth curve can be obtained by graphing absorbance (Y-axis) vs. time (X-axis). The rate of growth is indicated by the slope of the lines; faster growth produces a steeper (higher-number) slope (FIGURE 20.1).

Most bacteria grow within a particular temperature range (FIGURE 20.2). The *minimum growth temperature* is the lowest temperature at which a species will grow. A species grows fastest at its *optimum growth temperature*. The highest temperature at which a species can grow is its *maximum growth temperature*. Some heat is necessary for growth. Heat probably increases the rate of collisions between molecules, thereby increasing enzyme activity. At temperatures near the maximum growth temperature, growth ceases, presumably because of denaturation of enzymes.

In this exercise, we will plot growth curves for a bacterium at different temperatures to determine the optimum temperature range of this species.

MATERIALS

Flask containing 100 ml of nutrient broth
Spectrophotometer tubes (2)
Sterile 5-ml pipettes (11)
Sterile 10-ml pipette
Spectrophotometer

CULTURE

Escherichia coli

TECHNIQUES REQUIRED

Aseptic technique (Exercise 4)
Pipetting (Appendix A)
Spectrophotometry (Appendix C)
Graphing (Appendix D)

PROCEDURE

Wear goggles when pipetting the bacteria.

Each student group is assigned a temperature: 15°C, room temperature, 35°C, 45°C, or 55°C.

1. Aseptically transfer 2–4 ml of nutrient broth to one of the spectrophotometer tubes. This is your control to *standardize* the spectrophotometer. Read Appendix C. Also refer to FIGURE 20.3.
2. Aseptically inoculate a flask of nutrient broth with 10 ml of *E. coli*.
3. Swirl to mix the contents. Transfer 2–4 ml from the flask to the second spectrophotometer tube. Wipe the surface of the spectrophotometer tube with a low-lint, nonabrasive paper such as a Kimwipe. Place the tube in the spectrophotometer, wait about 45 sec, and measure the absorbance (Abs.) on the spectrophotometer. Discard the broth into the container of disinfectant.
4. Record all measurements.
5. Place the flask at your *assigned* temperature.
6. Record the absorbance every 10 minutes for 60 to 90 minutes. Remove the flask from the water bath

Time (min)	Temp A	Temp B
0	0.011	0.012
48	0.022	*a* 0.045
58	0.035	0.081
68	*a* 0.047	0.056
78	0.062	0.075
88	0.081	0.078
98	*b* 0.099	0.078
108	0.109	0.071
118	0.137	0.088
128	0.149	*b* 0.089
138	0.167	0.098
148	0.174	0.106
158	0.190	0.109
168	0.207	0.114
178	0.222	0.117
188	0.297	0.118
198	0.288	0.126
600	0.800	0.125
1470	0.864	0.130
Slope between *a* and *b*	0.0018	0.0004
Generation time (min)	29	80

(a) Generation times. To determine the generation time: first select two points, (*a*) and (*b*), in log phase growth during which the absorbance doubled. Determine the time required for the culture to double. Use your graphing application to calculate the slope of the line. A higher (steeper) slope indicates faster growth.

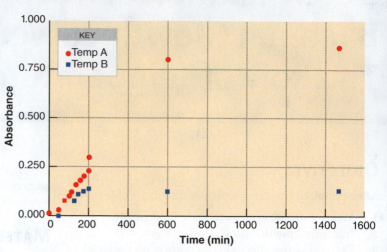

(b) A graph of all the data shows logarithmic and stationary phases.

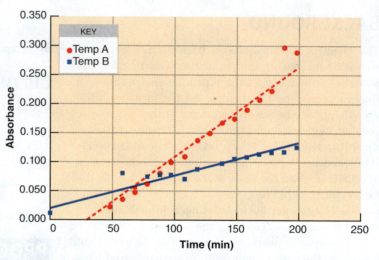

(c) A graph of the first 3 hours expands the log phase. You can compare the growth by comparing the slopes of the best-fit lines.

FIGURE 20.1 **Showing bacterial growth using absorbance values. (a)** The generation time is the number of minutes required for the absorbance to double. **(b)** Plotting all the data shows when the cultures reach stationary phase. **(c)** This graph is used to examine log phase. The best-fit lines show the rate at which the population is growing.

After taking a reading, empty the spectrophotometer tube into disinfectant. Rinse the spectrophotometer tube with distilled water, and discard the water into disinfectant.

Wear safety goggles when pipetting.

for the minimum time possible to take each sample. To take a sample, aseptically pipette 2–4 ml into the second spectrophotometer tube.

7. Graph the data you obtained. *Read* Appendix D before drawing your graph.

8. Determine the generation time. Select two points, (*a*) and (*b*), in the log phase during which the absorbance doubled, and determine the number of minutes required for the culture to go from (*a*) to (*b*). (See Figure 20.1.)

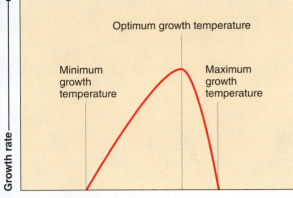

Growth rate →

Optimum growth temperature

Minimum
growth
temperature

Maximum
growth
temperature

Increasing temperature →

FIGURE 20.2 **Growth response of bacteria within their growing temperature range.**

2–4 ml

Nutrient broth

Spectrophotometer
tube (control)

Place tube in the spectrophotometer
and set absorption to zero.

10 ml

2–4 ml

E. coli

Inoculate the flask with *E. coli*
and swirl to mix contents.

Spectrophotometer
tube (2nd tube)

Place tube in the spectrophotometer
and record the absorbance.

2–4 ml

Place the flask at the
assigned temperature.

At 10-minute intervals for
60–90 minutes, remove the
flask and pipette 2–4 ml into
the second spectrophotometer
tube.

Place tube in the spectrophotometer
and record the absorbance.

FIGURE 20.3 Procedure for determining the absorbance at a particular time.

Name: _____ Date: _____ Lab Section: _____

Determination of a Bacterial Growth Curve: The Role of Temperature

PURPOSE _____

_____

_____

EXPECTED RESULTS

1. At what temperature would you expect *Escherichia coli* to grow best: 15°C, room temperature, 35°C, 45°C, or

    55°C? _____

2. If *Escherichia coli* has a generation time of 20 minutes, will you be able to construct a growth curve similar to

    the one on page 152 from your laboratory data? _____

RESULTS

Record your data below or in a computer application spreadsheet.
Temperature:

Time (min)	Absorbance	Time (min)	Absorbance
0			

Plot your data on graph paper or use a computer graphing application. Mark absorbance on the Y-axis and time on the X-axis. Obtain data for other temperatures from your classmates. Plot these data on the same graph, using different lines or colors. Label the phases of growth.

Use your data and those of your classmates to fill in this table:

Temperature	Doubling Time (min)

CONCLUSIONS

1. Summarize the effect of temperature on growth of *E. coli*. _____

_____

_____

2. From the class data, what is *E. coli*'s optimum growth temperature? _____ Does this result agree with the

result that you predicted? _____ If not, provide a brief explanation. _____

QUESTIONS

1. Why aren't you likely to get lag and death phases in this experiment? _____

_____

_____

2. What is the optimum growth temperature for human pathogens? _____

3. What is the effect of temperature on enzymes? _____

_____

Can you use any examples from your experiments? _____

CRITICAL THINKING

1. The following graph shows a likely relationship between bacterial growth and oxygen use in glucose broth. Explain the relationship.

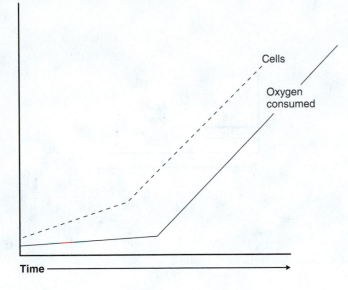

2. The following data were obtained for a log phase culture using plate counts and spectrophotometry.

No. of Cells/ml	Absorbance
5×10^4	0.04
5×10^5	0.26
6×10^7	0.72

Graph these data to answer the following question. Assume that you inoculated a flask yesterday, and now you need a broth culture containing 5×10^6 cells/ml to perform an experiment. The concentration of cells in the 24-hour culture is much too high; its absorbance is greater than 1. What is the easiest way to get the desired cell concentration from your 24-hour flask?

3. You have isolated a bacterium that produces an antibiotic. The bacterium grows well at 37°C but doesn't produce the antibiotic at that temperature. Antibiotic production is best at 30°C, but the cells grow so poorly at that temperature that you can't get enough antibiotic. The relationships are shown in the following graphs. Design a procedure to maximize the biomass (total cells) and the yield of the antibiotic.

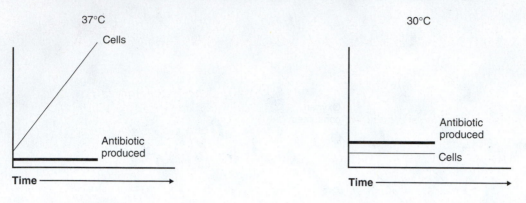

4. Why is turbidity *not* an accurate measurement of viable bacteria in a culture?

CLINICAL APPLICATION

Infection with *Salmonella enterica* initiates a strong immunological response shown by tumor necrosis factor (TNF). The data that follow show the body's response to *S. enterica*. The concentration of TNF was determined by spectrophotometry. Graph these data, and show their relationship to bacterial growth in the body.

Time (days)	TNF (as % Abs.)
0	0
1	10
4	12
7	15
9	15
11	13
14	12

Biofilms

OBJECTIVES

After completing this exercise, you should be able to:

1. Define the term *biofilm*.
2. Explain the importance of biofilms in clinical medicine.
3. Demonstrate biofilm growth.

BACKGROUND

Biofilms are populations or communities of microorganisms that attach to and grow on a solid surface that has been exposed to water. These microorganisms are usually encased in an extracellular polysaccharide that they synthesize. As the polysaccharide enlarges, other microbes may adhere to it, so several different species of bacteria and fungi may be present in one biofilm. Biofilms may be found on essentially any environmental surface in which sufficient moisture is present. Biofilms may also float. The pellicle that you've seen on some liquid culture media is an example (see Figure 3.4c on page 34). Biofilm at the air–water interface allows the bacteria to float near the surface and get oxygen while getting nutrients from the liquid. Individual cells would have to expend too much energy swimming to stay near oxygen.

As bacterial cells grow, gene expression in bacterial cells changes in a process called quorum sensing. **Quorum sensing** is the ability of bacteria to communicate and coordinate behavior. Bacteria that use quorum sensing produce and secrete a signaling chemical called an *inducer*. As the inducer diffuses into the surrounding medium, other bacterial cells move toward the source and begin producing inducer. The concentration of inducer increases with increasing cell numbers. This, in turn, brings more cells and causes synthesis of more inducer. Inducers at an increased concentration often induce gene expression, resulting in a product such as a polysaccharide, which helps the biofilm to form.

Biofilms are not just layers of cells. A biofilm is usually composed of pillars and channels through which water can flow, bringing nutrients and taking away wastes (**FIGURE 21.1**).

Increasing evidence indicates that biofilms play a key role in disease. Healthy bronchioles are usually sterile; however, *Pseudomonas aeruginosa* establishes

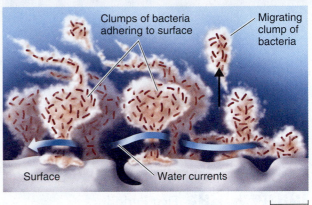

FIGURE 21.1 Biofilm. Water currents move through the pillars of slime formed by the growth of bacteria attached to a solid surface.

a permanent infection in cystic fibrosis patients by forming a biofilm. In healthy individuals, bacteria can form biofilms on implanted prostheses. Bacteria that break off biofilms of indwelling medical devices such as catheters or prostheses are a continuous source of infection in patients. The microbes in biofilms are generally more resistant to antibiotics and more difficult for the immune system to destroy. Biofilms form on teeth and provide the environment that leads to caries and eventually cavities.

MATERIALS

FIRST PERIOD

Microscope slides (3)
Coplin jar or beaker and slide rack

FOURTH PERIOD

Methylene blue

CULTURES

Hay infusion incubated 3 to 5 days in the light *or* pond or stream water

TECHNIQUES REQUIRED

Simple staining (Exercise 5)

FIGURE 21.2 In the environment. Set slides in a pond or stream where they won't be disturbed.

FIGURE 21.3 In the laboratory. Set slides in the Coplin jar. Fill the jar with liquid, leaving a few millimeters of the slides exposed above the liquid.

PROCEDURE First Period

1. Establish a biofilm by using one of the following methods:
   a. In the environment
      1. Place three slides in a horizontal staining dish or a plastic slide rack.
      2. Place the slide dish or rack in a pond or stream in an area that is easily accessible and where the slides will not be displaced (**FIGURE 21.2**).
      3. Mark the location, and note the location in your Laboratory Report.
   b. In the laboratory
      1. Place the three slides in a Coplin jar (**FIGURE 21.3**).
      2. Add enough of the culture to the jar to nearly cover the slides; leave a few millimeters of each slide exposed.
      3. Place the cap loosely on the jar, and incubate the jar at room temperature.

PROCEDURE Second Period

1. Collect your biofilm.
   a. Remove one slide, and label it "1."
   b. Wipe one side of the slide clean, leaving the biofilm on the other side.
   c. Air-dry and heat-fix the slide. Save the slide in your drawer for staining.

PROCEDURE Third Period

Repeat steps 1a–c performed during the second period, labeling the next slide "2."

PROCEDURE Fourth Period

1. Observe your biofilm.
   a. Remove and label the last slide "3," and repeat steps 1a–c from the second period.
   b. Compare the appearance of the slides without a microscope. Then stain the three slides with methylene blue (see Figure 5.3 on page 54).
   c. Examine the stained slides microscopically using the oil immersion objective. Record your observations.

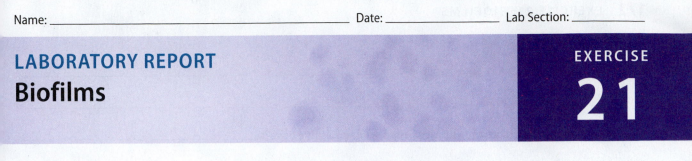

LABORATORY REPORT
Biofilms

PURPOSE _____

_____

_____

HYPOTHESIS

A biofilm will form on your slides. What result would indicate the presence of a biofilm? _____

_____

RESULTS

Where did you set your slides? _____

	Slide 1	Slide 2	Slide 3
Incubated (days)			
Appearance of the slide with the unaided eye			
Microscopic appearance of the slide			

CONCLUSIONS

1. Did your results support your hypothesis? _____

_____

2. Compare the slides. Did any changes occur over successive days? _____

_____

_____

3. What different types of microorganisms did you see? What was the most abundant? _____

_____

_____

QUESTIONS

1. Are the cells evenly distributed on your slides? Provide an explanation for their distribution. _____

_____

_____

2. Is a natural biofilm a pure culture? _____

_____

_____

3. How is a biofilm beneficial to bacteria? _____

_____

_____

4. What cell structures help bacteria attach to a solid surface? _____

_____

_____

5. In law, what is a quorum? _____

How is this related to quorum sensing in bacteria? _____

6. Why is quorum sensing important for biofilm formation? _____

_____

_____

CRITICAL THINKING

1. Bacteria are the second most common cause of artificial implant failures in humans. In these cases, why isn't an infection usually detected from cultures of blood or tissue?

2. Suggest a reason why dental researchers were the first to become aware of the importance of biofilms in disease.

3. Why are biofilms of major concern in industry when disinfectants are utilized?

CLINICAL APPLICATIONS

1. Infection control personnel in a hospital isolated *Mycobacterium chelonae* from 27 patients. All the patients had undergone an endoscopic procedure. Endoscopes were processed after each use with an automated disinfector, which washed the endoscopes with a detergent solution, disinfected them with 2% glutaraldehyde for 10 minutes, and rinsed them with sterile water. What was the source of the infection?

2. A 3-year-old boy was treated with amoxicillin for a middle-ear infection. After he completed antibiotic therapy, his condition improved for a few days, but the infection recurred. Amoxicillin was again prescribed because the *S. pneumoniae* isolated from ear fluid was amoxicillin sensitive. Again the boy was better after completing the antibiotic therapy, but the infection recurred. What is the underlying cause of the recurrent infections?

PART 7

Control of Microbial Growth

EXERCISES

In the 1760s, Lazzaro Spallanzani used heat to destroy microbes. He heated nutrient broth to kill preexisting life in his attempts to disprove the concept of spontaneous generation. In the 1860s, Pasteur heated broth in specially designed flasks and ended the debate over spontaneous generation (see the illustration on the next page). In Exercise 22, we will examine the effectiveness of heat for killing microbes.

The surgeon Joseph Lister was greatly influenced by Pasteur's demonstrations of the omnipresence of microorganisms and his proof that microorganisms cause decomposition of organic matter. Lister had observed the disastrous consequences of compound bone fractures (in which the skin is broken) compared to the relatively safe recovery from simple bone fractures. He had also heard of the treatment of sewage with carbolic acid (phenol) to prevent diseases in villages that had sewage spills. In 1867, Lister wrote,

> It appears that all that is requisite is to dress the wound with some material capable of killing these septic germs*

to prevent disease and death due to microbial growth.

---

*Quoted in W. Bulloch. *The History of Bacteriology.* New York: Dover Publications, 1974, p. 46.

In Part 7, we will examine current methods of controlling microbial growth, with special attention focused on the use of heat (Exercise 22), ultraviolet radiation (Exercise 23), disinfectants (Exercise 24), antimicrobials (Exercise 25), and hand scrubbing (Exercise 26).

Pasteur's experiment disproving spontaneous generation. The experiment involved boiling nutrient broth in an S-neck flask. Microorganisms did not appear in the cooled broth, even after a long period of time.

CASE STUDY: Double-Dipping in the School Cafeteria

At 7 a.m., Lucia went to the College Health Center saying she had been up all night with vomiting and diarrhea. The Health Center had seen 12 other students with the same symptoms since 8 p.m. the day before. All of the students had eaten at the Student Union and had the baked beans. An investigation into the preparation of the baked beans revealed that a new student worker had prepared them. He started work at noon on the previous day. The first thing he did was open the cans of commercially prepared baked beans and put them in the warming tray (50°C) to get ready for dinner. During the afternoon, the worker ate some of the beans directly from the warmer, taking several spoonfuls. Dinner service was from 5:30 to 8:30 p.m.

Questions

1. What two errors did the new worker make?
2. What is the most likely cause of the students' symptoms?

Physical Methods of Control: Heat

OBJECTIVES

After completing this exercise, you should be able to:

1. Compare the bactericidal effectiveness of dry heat and moist heat.
2. Evaluate the heat tolerance of microbes.
3. Define and provide a use for each of the following: *incineration, hot-air oven, pasteurization, boiling,* and *autoclaving*.

BACKGROUND

Extreme temperature is widely used to control the growth of microbes. Generally, heat kills microbes; cold temperatures inhibit microbial growth.

Heat Sensitivity

Bacteria exhibit different tolerances to the application of heat. Heat sensitivity is genetically determined and is partially reflected in the *optimal* growth ranges, which are **psychrophilic** (about 0°C to 20°C),

psychrotrophic (20°C to 30°C), **mesophilic** (25°C to 40°C), **thermophilic** (45°C to 65°C), and **hyperthermophilic** (about 80°C or higher), and by the presence of heat-resistant endospores (**FIGURE 22.1**). Overall, bacteria are more heat resistant than most other forms of life. Container size, cell density, moisture content, pH, and medium composition can all affect the heat sensitivity of organisms.

Types of Heating

Heat can be applied as dry or moist heat. **Dry heat,** such as that in hot-air ovens or incineration (for example, flaming loops), denatures enzymes, dehydrates microbes, and kills by oxidation effects. A standard application of dry heat in a hot-air oven is 170°C for 2 hours. The heat of hot air is not readily transferred to a cooler body such as a microbial cell. Moisture transfers heat energy to the microbial cell more efficiently than dry air, resulting in the denaturation of enzymes. **Moist heat** methods include pasteurization, boiling, and autoclaving. In **pasteurization,** the temperature

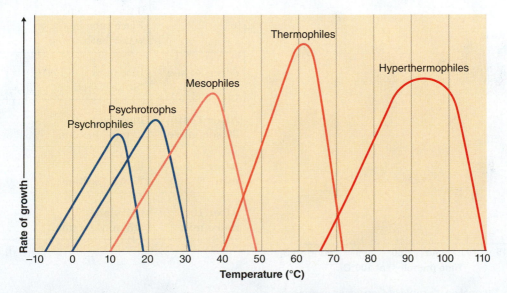

FIGURE 22.1 Typical growth responses of different types of microorganisms to temperature.

is maintained at 63°C for 30 minutes or 72°C for 15 seconds to kill designated organisms that are pathogenic or cause spoilage. **Boiling** (100°C) for 10 minutes will kill vegetative bacterial cells; however, boiling does not inactivate endospores. The most effective method of moist heat sterilization is **autoclaving,** the use of steam under pressure. Increased pressure raises the boiling point of water and produces steam with a higher temperature. Standard conditions for autoclaving are 15 psi, at 121°C, for 15 minutes. This is usually sufficient to kill endospores and render materials sterile.

Measuring Effectiveness

The effectiveness of heat against a specific microbe can be expressed as the thermal death time. **Thermal death time (TDT)** is the length of time required to kill all bacteria in a liquid culture at a given temperature. The less common **thermal death point (TDP)** is the temperature required to kill all bacteria in a liquid culture in 10 minutes. **Decimal reduction time (DRT,** or **D value**) is the time, in minutes, in which 90% of a population of bacteria at a given temperature will be killed (**FIGURE 22.2**).

MATERIALS

FIRST PERIOD

Petri plates containing nutrient agar (2)
Water bath *or*

> Thermometer
> Empty tube
> Beaker
> Hot plate or tripod and wire gauze
> Ice

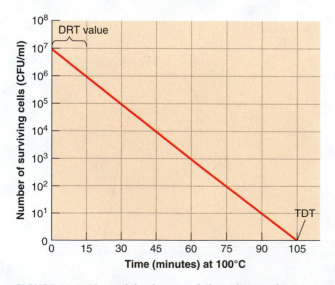

FIGURE 22.2 Thermal death curve of *Clostridium perfringens* endospores at 100°C. The DRT value is 15 minutes, and the thermal death time is 105 minutes.

CULTURES (AS ASSIGNED)

Group A
> *Staphylococcus epidermidis*
> *Escherichia coli*

Group B
> *Bacillus subtilis*
> *Escherichia coli*

Group C
> Mold *(Penicillium)* spore suspension
> *Bacillus subtilis*

DEMONSTRATION

Autoclaved and dry-heated soil

TECHNIQUES REQUIRED

Inoculating loop technique (Exercise 4)
Aseptic technique (Exercise 4)
Plate streaking (Exercise 11)
Graphing (Appendix D)

PROCEDURE First Period

Carefully read the steps before beginning. Remember to use aseptic techniques even though you must work quickly. A summary of the procedure is shown in the chart below.

	Inoculate Plate	Total Time Microbes Are in Water Bath (min)
Start	Inoculate 0 time section. Place tube in water bath.	
After 30 sec	Inoculate "30 sec" section.	
After 1 min, 30 sec more	Inoculate "2 min" section.	
After 3 more min	Inoculate "5 min" section.	
After 10 more min	Inoculate "15 min" section.	

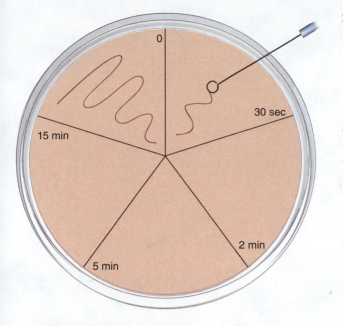

FIGURE 22.3 **Inoculating the nutrient agar plate.** Carefully streak an inoculum inside its section.

Each pair of students is assigned two cultures and a temperature.

Group	Group	Group
A: 63°C ____	A: 72°C ____	A: 100°C ____
B: 63°C ____	B: 72°C ____	B: 100°C ____
C: 63°C ____	C: 72°C ____	C: 100°C ____

You can share water baths as long as the effect of the same temperature is being evaluated.

1. Divide two plates of nutrient agar into five sections each. Label the sections "0," "30 sec," "2 min," "5 min," and "15 min."

2. Use a preset electric water bath, or set up your own water bath on a hot plate and a beaker. To set up a water bath in a beaker, add water to the beaker to a level higher than the level of the broth in the tubes. Do not put the broth tubes into the water bath at this time. Carefully put the thermometer in a test tube of water in the bath.

3. Streak the assigned organisms on the "0" time section of the appropriate plate (**FIGURE 22.3**).

4. Raise the temperature of the bath to the desired temperature, and maintain that temperature. Use ice to adjust the temperature. Why was 63°C selected as one of the temperatures? _____

_____

5. Place the broth tubes of your organism into the bath when the temperature is at the desired point. After 30 seconds, remove the tubes, resuspend the culture, streak a loopful on the corresponding sections, and return the tubes to the water bath. Repeat at 2, 5, and 15 minutes. What is the longest time period that any microbe is exposed to heat? _____

_____

6. When you are done, clean the beaker and return the materials. Incubate the plates, inverted, at 35°C until the next lab period.

PROCEDURE Second Period

1. Record your results and the results for the other organisms tested: (−) = no growth, (−) = minimum growth, (2+) = moderate growth, (3+) = heavy growth, and (4+) = maximum growth.

2. Examine the demonstration plates and record your observations (**FIGURE 22.4**). Collect results from your classmates to complete the data table in your Laboratory Report.

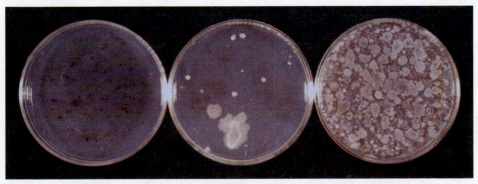

Autoclave   Dry heat   Control

FIGURE 22.4 **Comparing the effects of dry heat and autoclaving.** The control (right) is a pour plate inoculated with soil. The middle plate was inoculated with soil exposed to dry heat (121°C for 1 hour). The soil on the left was autoclaved (121°C for 15 min) and has no bacterial growth.

Name: _____ Date: _____ Lab Section: _____

PURPOSE _____

_____

_____

HYPOTHESIS

Vegetative bacterial cells will be killed at _____ °C in _____ minutes.

RESULTS

Record growth on a scale from (−) to (4+).

Organism	Temperature/Time														
	63°C					72°C					100°C				
	0	30 sec	2 min	5 min	15 min	0	30 sec	2 min	5 min	15 min	0	30 sec	2 min	5 min	15 min
Bacillus subtilis															
Staphylococcus epidermidis															
Escherichia coli															
Mold (Penicillium) spores															

Demonstration Plates

	Control	Autoclaved	Dry-Heated
Number of colonies			
Number of different colonies			

Use a computer graphing application to graph the effect of heating on each organism, or draw your graphs below.

Graph your cultures at _____ °C.

Organism _____   Organism _____

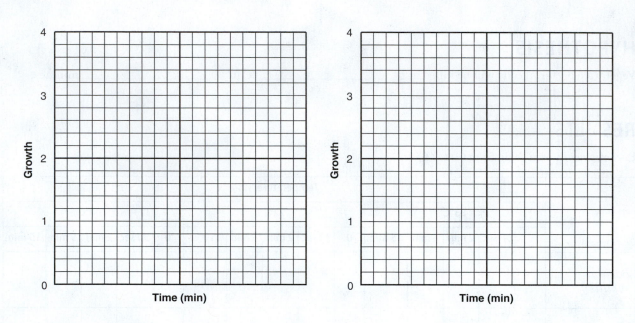

CONCLUSIONS

1. Do your results affirm your hypothesis? _____

_____

2. What conclusions can you draw from your data? _____

_____

3. In the exercise, was the thermal death time or the thermal death point determined? _____

_____

4. Compare the heat sensitivity of fungal spores to that of bacterial endospores. _____

_____

QUESTIONS

1. Compare the effectiveness of autoclaving and dry heat (see Figure 22.4). _____

    _____

    _____

2. Give an example of an application (use) of thermal death time. _____

    _____

    _____

3. Give an example of a nonlaboratory use of each of the following methods to control microbial growth.

    a. Incineration: _____

    b. Pasteurization: _____

    c. Autoclaving: _____

4. Define the term *pasteurization*. What is the purpose of pasteurization? _____

    _____

    _____

CRITICAL THINKING

1. Explain why fungi and *Bacillus* sometimes grow better after heat treatment.

2. Assume that a DRT value for autoclaving a culture is 1.5 minutes. How long would it take to kill all the cells if 10^6 cells were present? What would happen if you stopped the heating process at 9 minutes?

3. Indicators are used in autoclaving to ensure that sterilization is complete. One type of chemical indicator turns color when it has reached a specific temperature; the other type turns color when it has reached a specified temperature and been exposed to steam. Which type of indicator should be used?

4. A biological indicator used in autoclaving is a vial containing 10^9 *Geobacillus stearothermophilus* cells that is placed in the autoclave with the material to be sterilized. After autoclaving, the vial is incubated and examined for growth. Why is this species used rather than *E. coli* or *Bacillus subtilis*?

CLINICAL APPLICATION

The source of Legionellaceae causing hospital-acquired pneumonia has been linked to hospital water supplies. In healthcare settings, hot water coming out of the faucet should not be hotter than 40.6°C to prevent scalding skin. However, *Legionella* can grow if incubated at that temperature. The CDC recommendation for hot water storage (in a hot water heater) is significantly higher. Using the DRT values below, determine how long it would take to kill 10^7 *Legionella* cells at 45°C. Heating water is expensive, so what is the minimum temperature (from the data) that you would recommend for hospital water storage?

Temperature (°C)	DRT value (min)
45	2500
50	380
55	139
60	2.3–4.8
80	0.4–0.6

Physical Methods of Control: Ultraviolet Radiation

OBJECTIVES

After completing this exercise, you should be able to:

1. Examine the effects of ultraviolet radiation on bacteria.
2. Explain the method of action of ultraviolet radiation and light repair of mutations.

BACKGROUND

Radiant energy comes to Earth from the sun and other extraterrestrial sources, and some is generated on Earth from natural and human-made sources. The **radiant energy spectrum** is shown in **FIGURE 23.1**. Radiation differs in wavelength and energy. The shorter wavelengths have more energy. X rays and gamma rays are forms of **ionizing radiation.** Their principal effect is to ionize water into *highly reactive free radicals* (with unpaired electrons) that can break strands of DNA. Many variables, such as the age of the cells, media composition, and temperature, influence the effect of radiation.

Ultraviolet Radiation

Some **nonionizing** wavelengths are essential for biochemical processes. The main absorption wavelengths for green algae, green plants, and photosynthetic bacteria are shown in **FIGURE 23.1a**. Animal cells synthesize vitamin D in the presence of light around 300 nm. Nonionizing radiation between 15 and 400 nm is called **ultraviolet (UV).** Wavelengths below 200 nm are absorbed by air and do not reach living organisms. The most lethal wavelengths, sometimes called *biocidal,* are in the **UVC** range, 200 to 290 nm. These wavelengths correspond to the optimal absorption wavelengths of DNA (**FIGURE 23.1b**). **UVB** wavelengths (290–320 nm) can also cause damage to DNA. **UVA** wavelengths (320–400 nm) are not as readily absorbed and are therefore less active on living organisms.

Effects of Ultraviolet Radiation

Ultraviolet light induces *thymine dimers* in DNA, which result in a mutation. Mutations in critical genes may result in the death of the cell unless the damage is repaired. When thymine dimers are exposed to visible light, *photolyases* are activated; these enzymes split the dimers, restoring the DNA to its undamaged state. This is called **light repair** or **photoreactivation.** Another repair mechanism, called **dark repair,** is independent of light. Dimers are removed by endonuclease, DNA polymerase replaces the nucleotides, and DNA ligase seals the sugar-phosphate backbone.

As a sterilizing agent, ultraviolet radiation is limited by its poor penetrating ability. It is used to sterilize some heat-labile solutions, to decontaminate hospital operating rooms and food-processing areas, and to disinfect water.

In this exercise, we will investigate the penetrating ability of ultraviolet light and light repair by using lamps of the desired wavelength (**FIGURE 23.2**).

MATERIALS

FIRST PERIOD

Petri plates containing nutrient agar (3)
Sterile cotton swabs (3)
Covers (choose one): Gauze; 3, 6, or 12 layers of paper; cloth; aluminum foil; clear glass; sunglasses; or plastic
Ultraviolet lamp (260 nm)
Plastic safety glasses

CULTURES (AS ASSIGNED)

Bacillus subtilis
Serratia marcescens
Staphylococcus epidermidis
Micrococcus luteus

TECHNIQUE REQUIRED

Aseptic technique (Exercise 4)

PROCEDURE First Period

1. Swab the surface of each plate with *one* of the cultures; to ensure complete coverage, swab the surface in two directions. Label the plates "A," "B," and "C." Label one-half of each plate "control."
2. Remove the lid of an inoculated plate, and cover the control half with one of the covering materials (**FIGURE 23.3**). Cover one-half of each of the remaining plates with the same material.

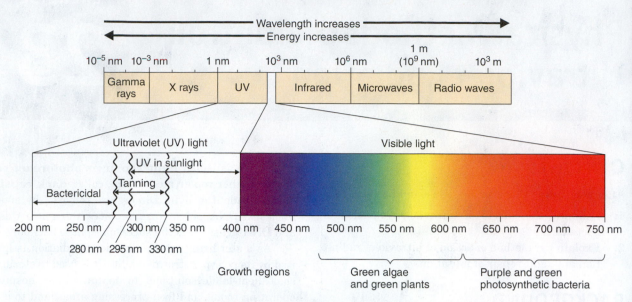

(a) Radiant energy spectrum and absorption of light for growth

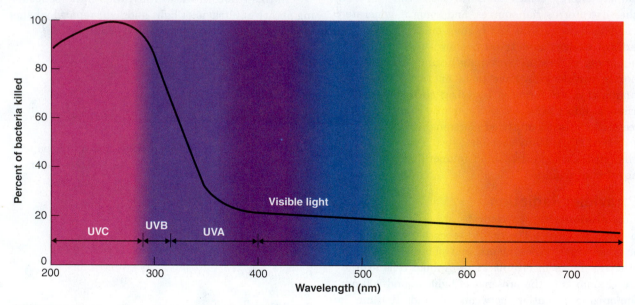

(b) Biocidal effectiveness of radiant energy between 200 and 700 nanometers (from UV to visible red light)

FIGURE 23.1 **Radiant energy.**

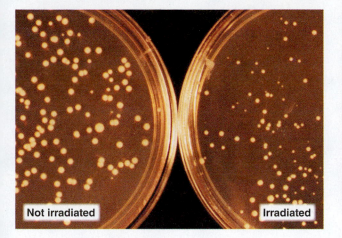

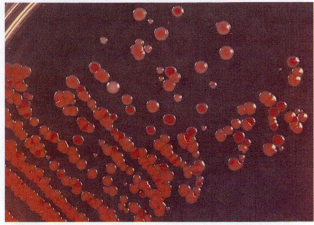

(a) *Saccharomyces cerevisiae* yeast

(b) *Serratia marcescens* bacteria colonies on nutrient agar after incubation at 25°C

FIGURE 23.2 **Effects of UV radiation. (a)** The plate on the right was irradiated for 25 seconds. **(b)** Note the color variations caused by mutations in prodigiosin production.

 Do not look at the ultraviolet light, and do not leave your hand exposed to it. Wear safety glasses.

3. Place each plate directly under the ultraviolet light 24 cm from the light with the *lid off,* agar-side up, with the covering material on one-half of the plate. Why should the lid be removed? _____

_____

_____

Plate A: Expose to UV light for 30 seconds. Remove the covering material, replace the lid, and incubate in a dark environment at room temperature.

Plate B: Expose to UV light for 30 seconds. Remove the covering material, and replace the Petri plate lid. Incubate in sunlight or under lights at room temperature.

Plate C: Expose to UV light for 60 seconds. Remove covering material, replace the lid, and incubate in a dark environment at room temperature.

4. Incubate all three plates at room temperature until the next period.

FIGURE 23.3 **With the lid removed, cover one-half of an inoculated Petri plate with one of the covering materials.**

PROCEDURE Second Period

Examine all plates and record your results. Observe the results of students using the other organisms. Record growth as (−) = no growth; (+) = minimum growth; (2+) = moderate growth; (3+) = heavy growth; and (4+) = maximum growth.

Name: _____ Date: _____ Lab Section: _____

LABORATORY REPORT
Physical Methods of Control:
Ultraviolet Radiation

PURPOSE _____

_____

_____

HYPOTHESES

1. UV radiation _____ bacteria. Incubation in light will increase/decrease/not affect survival.

2. The bacterial species that will survive UV radiation the best is _____

RESULTS

1. What organism did you use? _____

2. What did you use to cover one-half of each plate? _____

3. Sketch your results. Note any pigmentation. Identify the control half of each plate.

A

B

C

4. Record your results. Use your classmates' results to complete the table. Record growth on a scale from (−) to (4+).

	Covered Control	A: 30-sec UV Exposure; Incubated in Dark	B: 30-sec UV Exposure; Incubated in Light	C: 60-sec UV Exposure; Incubated in Dark
B. subtilis				
M. luteus				
S. marcescens				
St. epidermidis				

CONCLUSIONS

1. Do you accept your hypotheses? _____

_____

2. What effect did UV radiation have? _____

_____

3. Did UV radiation penetrate plastic? _____

_____

4. What evidence, if any, did you see of light repair? _____

_____

QUESTIONS

1. If the *Bacillus* had sporulated before exposure to radiation, would that have affected the results? _____

_____

_____

2. What are the variables in ultraviolet radiation treatment? _____

_____

_____

3. Many of the microorganisms found on environmental surfaces are pigmented. Of what possible advantage is

the pigment? _____

_____

_____

CRITICAL THINKING

1. Why are there still some colonies growing in the areas exposed to ultraviolet light?

2. How might the results differ if a UVA lamp were used? A UVB lamp?

3. Considering your results, discuss the possible effects of UV radiation on the ecology of a lake if the UV radiation has the same effect on the lake bacteria as it did on the bacteria in your experiment.

CLINICAL APPLICATION

Legionella pneumophila in hospital water-distribution systems has been linked to healthcare-associated infections. As a result, hospitals are investigating methods for eradicating *Legionella* in their water systems. Plate counts were used to determine the effects of heat (50°C), ozone, and UV light. Data are shown below. Which is the most effective method? If the rates continue, which method(s) will kill all the bacteria in 320 minutes?

Time (min)	*L. pneumophila* (CFU/ml)			
	Control at 23°C	50°C	Ozone at 23°C	UV light at 23°C
0	1×10^7	1×10^7	1×10^7	1×10^7
20	2×10^7	1×10^6	1×10^7	1×10^3
40	5×10^7	1×10^5	1×10^6	1×10^3
80	6×10^7	1×10^4	5×10^5	1×10^3
160	9×10^7	1×10^3	1×10^4	1×10^3
320	1×10^8			

Chemical Methods of Control: Disinfectants and Antiseptics

One nineteenth-century method of avoiding cholera: Wear a pouch of FOUL-SMELLING HERBS around your neck. If the odor is bad enough, disease carriers will spare you the trouble of avoiding them. – ANONYMOUS

OBJECTIVES

After completing this exercise, you should be able to:

1. Define the following terms: *disinfectant* and *antiseptic*.
2. Describe the use-dilution test.
3. Evaluate the relative effectiveness of various chemical substances as antimicrobial agents.

BACKGROUND

A wide variety of chemicals called **antimicrobial agents** are available for controlling the growth of microbes. Chemotherapeutic agents are used internally and will be evaluated in another exercise. **Disinfectants** are chemical agents used on inanimate objects to lower the level of microbes on their surfaces; **antiseptics** are chemicals used on living tissue to decrease the number of microbes. Disinfectants and antiseptics affect bacteria in many ways. Those that result in bacterial death are called **bactericidal agents.** Those causing temporary inhibition of growth are **bacteriostatic agents.**

No single chemical is the best to use in all situations. Antimicrobial agents must be matched to specific organisms and environmental conditions. Additional variables to consider in selecting an antimicrobial agent include pH, solubility, toxicity, organic material present, and cost. In evaluating the effectiveness of an antimicrobial agent, the concentration, length of contact, and whether it is lethal (*-cidal*) or inhibiting (*-static*) are the important criteria. Before a disinfectant is selected, the decimal reduction time (DRT) for the most common and persistent microbes identified at a health care facility should be determined. The **DRT** is the time it takes to kill 90% of the test microbial population. The standard method for measuring the effectiveness of a chemical agent is the **American Official Analytical Chemist's use-dilution test.** For most purposes, three strains of bacteria are used in this test: *Salmonella enterica* Choleraesuis, *Staphylococcus aureus,* and *Pseudomonas aeruginosa.* To perform a use-dilution test, metal rings are dipped into standardized cultures of the test bacteria grown in liquid media, removed, and dried. The rings are next placed into a solution of the disinfectant at the concentration recommended by the manufacturer for 10 minutes at 20°C. The rings are then transferred to a nutrient medium to permit the growth of any surviving bacteria. The effectiveness of the disinfectant can then be determined by the amount of resulting growth. The use-dilution test is limited to bactericidal compounds and cannot be used to evaluate bacteriostatic compounds.

In this exercise, we will perform a modified use-dilution test.

MATERIALS

FIRST PERIOD

Petri plates containing nutrient agar (2)
Sterile water
Sterile tubes (3)
Sterile 5-ml pipettes (2)
Sterile 1-ml pipettes (2)
Test substance: chemical agents such as bathroom cleaner, floor cleaner, mouthwash, lens cleaner, and acne cream. Bring your own.

CULTURES (AS ASSIGNED)

Escherichia coli
Staphylococcus epidermidis
Pseudomonas aeruginosa `BSL-2`
Staphylococcus aureus `BSL-2`

TECHNIQUES REQUIRED

Inoculating loop technique (Exercise 4)
Aseptic technique (Exercise 4)
Pipetting (Appendix A)

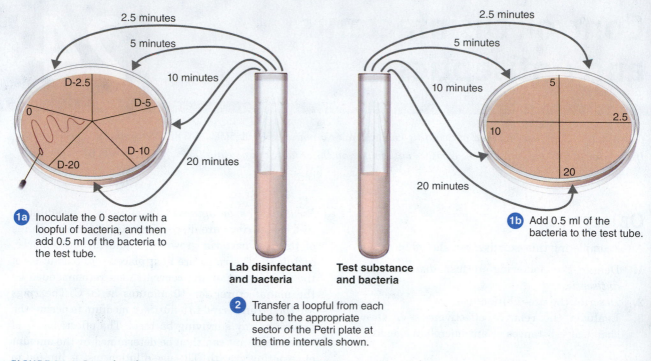

1a Inoculate the 0 sector with a loopful of bacteria, and then add 0.5 ml of the bacteria to the test tube.

Lab disinfectant and bacteria

Test substance and bacteria

2 Transfer a loopful from each tube to the appropriate sector of the Petri plate at the time intervals shown.

1b Add 0.5 ml of the bacteria to the test tube.

FIGURE 24.1 Procedure for testing the effectiveness of a disinfectant or antiseptic.

PROCEDURE First Period

Work with another student group so that you test the effects of the same disinfectant or antiseptic against the two bacteria available.

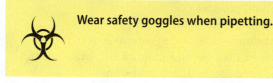

Wear safety goggles when pipetting.

1. Using sterile water, prepare a dilution of the test substance in a sterile tube, diluted to the strength at which it is normally used. If it is a paste, it must be suspended in sterile water. If the test substance is normally used at full strength, then don't dilute it for this experiment.

2. Transfer 5 ml of the test substance prepared in step 1 to a sterile tube. Label the tube. Add 5 ml of your laboratory disinfectant to another sterile tube. What is the disinfectant you use to disinfect your lab bench?_____

   Label the tube.

3. Divide one plate of nutrient agar into five sections. Label the sections "0," "D-2.5," "D-5," "D-10," and "D-20." The D stands for laboratory disinfectant.

4. Label the other nutrient agar plate for the other chemical, and divide it into four sections. Label the sections "2.5," "5," "10," and "20."

5. Inoculate the 0 sector with a loopful of your assigned bacteria.

6. Aseptically add 0.5 ml of the assigned culture to each tube prepared in step 2.

7. Transfer one loopful from each tube to a corresponding sector at 2.5 minutes, 5 minutes, 10 minutes, and 20 minutes (**FIGURE 24.1**).

8. Incubate the plates, inverted, at 35°C until the next lab period. (Discard the chemical/bacteria mixtures in the To Be Autoclaved area.)

PROCEDURE Second Period

Observe the plates for growth. Record the growth as (−) = no growth, (−) = minimum growth, (2+) = moderate growth, (3+) = heavy growth, and (4+) = maximum growth. Observe the results of students using the other organism.

LABORATORY REPORT

Chemical Methods of Control: Disinfectants and Antiseptics

PURPOSE _____

_____

_____

HYPOTHESIS

The disinfectants tested will kill all bacteria in _____ minutes.

RESULTS

1. What organism did you use? _____

    Be sure to record the chemical or product you tested.

Times of Exposure (min)	Amount of Growth (Record growth on a scale of − to 4+.)		
	Control	Lab Disinfectant	Test Substance
0			
2.5			
5			
10			
20			

2. Classmates' results for: _____ bacteria

Time of Exposure (min)	Amount of Growth		
	Control	Lab Disinfectant	Test Substance
0			
2.5			
5			
10			
20			

CONCLUSIONS

1. Do your results confirm your hypothesis? _____

_____

2. Was your test chemical effective? Briefly explain. _____

_____

_____

3. Was it more effective than the lab disinfectant? Briefly explain. _____

_____

_____

4. Was your test chemical bactericidal or bacteriostatic? _____

QUESTIONS

1. Was this a fair test? Is it representative of the effectiveness of the test substance? _____

_____

_____

2. Read the label of the preparation you tested. What is (are) the active ingredient(s)? _____

_____

3. Using your textbook or another reference, find the method of action of the active ingredient(s) in the test

    substance. _____

    _____

    _____

CRITICAL THINKING

1. How could the procedures used in this experiment be altered to measure bacteriostatic effects?

2. In the use-dilution test, a chemical is evaluated by its ability to kill 10^6 to 10^8 dried *Clostridium sporogenes* or *Bacillus subtilis* endospores. Why is this considered a stringent test?

CLINICAL APPLICATION

The effectiveness of disinfectants can be measured in DRT values. The DRT values for contact lens disinfectants against *Serratia marcescens* are as follows:

Disinfectant	DRT Value (min)	Disinfectant	DRT Value (min)
Chlorhexidine, 0.005%	2.8	Thimerosal, 0.002%	138.9
Hydrogen peroxide, 3%	3.1	Polyquaternium-1, 0.001%	383.3

Which disinfectant is most effective? _____ What is the minimum time that lenses with 10^2

bacteria should be soaked in chlorhexidine? _____

In polyquaternium-1? _____ What if the lenses are contaminated with *Staphylococcus* or

Acanthamoeba? _____

_____

Why isn't a higher concentration of disinfectant used? _____

Chemical Methods of Control: Antimicrobial Drugs

The aim of medicine is to prevent disease and PROLONG *life; the ideal of medicine is to* ELIMINATE *the need of a physician.* – W I L L I A M J A M E S M A Y O

OBJECTIVES

After completing this exercise, you should be able to:

1. Define the following terms: *antibiotic, antimicrobial drug,* and *MIC*.
2. Perform an antibiotic sensitivity test.
3. Provide the rationale for the agar diffusion technique.

BACKGROUND

The observation that some microbes inhibited the growth of others was made as early as 1874. Pasteur and others observed that infecting an animal with *Pseudomonas aeruginosa* protected the animal against *Bacillus anthracis.* Later investigators coined the word **antibiosis** (against life) for this inhibition and called the inhibiting substance an **antibiotic,** a substance produced by a microorganism that inhibits other microorganisms. In 1928, Alexander Fleming observed antibiosis around a *Penicillium* mold growth on a culture of staphylococci. He found that culture filtrates of *Penicillium* inhibited the growth of many gram-positive cocci and *Neisseria* spp. In 1940, Selman A. Waksman isolated the antibiotic streptomycin, produced by an actinomycete. This antibiotic was effective against many bacteria that penicillin did not affect. Actinomycetes remain an important source of antibiotics. *Streptomyces* bacteria produce nearly 70% of all antibiotics. Today, research investigators look for antibiotic-producing actinomycetes and fungi in soil and antimicrobial compounds made by plants and animals. Antimicrobial chemicals used internally, whether natural (antibiotics) or synthetic, are called **antimicrobial drugs.**

To treat an infectious disease, a physician or dentist needs to select the correct antimicrobial agent intelligently and administer the appropriate dose; then the practitioner must follow that treatment to be aware of resistant forms of the organism that might occur. The clinical laboratory isolates the **pathogen** (disease-causing organism) from a clinical sample and determines its sensitivity to antimicrobial agents.

Disk-Diffusion Method

In the **disk-diffusion method,** a Petri plate containing an agar growth medium is inoculated uniformly over its entire surface. Paper disks impregnated with various antimicrobial agents are placed on the surface of the agar. During incubation, the antimicrobial agent *diffuses* from the disk, from an area of high concentration to an area of lower concentration. An effective agent will inhibit bacterial growth, and measurements can be made of the size of the **zones of inhibition** around the disks (**FIGURE 25.1**). The zone size is affected by such factors as the diffusion rate of the antimicrobial agent and the growth rate of the organism. To minimize the variance between laboratories, the standardized **Kirby-Bauer test** for agar diffusion methods is performed in many clinical laboratories with strict quality controls. This test uses *Mueller-Hinton agar.* Mueller-Hinton agar allows the antimicrobial agent to diffuse freely.

Minimum Inhibitory Concentration

The **minimum inhibitory concentration (MIC)** of an antibiotic is determined by testing for bacterial growth in dilutions of the antibiotic in nutrient broth. When the MIC is determined, inhibition zones can be correlated with MICs. Then, inhibition zones can be compared to a standard table (**TABLE 25.1**) to use the disk-diffusion method to determine susceptibility.

In this exercise, we will evaluate antimicrobial agents by the disk-diffusion method.

MATERIALS

FIRST PERIOD

Petri plate containing Mueller-Hinton agar
Sterile cotton swab
Dispenser and antimicrobial disks
Forceps
Alcohol

SECOND PERIOD

Ruler

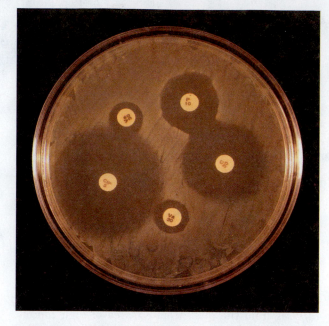

FIGURE 25.1 Zones of inhibition. Antimicrobial sensitivity of this *Staphylococcus aureus* culture is demonstrated by the Kirby-Bauer disk-diffusion method.

CULTURES (AS ASSIGNED)

Staphylococcus epidermidis broth
Escherichia coli broth
Pseudomonas aeruginosa broth **BSL-2**

TECHNIQUE REQUIRED

Aseptic technique (Exercise 4)

PROCEDURE First Period

1. Aseptically swab the assigned culture onto the appropriate plate. Swab in three directions to ensure complete plate coverage (**FIGURE 25.2**). Why is complete coverage essential? _____

   _____

   Let stand at least 5 minutes.
2. Follow procedure a or b.
   a. Place the antimicrobial-impregnated disks by pushing the dispenser over the agar (**FIGURE 25.3a**). Sterilize your loop, and touch each disk with the sterile inoculating loop to ensure better contact with the agar. Record the agents and the disk codes in your Laboratory Report. (Circle the corresponding chemicals in Table 25.1.)
   b. Sterilize forceps by dipping them in alcohol and burning off the alcohol.

   > ⚠️ **While the alcohol is burning off, hold the forceps pointed down. Keep the beaker of alcohol away from the flame.**

   Obtain a disk impregnated with an antimicrobial agent, and place it on the surface of the agar (**FIGURE 25.3b**). Gently tap the disk with the forceps to ensure better contact with the agar. Repeat, placing five different disks the same distance apart on the Petri plate. See the location of the disks in **FIGURE 25.3c**. Record the agents and the disk codes in your Laboratory Report. (Circle the corresponding chemicals in Table 25.1.)
3. Incubate the plate, inverted, at 35°C until the next period.

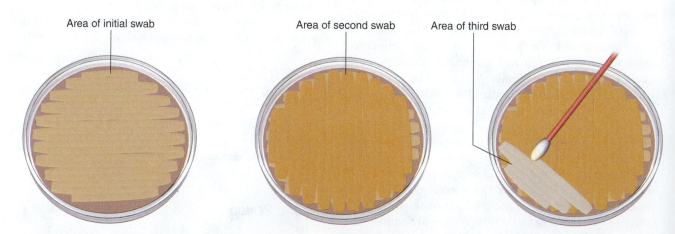

FIGURE 25.2 Inoculating the Mueller-Hinton agar plate. Dip a cotton swab in the culture to be tested, and swab across the surface of the agar *without leaving any gaps.* Using the same swab, swab the agar in a direction perpendicular to the first inoculum. Repeat, swabbing the agar at a 45° angle to the first inoculum.

Area of initial swab   Area of second swab   Area of third swab

TABLE 25.1 INTERPRETING INHIBITION ZONES OF TEST CULTURES

Disk Symbol	Antimicrobial Agent	Disk Content	Diameter of Zones of Inhibition (mm)		
			Resistant	Intermediate	Susceptible
AM	Ampicillin when testing gram-negative bacteria	10 µg	≤13	14–16	≥17
	Ampicillin when testing gram-positive bacteria	10 µg	≤28	—	≥29
C	Chloramphenicol	30 µg	≤12	13–17	≥18
CAZ	Ceftazidime	30 µg	≤14	15–17	≥18
CB	Carbenicillin	100 µg	≤19	20–22	≥23
	Carbenicillin when testing *Pseudomonas*	100 µg	≤13	14–16	≥17
CF	Cephalothin	30 µg	≤14	15–17	≥18
CIP	Ciprofloxacin	5 µg	≤15	16–20	≥21
E	Erythromycin	15 µg	≤13	14–22	≥23
FOX	Cefoxitin	30 µg	≤14	15–17	≥18
G	Sulfisoxazole	25 µg	≤12	13–16	≥17
GM	Gentamicin	10 µg	≤12	13–14	≥15
IPM	Imipenem	10 µg	≤13	14–15	≥15
P	Penicillin when testing staphylococci	10 units	≤28	—	≥29
	Penicillin when testing other bacteria	10 units	≤14	—	≥15
PB	Polymyxin	300 units	≤8	9–11	≥12
R	Rifampin	5 µg	≤16	17–19	≥20
S	Streptomycin	10 µg	≤11	12–14	≥15
SXT	Trimethoprim-sulfamethoxazole (TMP-SMZ)	1.25 µg 23.75 µg	≤10	11–15	≥16
Te	Tetracycline	30 µg	≤14	15–18	≥19
Va	Vancomycin when testing *Staphylococcus* spp.	30 µg	—	—	≥15
	Vancomycin when testing enterococci	30 µg	≤14	15–16	≥17

Clinical and Laboratory Standards Institute. *Performance Standards for Antimicrobial Susceptibility Tests,* 2014.

PROCEDURE Second Period

1. Measure the zones of inhibition in millimeters, using a ruler on the underside of the plate (see Figure 25.3c). If the diameter is difficult to measure, measure the radius from the center of the disk to the edge of the zone. Multiply the radius by 2 to get the diameter of the zone.

2. Record the zone size and, based on the values in Table 25.1, indicate whether the organism is susceptible, intermediate, or resistant. Record the results of students using the other two bacteria. Colonies within a zone of inhibition are resistant to that antibiotic (**FIGURE 25.4**).

(a) Place the disk dispenser over the agar, and push the plunger to dispense the disk.

(b) Place disks impregnated with antimicrobial agents on an inoculated culture medium with sterile forceps to get the pattern shown in **(c)**.

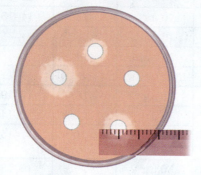

(c) After incubation, measure the diameters of zones of inhibition.

FIGURE 25.3 Disk-diffusion method.

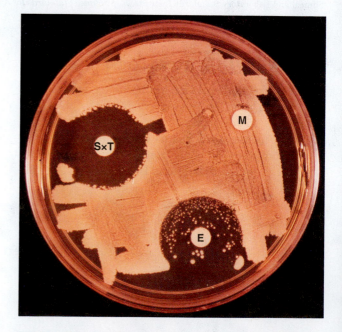

FIGURE 25.4 Antibiotic resistance. This strain of *Staphylococcus aureus* is resistant to methicillin (M) and sensitive to Trimethoprim-sulfamethoxazole (SxT). Some erythromycine-resistant mutant cells grew around the erythromycin (E) disk.

Chemical Methods of Control: Antimicrobial Drugs

PURPOSE

_____

_____

_____

HYPOTHESES

1. All antibiotics will be equally effective against gram-positive and gram-negative bacteria. Agree/disagree

2. All antibiotics will be equally effective against gram-negative *E. coli* and *P. aeruginosa* bacteria. Agree/disagree

RESULTS

Antimicrobial Agent	Disk Code	*Staphylococcus epidermidis*		*Escherichia coli*		*Pseudomonas aeruginosa* BSL-2	
		Zone Size	S, I, or R*	Zone Size	S, I, or R*	Zone Size	S, I, or R*
1.							
2.							
3.							
4.							
5.							
6.							
7.							
8.							

*S = susceptible; I = intermediate; R = resistant.

CONCLUSIONS

1. Do you accept or reject your hypotheses? _____

    _____

2. Which antimicrobial agents were most effective against each organism? _____

    _____

    _____

3. Why isn't one antimicrobial agent equally effective against all three bacteria? _____

    _____

    _____

QUESTIONS

1. Is the disk-diffusion technique measuring bacteriostatic or bactericidal activity? Briefly explain. _____

    _____

    _____

2. In which growth phase is an organism most sensitive to an antimicrobial agent? _____

3. Why is the disk-diffusion technique not a perfect indication of how the drug will perform in vivo? What other

    factors are considered before using the antimicrobial agent in vivo? _____

    _____

    _____

CRITICAL THINKING

1. What effect would the presence of tetracycline in the body have on penicillin therapy?

2.   The following results were obtained from a disk-diffusion test against a bacterium.

Antibiotic	Zone of Inhibition (mm)
A	6
B	18
C	11
D	18

Which drug should be used to treat an infection caused by this bacterium? Briefly explain.

CLINICAL APPLICATION

The broth dilution test can be used to determine the effectiveness of an antibiotic. In this test, serial dilutions of the antibiotic were set up in the wells of a microtiter plate. Equal amounts of broth culture of *Staphylococcus aureus* were added to each well. After incubation, the wells were examined for bacterial growth. Wells with no growth were subcultured in nutrient broth without the antibiotic. Results were recorded as (+) for growth and (−) for no growth.

Antibiotic	Dilution	Growth	Growth in Subculture
A	1:10 through 1:70 1:80 1:90 1:100 1:200 through 1:500	− − − − +	− − + + +
B	1:10 through 1:150 1:160 1:170 1:180 1:190 through 1:500	− − + + +	− + + + +

What is the minimum bactericidal concentration of each antibiotic? _____

What is the minimum bacteriostatic concentration? _____

Which antibiotic is more effective against *Staphylococcus aureus*? _____

Which antibiotic is more effective against *Salmonella enterica*? _____

Effectiveness of Hand Scrubbing

OBJECTIVES

After completing this exercise, you should be able to:

1. Evaluate the effectiveness of handwashing and a surgical scrub.
2. Explain the importance of aseptic technique in the hospital environment.

BACKGROUND

The skin is sterile during fetal development. After birth, a baby's skin is colonized by many bacteria for the rest of its life. As an individual ages and changes environments, the microbial population changes to match the environmental conditions. The microorganisms that are more or less permanent are called **normal microbiota.** Microbes that are present only for days or weeks are referred to as **transient microbiota.**

Importance of Handwashing

Discovery of the importance of handwashing in preventing disease is credited to Ignaz Semmelweis at Vienna General Hospital in 1846. He noted that the lack of aseptic methods was directly related to the incidence of puerperal fever and other diseases. Medical students would go directly from the autopsy room to the patient's bedside and assist in child delivery without washing their hands. Less puerperal sepsis occurred in patients attended by midwives, who did not touch cadavers. Semmelweis ordered the medical students to wash their hands with a chloride of lime solution, a policy that caused the death rate due to puerperal sepsis to drop from 12% to 1.2% in one year. Guidelines from the Centers for Disease Control and Prevention (CDC) state that "handwashing is the single most important procedure for preventing healthcare-associated infections," yet studies in hospitals show handwashing rates average only 40%.*

*CDC, *MMWR* 51:RR 16. October 25, 2002.

Clinical Handwashing

A layer of oil and the overlapping layers of cells of the skin prevent handwashing from removing all bacteria. Soap helps remove the oil, however, and scrubbing will maximize the removal of bacteria. Hospital procedures require personnel to wash their hands before attending a patient and to perform a complete surgical scrub—removing the transient and many of the resident microbiota—before surgery. Usually, 3 to 5 minutes of scrubbing with an antimicrobial soap will remove transient microbiota. The surgeon's skin is never sterilized. Only burning or scraping it off would achieve that.

In this exercise, we will examine the effectiveness of washing skin with soap and water. Only organisms capable of growing aerobically on nutrient agar will be observed. Because organisms with different nutritional and environmental requirements will not grow, this procedure will involve only a minimum number of the skin microbiota.

MATERIALS

FIRST PERIOD

Petri plates containing nutrient agar (2)
Scrub brush
Bar soap or liquid soap (bring one from home)
Waterless hand cleaner

TECHNIQUE REQUIRED

Colony morphology (Exercise 3)

PROCEDURE First Period

1. Select two nutrient agar plates.
   a. Divide one nutrient agar plate into four quadrants. Label the sections 1 through 4. Label the plate "Water."
   b. Divide the other nutrient agar plate into five sections. Label the sections 1 through 5. Label the plate "Soap." Which soap will you use? ____

2. Use the "Water" plate first. Touch section 1 with your fingers, and wash well *without* soap for your normal handwashing time (**FIGURE 26.1**). Don't touch the faucet with your "test" hand to turn the water off. Shake off excess water, and, while your hands are still wet, touch section 2. Do not dry your fingers with a towel. Wash again, and, while your hands are still wet, touch section 3. Wash a final time, and touch section 4. Touch the same fingers to the plate each time.

3. Use your same hand on the plate labeled "Soap." Wash well with soap, rinse, shake off the excess water, and then touch section 1.

4. Wash again with soap, rinse, shake off the excess water, and then touch section 2.

5. Using a brush and soap, scrub your hand for 2 minutes, rinse, and shake off the excess water; then touch section 3.

6. Repeat the soap-and-brush scrub for 4 minutes, rinse, and shake off the excess water; then touch section 4.

7. Let your hands air dry, then use a waterless hand-cleaning product, rinse with water, and then touch section 5. What are the active ingredients in the product? _____

_____

8. Incubate the plates, inverted, at 35°C until the next period.

9. Speculate on your expected results, and record them in your Laboratory Report.

PROCEDURE Second Period

Record the growth as (−) = no growth, (+) = minimum growth, (2+) = moderate growth, (3+) = heavy growth, and (4+) = maximum growth.

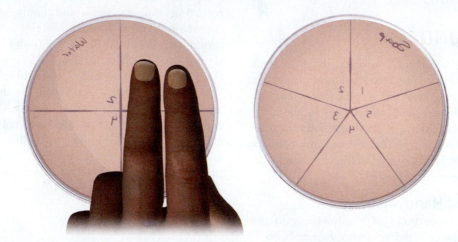

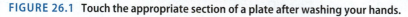

FIGURE 26.1 Touch the appropriate section of a plate after washing your hands.

LABORATORY REPORT
Effectiveness of
Hand Scrubbing

PURPOSE _____

_____

_____

EXPECTED RESULTS

Indicate the relative amounts of growth you *expect* in each quadrant on a scale of (−) to (4+).

Section	Water Alone	Soap
1.	(No washing)	
2.		
3.		
4.		
5.		(Waterless hand cleaner)

RESULTS

Indicate the relative amounts of growth in each quadrant.

Section	Water Alone	Soap (type: _____)
1.	(No washing)	
2.		
3.		
4.		
5.		(Waterless hand cleaner)

CONCLUSIONS

1. Did your results differ from your expected results? _____ Briefly explain why or why not. _____

_____

2. Using your classmates' data, compare the results from bar soap and liquid soap. _____

_____

_____

QUESTIONS

1. What is a surgeon trying to accomplish with a 5-minute scrub with a brush followed by an antiseptic?

_____

_____

2. How do normal microbiota and transient microbiota differ? _____

_____

CRITICAL THINKING

If most of the normal microbiota and transient microbiota aren't harmful on skin, then why must hands be scrubbed before surgery?

CLINICAL APPLICATION

The following data were collected from soaps after 1 week of use at a hospital nurses' handwashing station. Neither bacteria nor fungi were isolated from any of the products before use.

Aerobic bacteria were isolated from 25 soap products. Data are expressed as percentage of soap products contaminated.

Organisms	Bar Soap	Liquid Soap: Type of Closure			
		Screw Top	Slit/Flip	Flip/Pump	Pump
Total bacteria	95%	71%	39%	10%	0%
Gram-positive cocci	95%	71%	39%	10%	0%
Gram-negative rods	12%	1%	1%	1%	0%

What conclusions can you draw from these data?

PART 8

Microbial Genetics

EXERCISES

All of the characteristics of bacteria—including growth patterns, metabolic activities, pathogenicity, and chemical composition—are inherited. These traits are transmitted from parent cell to offspring through genes. **Genetics** is the study of genes: how they carry information, how they are replicated and passed to the next generation or to another organism, and how their information is expressed within an organism to determine the particular characteristics of that organism (see the illustration on the next page).

The information stored in genes is called the **genotype.** Not all of the genes may be expressed. **Phenotype** refers to the actual expressed characteristics, such as the ability to perform certain biochemical reactions (the synthesis of a capsule or pigment, for example).

The control of gene expression is explained by the operon model (Exercise 27). This model was first proposed by François Jacob and Jacques Monod to explain why *E. coli* makes lactose-catabolizing enzymes in the presence, but not in the absence, of lactose. Bacteria are invaluable to the study of genetics because large numbers can be cultured inexpensively, and their relatively simple genetic composition facilitates studying the structure and function of genes.

Bacteria undergo genetic change because of mutation or recombination. These genetic changes result in a change in the genotype. When a change occurs, in most instances we can expect that the product encoded by certain genes will be changed. An enzyme encoded by a changed gene may become inactive. This might be disadvantageous or even lethal if the cell loses a phenotypic trait it needs. Some genetic changes may be beneficial. For instance, if an altered enzyme encoded by a changed gene has new enzymatic activity, the cell may be able to grow in new environments.

Genetic change in microbial populations is relatively easy to observe. In 1949, Howard B. Newcombe wrote,

Numerous bacterial variants are known which will grow in environments unfavorable to the parent strain, and to explain their occurrence two conflicting hypotheses have been advanced. The first assumes that the particular environment produces the observed change in some bacteria exposed to it, whereas the second assumes that the variants arose spontaneously during growth under normal conditions. *

Salvador Luria and Max Delbrück first demonstrated such **spontaneous mutation** in 1943.

It is often possible to use mutated bacteria to get a different product from an altered gene (Exercise 28). New genes can be inserted into bacteria by transformation (Exercise 29). The desired gene can be isolated and identified by DNA fingerprinting (Exercise 30). Artificial manipulation of genes is known as **genetic engineering.** In genetic engineering, genes can be inserted into a vector and incorporated into a bacterium by transformation (Exercise 31).

Cancer is usually the result of mutations in the nucleic acid of a chromosome. In Exercise 32, we will perform a test using principles of microbial genetics to identify possible cancer-inducing substances.

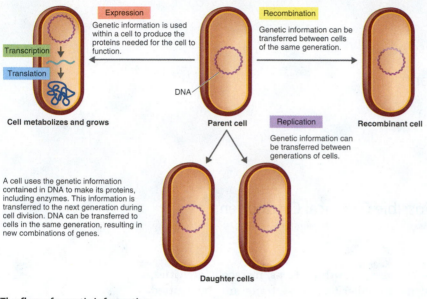

Expression
Genetic information is used within a cell to produce the proteins needed for the cell to function.

Transcription

Translation

DNA

Recombination
Genetic information can be transferred between cells of the same generation.

Cell metabolizes and grows

Parent cell

Replication
Genetic information can be transferred between generations of cells.

Recombinant cell

A cell uses the genetic information contained in DNA to make its proteins, including enzymes. This information is transferred to the next generation during cell division. DNA can be transferred to cells in the same generation, resulting in new combinations of genes.

Daughter cells

The flow of genetic information.

*Quoted in H. A. Lechevalier and M. Solotorovsky. *Three Centuries of Microbiology.* New York: Dover Publications, 1974, p. 502.

CASE STUDY: A Mouse in the House

As an epidemiologist at the Centers for Disease Control and Prevention, you are notified by the New Mexico Health Department (NMHD) that two people living in the same household died within 5 days of each other. Their illnesses were characterized by abrupt onset of fever, muscle pain, headache, and cough, followed by the rapid development of respiratory failure. The NMHD tests for bacterial and viral pathogens were negative. The NMHD has notified you because an additional 24 cases with the same symptoms occurred during the following month. You fly to New Mexico and, wearing protective clothing, collect patient and environmental samples. You think the disease symptoms are similar, although not identical, to those of a viral disease that was discovered in Asia in the 1950s. Back at your BSL-4 lab in Atlanta, you find a virus in the samples. The virus looks similar to the Asian virus, and both viruses have RNA.

Questions

1. How would you determine whether these are the same virus?
2. How would you explain differences if your resulting gels were similar but not identical?
3. Your CDC colleagues trace the disease to a virus carried by the numerous local deer mice. How was the disease traced to deer mice?

Regulation of Gene Expression

OBJECTIVES

After completing this exercise, you should be able to:

1. Define the terms *operon, induction,* and *repression*.
2. Determine the inducers for the enzymes tested in this exercise.

BACKGROUND

Many genes are expressed constantly throughout the life of a cell. However, some proteins may be needed only during a particular growth phase or in a particular environment. A cell can conserve energy by making only those proteins needed at a particular time.

The Operon Model

The operon model was proposed to explain why *Escherichia coli* makes the enzyme β-galactosidase in the presence of lactose but not in its absence. An **operon** consists of a promoter, the region where RNA polymerase binds; the operator, which signals whether transcription will occur; and the structural genes they control. The structural genes code for the peptides. A regulatory gene encodes a regulatory protein, which binds to the operator to activate or inhibit transcription. In an **inducible operon,** transcription is activated or induced by the presence of a particular substance—in this case, lactose. The regulatory protein has an allosteric site to which the inducer can bind. The inducer then inactivates the repressor. A **repressible operon** is inhibited when a specific small molecule is present. The synthesis of the amino acid tryptophan in *E. coli* is inhibited by tryptophan. If tryptophan binds to the regulatory protein, the tryptophan–protein complex acts as a repressor to inhibit synthesis of more tryptophan.

nar Operon and *ara* Operon

In *E. coli,* the genes for nitrate reductase are encoded by the **nar** operon (FIGURE 27.1a), and the enzymes for the catabolism of arabinose are encoded by the **ara operon** (FIGURE 27.1b). The regulatory genes for these operons produce a protein that binds to the operator

to inhibit transcription or to promote binding of RNA polymerase, depending on the shape of the regulatory proteins (FIGURE 27.2).

In this experiment, the *ara* structural genes have been replaced with a gene called pGLO* that encodes green fluorescent protein (FIGURE 27.3). The pGLO gene, isolated from the jellyfish *Aequoria victoria,* is widely used to study gene expression.

You will investigate the *ara* or *nar* operons in *E. coli* in this exercise. Before beginning, write your hypothesis in your Laboratory Report.

MATERIALS

FIRST PERIOD

Small test tubes (3)
1-ml pipettes (3)
Nitrate solution
Brewer anaerobic jar
Petri plate containing glucose-nutrient agar
Petri plate containing glucose-arabinose-nutrient agar
Petri plate containing arabinose-nutrient agar
Spreading rod
Alcohol
Nitrate reagents A and B

SECOND PERIOD

Ultraviolet lamp

CULTURES

Escherichia coli for nitrate reductase
Escherichia coli pGLO for arabinose catabolism

TECHNIQUES REQUIRED

Anaerobic culture techniques (Exercise 19)
Nitrate reduction test (Exercise 17)
Pipetting (Appendix A)

---

pGLO is a trademark of Bio-Rad Laboratories.

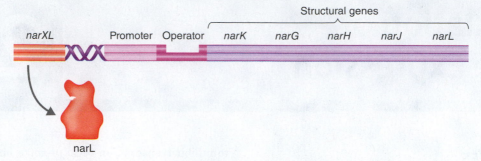

(a) The *nar* operon. The regulatory gene (*narXL*) encodes the regulatory protein narL.

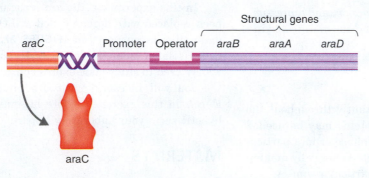

(b) The *ara* operon. The regulatory gene (*araC*) encodes the regulatory protein araC.

FIGURE 27.1 **Operons.** An operon consists of a promoter, an operator, and the structural genes they control.

PROCEDURE First Period

Nitrate Reductase

 Wear safety goggles when working with liquids.

1. Label two small tubes "Aerobic" and "Anaerobic." Add 0.5 ml *E. coli* to each tube. Then add 0.2 ml nitrate (NO_3^-) solution to each tube.
2. Immediately place the "anaerobic" tube in the anaerobic Brewer jar and add the activated charcoal sachet (see Figure 19.2 on page 154). Begin shaking the "aerobic" tube. Do not splash the contents out of the tube, but shake it enough to aerate the entire culture.
3. After 15 minutes, test for the presence of nitrite ions (NO_2^-) by adding 5 drops of nitrate reagent A and 5 drops of nitrate reagent B to each tube. Be sure to test an uninoculated nitrate broth tube for nitrite ion. Let the tubes stand at room temperature for 15 to 20 minutes. The presence of a pink color indicates nitrite ions. Observe the tubes for a color change, and record your results (see Figure 17.2 on page 138).

Arabinose Catabolism

1. Aseptically transfer 0.1 ml of the *E. coli* pGLO culture to the surface of the glucose-nutrient agar, the glucose-arabinose-nutrient agar, and the arabinose-nutrient agar.
2. Disinfect a spreading rod by dipping it in alcohol, quickly igniting the alcohol in a Bunsen burner flame, and letting the alcohol burn off.

 While the alcohol is burning off, hold the spreading rod pointed *down*. Keep the beaker of alcohol away from the flame.

Let the spreading rod cool.
3. Spread the liquid on the surface of each plate. Let the Petri plates sit undisturbed until the liquid diffuses into the agar (**FIGURE 27.4**). Disinfect the spreading rod and return it.
4. Incubate the plates in an inverted position at 35°C for 24 to 48 hours.

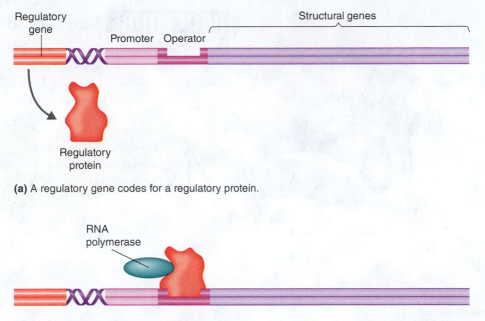

Regulatory gene

Promoter Operator Structural genes

Regulatory protein

(a) A regulatory gene codes for a regulatory protein.

RNA polymerase

(b) The regulatory protein binds to the operator and prevents transcription of the structural genes.

(c) A change in shape of the regulatory protein activates transcription of the structural genes.

FIGURE 27.2 Regulatory proteins. The regulatory proteins that control the *nar* and *ara* operons can inhibit or activate transcription.

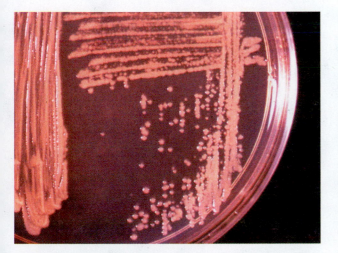

(a) Viewed with visible light

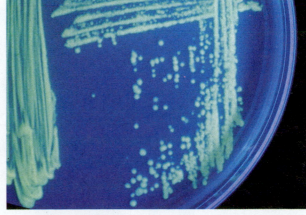

(b) Viewed with UV light

FIGURE 27.3 *E. coli* **containing pGLO.** The colonies of *E. coli* fluoresce under UV light because the cells contain the jellyfish gene for green fluorescent protein.

PROCEDURE Second Period

Observe the plates using an ultraviolet lamp. Are any of the colonies producing a fluorescent protein?

_____

_____

 Do not look at the ultraviolet light, and do not leave your hand exposed to it. Wear safety goggles.

FIGURE 27.4 Inoculation using a spreading rod.

LABORATORY REPORT
Regulation of Gene Expression

PURPOSE _____

_____

_____

HYPOTHESES

1. Nitrate reduction will occur with/without air.

2. Green fluorescent protein will be made with/without arabinose.

RESULTS

Nitrate Reductase

	Color after NO_3^- Reagents	NO_3^- Reduction
Control		
E. coli, aerobic		
E. coli, anaerobic		

Arabinose Catabolism

	Growth	Fluorescence
Glucose-nutrient agar		
Arabinose-nutrient agar		
Glucose-arabinose-nutrient agar		

CONCLUSIONS

1. Did you accept or reject your hypothesis? Briefly explain. _____

_____

_____

_____

2. What is the inducer for *nar*? _____

3. What is the inducer for *ara*? _____

4. What effect did the presence of glucose have on *ara*? Briefly explain. _____

_____

_____

QUESTIONS

1. Write the chemical reaction catalyzed by nitrate reductase.

2. Of what value is nitrate reductase to *E. coli*? _____

_____

_____

3. Chemically, what is arabinose? _____

CRITICAL THINKING

1. Several bacteria, including *Bacillus subtilis,* reduce nitrate to ammonia as the first step in amination. This is called assimilatory nitrate reduction. *E. coli*'s transformation of nitrate is called dissimilatory nitrate reduction. How do assimilatory and dissimilatory nitrate reduction differ?

2. What is the valence of N in NO_3^-? In NO_2^-? What is gaining electrons in nitrate reduction?

3. *E. coli* cells do not normally fluoresce in the presence of arabinose. What do they normally do?

4. Why did the *E. coli* cells fluoresce in the presence of arabinose in this experiment? Why didn't they do so in the absence of arabinose?

CLINICAL APPLICATION

Harvard researchers replaced the pilus gene with the *gfp* gene in *Vibrio cholerae* bacteria. The bacteria were then inoculated into nutrient broth at 37°C or into mice. After 12 hours, bacteria were collected and analyzed for fluorescence. The results are shown below.

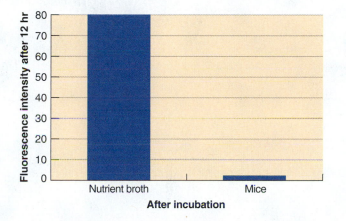

a. What can you conclude from these data?

b. What disease does *V. cholerae* cause?

c. Of what value is the experiment?

Isolation of Bacterial Mutants

Though coli may **BOTHER AND VEX US,**
It's hard to believe they outsex us.
They accomplish seduction
by **VIRAL TRANSDUCTION**
Forgoing the joys of amplexus. – S E Y M O U R   G I L B E R T

OBJECTIVES

After completing this exercise, you should be able to:

1. Define the following terms: *prototroph, auxotroph,* and *mutation*.
2. Differentiate direct from indirect selection.
3. Isolate bacterial mutants by replica plating.

BACKGROUND

Metabolic Mutants

For practical purposes, genes and the characteristics for which they code are stable. However, when millions of bacterial progeny are produced in just a few hours of incubation, genetic variants may be found. A variant is the result of a **mutation**—that is, a change in the sequence of nucleotide bases in the cell's DNA.

Metabolic mutants can be easily identified and isolated. Catabolic mutations could result in a bacterium that is deficient in the production of an enzyme needed to utilize a particular substrate. Anabolic mutations result in bacteria that are unable to synthesize an essential organic chemical. "Wild-type," or nonmutated, bacteria are called **prototrophs,** and mutants are called **auxotrophs.** Auxotrophs have been isolated that either cannot catabolize certain organic substrates or cannot synthesize certain organic chemicals, such as amino acids, purine or pyrimidine bases, or sugars.

The cultures used in this exercise are capable of synthesizing all of their growth requirements from glucose–minimal salts medium (**TABLE 28.1**). Mutations will be induced by exposing the cells to the mutagenic wavelengths of ultraviolet light (Exercise 23). After exposure to ultraviolet radiation, auxotrophs that cannot grow on the glucose–minimal salts medium will be identified.

Replica Plating

Because of the low rate of appearance of recognizable mutants, special techniques have been developed to

TABLE 28.1   COMPOSITION OF GLUCOSE–MINIMAL SALTS AGAR

Glucose	0.1 g
Dipotassium phosphate	0.7 g
Monopotassium phosphate	0.2 g
Sodium citrate	0.05 g
Magnesium sulfate	0.01 g
Ammonium sulfate	0.15 g
Agar	1.5 g
Water	100 ml

select for desired mutants. **Direct selection** is used to pick out mutant cells while rejecting the unmutated parent cells. (Direct selection will be used in Exercises 29 and 31.) We will use **indirect selection** with the **replica plating technique** in this exercise.

In replica plating, mutated bacteria are grown on a nutritionally complete solid medium. An imprint of the colonies is made on velveteen-covered or rubber-coated blocks and transferred to a glucose–minimal salts solid medium and to a complete solid medium. Colonies that grow on the complete medium but not on the minimal medium are auxotrophs. Auxotrophs are identified *indirectly* because they will not grow.

MATERIALS

FIRST PERIOD

Petri plates containing nutrient agar (complete medium) (3)
99-ml water dilution blanks (3)
Sterile 1-ml pipettes (4)
Ultraviolet lamp, 260 nm
Spreading rod

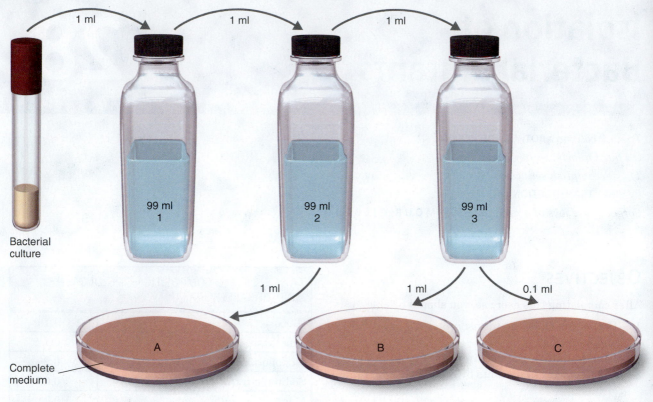

FIGURE 28.1 Dilution procedure.

Alcohol
Safety goggles

SECOND PERIOD

Petri plate containing glucose–minimal salts agar (minimal medium)
Petri plate containing nutrient agar (complete medium)
Sterile replica-plating block

CULTURE

Serratia marcescens

TECHNIQUES REQUIRED

Spreading rod technique (Exercise 27)
Pipetting (Appendix A)
Serial dilution technique (Appendix B)

PROCEDURE* First Period

1. Label the nutrient agar plates "A," "B," and "C" and the dilution blanks "1," "2," and "3."

---

*Adapted from C. W. Brady. "Replica Plate Isolation of an Auxotroph." Unpublished paper. Whitewater: University of Wisconsin, n.d.

 Wear safety goggles when pipetting.

2. Aseptically pipette 1 ml of *Serratia* to dilution blank 1 and mix well.
3. Using another pipette, transfer 1 ml from dilution blank 1 to dilution blank 2, as shown in **FIGURE 28.1**, and mix well. What dilution is in bottle 2? _____
4. Using another pipette, transfer 1 ml from dilution blank 2 to bottle 3 and 1 ml to the surface of plate A.
5. Mix bottle 3, and transfer 1 ml to the surface of plate B with a sterile pipette and 0.1 ml to the surface of plate C.
6. Disinfect a spreading rod by dipping it in alcohol, quickly igniting the alcohol in a Bunsen burner flame and letting the alcohol burn off.
   Let the spreading rod cool.

⚠ **While the alcohol is burning off, hold the spreading rod pointed *down*. Keep the beaker of alcohol away from the flame.**

7. Spread the liquid on the surface of plate C over the entire surface. Do the same with plates B and A

(see Figure 27.4 on page 216). Let the Petri plates sit undisturbed until the liquid diffuses into the agar.

8. Disinfect the spreading rod and return it. Why is it not necessary to disinfect the spreading rod between plates when proceeding from plate C to B to A? _____

_____

9. Put the plates with the covers off, agar-side up, about 30 cm from the ultraviolet lamp. Position the plates directly under the lamp. Turn the lamp on for 30 to 60 seconds, as assigned by your instructor.

> ⚠ **Do not look at the ultraviolet light, and do not leave your hand exposed to it. Wear safety goggles.**

10. Incubate the plates at 25°C until the next period.

PROCEDURE Second Period

1. Select a plate with 25 to 50 isolated colonies. Mark the bottom of the plate with a reference mark (**FIGURE 28.2a**). This is the master plate.
2. Mark the uninoculated complete and minimal media with a reference mark on the bottom of each plate.
3. If your replica-plating block is already assembled, proceed to step 4. If not, follow these instructions. Carefully open the package of velveteen. Place the replicator block on the center of the velveteen. Pick up the four corners of the cloth, and secure them tightly on the handle with a rubber band.

4. Inoculate the sterile media using *either* step a *or* step b, as follows:
   a. Hold the replica-plating block by resting it on the table with the rough surface, or velveteen, up (**FIGURE 28.2b**). Invert the master plate selected in step 1 on the block, and allow the master plate agar to lightly touch the block. Remove the lid from the minimal medium, align the reference marks, and touch the uninoculated minimal agar with the inoculated replica-plating block. Replace the lid. Remove the lid from the uninoculated complete medium and inoculate with the replica-plating block, keeping the reference marks the same.
   b. Place the master plate and the uninoculated plates on the table. Place the reference marks in a 12 o'clock position, and remove the lids. Touch the replica-plating block to the master plate; then, without altering its orientation, gently touch it to the minimal medium, then to the complete medium (**FIGURE 28.2c**).
5. Replace the lids, and incubate as before. Refrigerate the master plate.

PROCEDURE Third Period

After incubation, compare the plates. Record your results and those of your classmates.

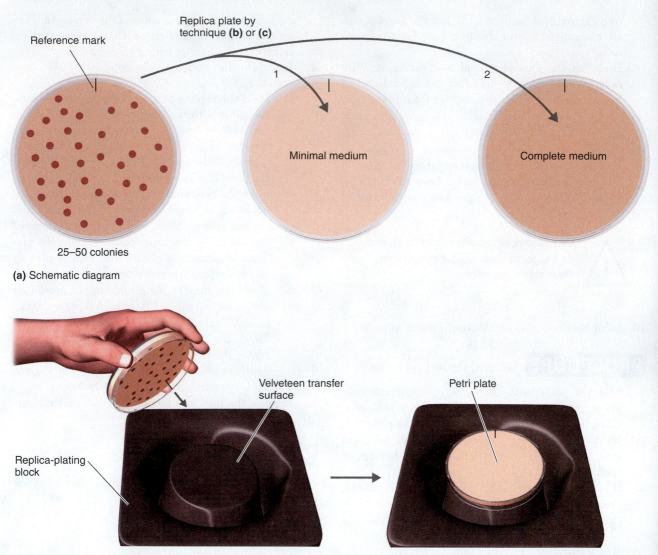

Reference mark

Replica plate by technique **(b)** or **(c)**

1

2

Minimal medium

Complete medium

25–50 colonies

(a) Schematic diagram

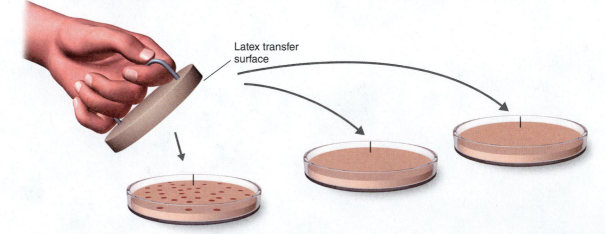

Velveteen transfer surface

Petri plate

Replica-plating block

(b) Transfer with a stationary replica-plating block

Latex transfer surface

(c) Transfer by moving the latex transfer surface from plate to plate

FIGURE 28.2 **Replica plate.**

LABORATORY REPORT

Isolation of Bacterial Mutants

PURPOSE _____

_____

_____

HYPOTHESIS

Exposure to UV radiation will increase/decrease the number of auxotrophs.

RESULTS

Mark the location of colonies, and note any changes in pigmentation. Circle the auxotrophs on the diagram of the complete medium.

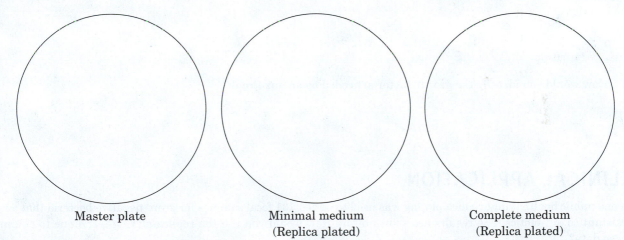

| Master plate | Minimal medium (Replica plated) | Complete medium (Replica plated) |

Class Data

	S. marcescens	
Students	UV Exposure (seconds)	Number of Auxotrophs

CONCLUSION

Based on your results, do you accept or reject your hypothesis? _____

_____

QUESTIONS

1. Why was the complete medium, as well as the minimal medium, replica plated? _____

_____

_____

2. Why is replica plating usually used for indirect selection of mutants? _____

_____

_____

CRITICAL THINKING

1. How would you isolate a mutant from the replica plate? Does your technique involve a catabolic or anabolic mutant?

2. How would you identify the growth factor(s) needed by an auxotroph?

CLINICAL APPLICATION

In one public health study, replica plating was used to screen 131 fecal samples for gram-negative bacteria that are resistant to antibiotics. The results are shown below. How would you use the replica-plating technique to obtain these data?

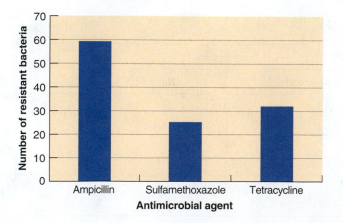

Transformation of Bacteria

OBJECTIVES

After completing this exercise, you should be able to:

1. Define the term *transformation*.
2. Transform *E. coli* bacteria.

BACKGROUND

Isolation of DNA

Transformation is a rare event involving acquisition by a *recipient* bacterium of small pieces of DNA released from a dead *donor* bacterium. The acquired pieces of DNA can give the recipient bacterium new characteristics.

In this experiment, you will transform *Escherichia coli* bacteria with a gene encoding green fluorescent protein (GFP; see Figure 27.3). We will use *E. coli* K-12, which has a long history of safe commercial use. *E. coli* K-12 does not normally colonize the human intestine and survives poorly in the environment.

The *gfp* gene is from the jellyfish *Aequorea victoria*. GFP causes the jellyfish to fluoresce under ultraviolet light. To genetically modify a bacterium by transformation, the desired gene is put into a plasmid (Exercise 31). You will use the pGLO plasmid that has been genetically modified to carry the *gfp* and ampicillin-resistance genes. The ampicillin-resistance gene makes possible the direct selection of transformed cells (Exercise 28).

E. coli bacteria are not naturally competent; that is, they don't readily take up foreign DNA. Treatment with Ca^{2+} neutralizes the negatively charged phosphates on DNA and the phospholipid membrane. Exposure to heat for a short time, called **heat shocking,** increases the fluidity of the plasma membrane, allowing DNA to enter. We will use **LB agar,** which is similar to nutrient agar (see Table 12.1 on page 103) but nutritionally rich to grow *E. coli* K-12. LB agar contains 0.1% tryptose and 0.5% yeast extract.

MATERIALS

FIRST PERIOD

Petri plate containing LB agar

SECOND PERIOD

Microcentrifuge tubes (2)
Foam floating rack for tubes
Micropipette, 2–20 μl
Micropipette, 200–1000 μl
Sterile small micropipette tips (3)
Sterile large micropipette tips (2)
50 mM $CaCl_2$ (500 μl)
pGLO plasmid, 0.08 μg DNA/μl (10 μl)
LB broth, 500 μl
Petri plates containing LB agar (2)
Petri plates containing LB agar + 0.01% ampicillin (2)
Ice
42°C water bath
Spreading rod
Alcohol
UV lamp

CULTURE

Escherichia coli K-12

TECHNIQUES REQUIRED

Aseptic technique (Exercise 4)
Streak plate technique (Exercise 11)
Spreading rod technique (Exercise 27)
Pipetting (Appendix A)

PROCEDURE First Period

Streak an LB agar plate with *E. coli,* and incubate the plate for 24 hours at 35°C.

PROCEDURE Second Period

Preparing Competent Cells (Figure 29.1a)

1. Label two microcentrifuge tubes: +pGLO and −pGLO.
2. Aseptically transfer 250 μl $CaCl_2$ into each tube. Place the tubes in ice.
3. Using a sterile loop, transfer two to four large colonies of *E. coli* to one tube. Mix with the loop until the entire colony is dispersed in the $CaCl_2$ solution.

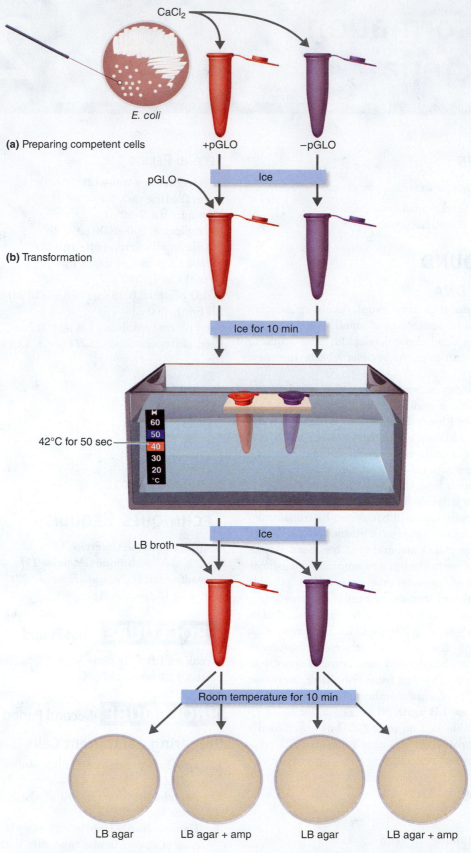

FIGURE 29.1 Transformation of *E. coli*.

Place the tube back on ice. Sterilize your loop, and transfer two to four large colonies to the other tube.

Transformation (Figure 29.1b)

1. Label one LB agar plate "+pGLO" and another LB plate "−pGLO." Label two LB + ampicillin plates: "+pGLO" and "−pGLO."
2. Examine the tube containing pGLO with a UV light. Does it fluoresce? _____
3. Aseptically pipette 10 µl of pGLO plasmid into the +pGLO tube and mix. Incubate the tubes in ice for 10 minutes.
4. Place both tubes in a foam floating rack. Float both tubes in a 42°C water bath for exactly 50 seconds. Then quickly move the tubes to ice. Incubate the tubes in ice for 2 minutes.
5. Aseptically add 250 µl of LB broth to each tube. Incubate the tubes for 10 minutes at room temperature.
6. Gently flick the tubes to resuspend the bacteria. Using a different pipette tip for each tube, transfer 100 µl from each tube onto the appropriate nutrient agar plate.
7. Disinfect a spreading rod by dipping it in alcohol, quickly igniting the alcohol, and letting the alcohol burn off.

 While the alcohol is burning off, hold the spreading rod pointed *down*. Keep the beaker of alcohol away from the flame.

Let the spreading rod cool.

8. Spread cells evenly over one plate with the sterile spreading rod (see Figure 27.4 on page 216). Disinfect the spreading rod.
9. Repeat steps 7 and 8 to spread bacteria on the other plates.
10. Incubate the plates at 35°C for 24 to 48 hours. Which plate is the control? _____

_____

PROCEDURE Third Period

 Do not look at the ultraviolet light, and do not leave your hand exposed to it.

Examine your plates in visible light and under UV light. Count the colonies, and record your results.

LABORATORY REPORT
Transformation of Bacteria

PURPOSE _____

_____

_____

EXPECTED RESULTS

Before you complete the experiment, indicate where you expect growth (+), where no growth (−), and where fluorescence (F).

	Inoculum	
	E. coli − pGLO	*E. coli* + pGLO
LB agar		
LB agar + ampicillin		

RESULTS

	E. coli − pGLO		*E. coli* + pGLO	
	Number of Colonies	Fluorescence?	Number of Colonies	Fluorescence?
LB agar				
LB agar + ampicillin				

CONCLUSION

Did you transform *E. coli*? How do you know? _____

_____

QUESTIONS

1. What new gene(s) were you looking for in the transformed *E. coli*? _____

2. Why do you expect colonies on the ampicillin agar to fluoresce? _____

_____

3. Using the following formulas, calculate transformation efficiency.

Total DNA used = Concentration of DNA × Volume of DNA

$$\text{Fraction of DNA spread on plate} = \frac{\text{Volume spread on plate}}{\text{Total sample in tube}}$$

Amount of DNA spread on plate = Total amount of DNA × Fraction of DNA used

$$\text{Efficiency} = \frac{\text{Number of transformed colonies}}{\text{Amount of DNA spread on plate}}$$

This transformation procedure has a transformation efficiency of 8.0×10^2 to 7.0×10^3 transformants/μg DNA.

How do your results compare with this? _____

CRITICAL THINKING

1. From your results, how could you rule out contamination? Mutation?

2. Osamu Shimomura discovered GFP in 1962. Later, Martin Chalfie inserted the *gfp* gene into the *Caenorhabditis* worm, and Roger Tsien inserted *gfp* into yeast. The 2008 Nobel Prize in Chemistry was awarded to Shimomura, Chalfie, and Tsien. There are many fluorescent compounds in nature—even human skin fluoresces because of NADH, tryptophan, and collagen. Why was GFP so noteworthy?

CLINICAL APPLICATION

Researchers inserted the gene for green fluorescent protein into influenza A virus (IAV) and inoculated mice with the modified virus. They tracked the virus in the mice using fluorescence molecular tomography, an in vivo imaging technique that provides deep-tissue images of small animals. The researchers also watched the infection after treatment with oseltamivir, an IAV inhibitor. The results are shown below.

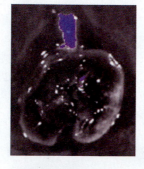 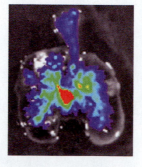 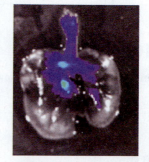

(a) Uninfected **(b)** No treatment **(c)** Oseltamivir treated

a. Describe the virus progress in a mammal. What was the effect of oseltamivir?

b. What is the value of this experiment?

DNA Fingerprinting

OBJECTIVES

After completing this exercise, you should be able to:

1. Define the terms *restriction enzyme* and *RFLP*.
2. Perform agarose gel electrophoresis.
3. Identify your unknown.

BACKGROUND

Restriction enzymes recognize specific short nucleotide sequences in double-stranded DNA and cleave both strands of the molecule (**TABLE 30.1**). Restriction enzymes are called **endonucleases** because they hydrolyze nucleic acids *in* the nucleic acid polymer; an **exonuclease** removes nucleotides from the end of a nucleic acid. Restriction enzymes are bacterial enzymes that restrict the host range of bacteriophages. The enzymes were discovered in laboratory experiments when phages were used to infect bacteria other than their usual hosts. Restriction enzymes in the new host destroyed the phage DNA. Today, more than 300 restriction enzymes have been isolated and purified for use in DNA research. Each enzyme recognizes a different nucleotide sequence in DNA. The enzymes are named with three-letter abbreviations for the bacteria from which they were isolated.

RFLPs

DNA cleaved with restriction enzymes produces **restriction fragment length polymorphisms**, or **RFLPs** (pronounced "rif-lips"). The size and number of the pieces is determined by **agarose gel electrophoresis** (Appendix G). The digested DNA is placed near one end of a thin slab of agarose and immersed in a buffer to allow current to flow through the agarose. Electrodes are attached to both ends of the gel, and a current is applied. Each piece of DNA then migrates toward the positive electrode at a rate determined by its size. The DNA fragments are visualized by staining (**FIGURE 30.1**).

Applications of RFLPs

These enzymes can be used to characterize DNA because a specific restriction enzyme will cut a molecule of DNA everywhere a specific base sequence occurs. When the DNA molecules from two different microorganisms are treated with the same restriction enzyme, the restriction fragments that are produced can be separated by electrophoresis. A comparison of the number and sizes of restriction fragments produced from different organisms provides information about their genetic similarities and differences.

In forensic science, analyses of DNA can be used to determine the father of a child or the perpetrator of a crime. DNA samples are digested to produce RFLPs that are separated by gel electrophoresis. The DNA fragments are transferred to a membrane filter by a process called Southern blotting so the fragments on the paper are in the same positions as the fragments on the gel. The paper is then heated to produce single strands of DNA. Small pieces of enzyme- or fluorescence-labeled single-stranded DNA called *probes* are used to distinguish core sequences of nucleotides. When the probe is added to the filter, it will anneal to its complementary strand. The filter can then be stained with the enzyme-substrate or viewed with UV light to look for specific RFLPs.

In this exercise, we will compare plasmids from an unknown bacterial species to plasmids from known bacteria to identify the unknown.

TABLE 30.1	RECOGNITION SEQUENCES OF SOME RESTRICTION ENDONUCLEASES	
Enzyme	**Bacterial Source**	**Recognition Sequence**
*Eco*RI	*Escherichia coli*	G↓AATTC CTTAA↑G
*Hae*II	*Haemophilus aegyptius*	GG↓CC CC↑GG

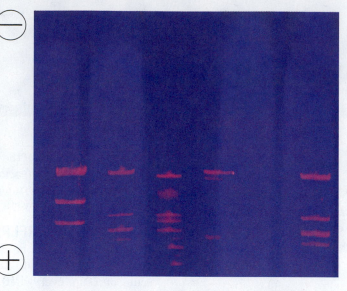

FIGURE 30.1 Restriction enzyme digests of a plasmid. To make the bands visible, the DNA was stained with ethidium bromide, and the gel was illuminated with UV light.

MATERIALS

FIRST PERIOD: DIGESTION OF DNA

DNA samples
Restriction enzyme
Restriction buffer
Micropipette, 1–20 μl
Micropipette tips (7)
Microcentrifuge tubes (5)

SECOND PERIOD: ELECTROPHORESIS OF DNA SAMPLES

Electrophoresis buffer
Casting tray and comb
Agarose, 0.8%
Tracking dye
Micropipette, 1–20 μl
Micropipette tips (6)
Electrophoresis chamber and power supply
SYBR Safe dye, ethidium bromide, or methylene blue
Transilluminator or light box
Safety goggles

TECHNIQUES REQUIRED

Pipetting (Appendix A)
Electrophoresis (Appendix G)

PROCEDURE First Period

Every pair of students will perform five digests: one from the unknown, and one each from four known bacteria.

Digestion of DNA

1. Label five microcentrifuge tubes "U," "1," "2," "3," and "4."
2. To each tube add 5 μl of restriction buffer.
3. Using a new pipette tip, add 4 μl of restriction enzyme to each tube.
4. Add the DNA samples to each tube as follows. Use a different pipette tip for each sample. Why? _____

Tube	DNA
U	Unknown, 4 μl
1	Species 1, 4 μl
2	Species 2, 4 μl
3	Species 3, 4 μl
4	Species 4, 4 μl

5. Centrifuge the tubes for 1 to 2 seconds to mix. Be sure to balance the centrifuge. Then incubate the tubes in a 37°C water bath for 45 minutes. After incubation, the tubes can be frozen if electrophoresis will be done during another lab period.

PROCEDURE Second Period

Electrophoresis of DNA Samples

1. Read about electrophoresis in Appendix G. Use the melted agarose to pour a gel, as described in Appendix G.

2. If necessary, defrost the tubes of digested DNA by holding them in your hand or placing them in a 37°C water bath.

3. Add 1 µl of tracking dye to each tube. Centrifuge the tubes for 1 to 2 seconds to mix. *Be sure the centrifuge is balanced.*

4. Load 14 µl of one of your previously prepared samples into a well. Using a different pipette tip for each sample, load your other samples.

5. Once the wells have been filled, apply power (125 V) to the chamber. During the run, you will see the tracking dye migrate. The tracking dye will separate into its two component dyes during electrophoresis. Turn off the power before the faster dye runs off the gel. Good separation of the DNA fragments occurs when the two dyes have separated by 4 or 5 cm. Why do the two dyes separate?_____ Run the gel until the dye is near the end of the gel; then turn off the current and remove the gel.

6. To stain the gel, use *either* step a, step b, *or* step c as follows (see Figure 30.1).

    a. Transfer the gel to the ethidium bromide staining tray for 5 to 10 minutes.

Wear safety goggles and gloves. Do not touch the ethidium bromide. It is a mutagen.

Transfer the gel to tap water to destain for 5 minutes. (Chlorine in tap water will inactivate residual ethidium.) Ethidium bromide that is bound to DNA does not wash off and will fluoresce with UV light. Place your gel on the transilluminator, and close the plastic lid.

Do not look directly at the UV light. Do not turn on the UV light until the plastic lid is down.

    b. Transfer the gel to the SYBR Safe staining tray. Cover the tray with aluminum foil. Gently and continuously agitate the gel at room temperature (for example, on an orbital shaker at 50 rpm). Observe the DNA bands on a UV transilluminator.

    c. Transfer the gel to the methylene blue staining tray for 30 minutes to 2 hours. Destain by placing the gel in water for 30 minutes to overnight. Place your gel on a light box. Methylene blue that is bound to DNA does not wash off.

7. Carefully draw the location of the bands, or photograph your gel.

LABORATORY REPORT
DNA Fingerprinting

PURPOSE _____

_____

_____

HYPOTHESIS

The restriction enzymes will/will not produce different RFLPs for species 1 through 4.

RESULTS

Draw your gel so that you have all five digests represented. Label the contents of each lane.

CONCLUSIONS

1. Do you accept or reject your hypothesis? _____

_____

2. What is the identity of your unknown? How can you tell? _____

_____

_____

QUESTIONS

1. What restriction enzyme did you use? _____

What nucleotide sequence does it recognize? _____

2. Using this map of pLAB30 (18 kb), give the number and lengths of the restriction fragments that would result from digesting pLAB30 with *Eco*RI, with *Bam*HI, and with both enzymes together.

	Fragments	
Enzyme	Number	Size(s)
*Eco*RI	_____	_____
*Bam*HI	_____	_____
*Eco*RI + *Bam*HI	_____	_____

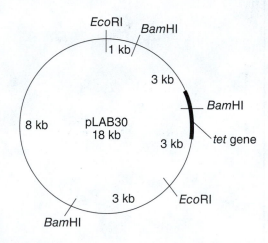

Which enzyme will give the smallest piece containing the tetracycline-resistance gene? _____

CRITICAL THINKING

1. Differentiate a gene from a RFLP.

2. *Nocardia* species can be identified with RFLP analysis of the 16S rRNA gene. Following extraction of the DNA, the gene is amplified by PCR and digested with a restriction enzyme. Briefly discuss the advantages and disadvantages of RFLP analysis compared with conventional biochemical testing.

CLINICAL APPLICATIONS

1. How could you use this technique to trace the source of *E. coli* O157:H7?

   Why would you want to trace *E. coli*?

2. In March, carbapenem-susceptible *Escherichia coli* (Eco1) and carbapenem-resistant *Klebsiella pneumoniae* (Kpn) were isolated from rectal swabs of a patient. In April, carbapenem-susceptible *E. coli* (Eco2) was isolated from gallbladder fluid from the same patient. Bacterial DNA was extracted from each bacterium and digested with *spe*1 restriction enzyme.

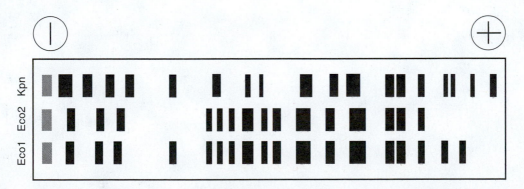

   What can you conclude from these data? Briefly explain how you arrived at your conclusion.

Genetic Engineering

OBJECTIVES

After completing this exercise, you should be able to:

1. Define the terms *transformation, restriction enzyme,* and *ligation.*
2. Analyze DNA by electrophoresis.
3. Accomplish genetic change through transformation.

BACKGROUND

Scientists from many different disciplines have used the technique of genetic engineering to genetically alter bacteria and eukaryotic organisms. In **genetic engineering,** genes are isolated from one organism and incorporated by manipulation in a laboratory into other bacterial or eukaryotic cells.

To genetically engineer a bacterium by using transformation, the desired gene is first inserted into a plasmid. Next, the recombinant plasmid is acquired by the bacterium. If the gene is expressed in the cell, the cell will produce a new protein product.

Cutting DNA

Genetic engineering is made possible by restriction enzymes. A **restriction enzyme** is an enzyme that recognizes and cuts only one particular sequence of nucleotide bases in DNA, and it cuts this sequence in the same way each time. These enzymes are divided into two major classes, based on their manner of cleavage. Class I enzymes bind to specific sequences but catalyze cleavage at sites up to thousands of base pairs from the binding sequence. Class II enzymes recognize specific sequences and catalyze cleavage within or next to these sequences (see Table 30.1). The recognition sequences usually include inverted repeats (*palindromes*) so that the same sequence is cut at the same point in both strands. Some of these enzymes catalyze cleavage of the two strands so that the ends of the resulting fragments are fully base-paired (**blunt ends**). Other enzymes catalyze a staggered cleavage of the two DNA strands so that resulting fragments have unpaired single strands (**sticky ends**) (FIGURE 31.1).

We will use the restriction enzymes *Hind*III and *Bam*HI. *Hind*III, an enzyme isolated from *Haemophilus influenzae,* cuts double-stranded DNA at the arrows in the sequence

A↓AGCTT

TTCGA↑A

*Bam*HI, isolated from *Bacillus amyloliquefaciens,* cuts double-stranded DNA at the arrows in the sequence

G↓GATCC

CCTAG↑G

Recombining DNA

In this exercise, we will use a plasmid containing an ampicillin-resistance gene (pAMP) and another plasmid with a kanamycin-resistance gene (pKAN) to transform an antibiotic-sensitive strain of *E. coli* (amp^s/kan^s). We will cut the two circular plasmids using restriction enzymes (FIGURE 31.2). The DNA fragments will be ligated using DNA ligase, and the ligated DNA will be used to transform the *E. coli*. *E. coli* does not normally take up new DNA; its cell wall will be damaged by treatment with $CaCl_2$ so DNA can enter after heat shocking (Exercise 29 on page 227). We will use agarose gel electrophoresis to determine whether the restriction enzymes cut the plasmids.

C-C-G-**G-A-A-T-T**-C-C-C-C-A-C-**G-G-C-C**-G-A
G-G-C-**C-T-T-A-A**-G-G-G-G-T-G-**C-C-G-G**-C-T

(a) Before cleavage

C-C-G-**G**   **A-A-T-T**-C-C-C-C-A-C-**G-G**   **C-C**-G-A
G-G-C-**C-T-T-A-A**   G-G-G-G-T-G-**C-C**   **G-G**-C-T

(b) After cleavage

FIGURE 31.1 Restriction digests. *Eco*RI recognizes G↓AATTC and leaves sticky ends. *Hae*III recognizes GG↓CC to produce blunt ends.

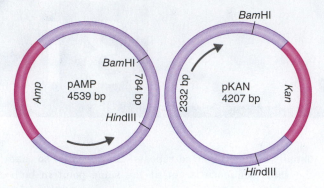

FIGURE 31.2 Restriction maps of plasmids showing locations of antibiotic-resistance markers. DNA is measured in the number of base pairs (bp).

Recombinant cells will be cultured on an enriched medium called LB agar (see Exercise 29).

MATERIALS

PLASMID DIGESTION

Sterile microcentrifuge tubes (2)
Micropipette, 1–10 µl
Sterile micropipette tips (4)
pAMP plasmid
pKAN plasmid
Restriction enzyme buffer
*Bam*HI and *Hind*III enzymes
35°C water bath

ELECTROPHORESIS OF PLASMIDS

Microcentrifuge tubes (3)
Micropipette, 1–10 µl
Sterile micropipette tips (7)
Casting tray and comb
Agarose, 0.8%
pAMP plasmid
Electrophoresis buffer
Tracking dye
Electrophoresis chamber and power supply
SYBR Safe dye, ethidium bromide, or methylene blue stain
Transilluminator or light box
Safety goggles

LIGATION OF DNA PIECES

Sterile microcentrifuge tubes (2)
Micropipette, 1–10 ml
Sterile micropipette tips (5)
Ligation buffer
65°C water bath
Ice
Sterile distilled water (dH$_2$O)
DNA ligase

TRANSFORMATION OF COMPETENT AMP^s/KAN^s *E. COLI* CELLS

Plate containing LB agar
Sterile microcentrifuge tubes (2)
Micropipettes, 1–10 µl and 200–1000 µl
Sterile micropipette tips (4)
50 mM CaCl$_2$
Ice
43°C water bath
35°C water bath
Sterile LB broth

CULTURING OF TRANSFORMED CELLS

Micropipette, 10–100 µl
Sterile micropipette tips (2)
Plates containing LB agar (3)
Plates containing 0.01% ampicillin–LB agar (3)
Plates containing 0.01% kanamycin–LB agar (3)
Plates containing 0.01% ampicillin–0.01% kanamycin–LB agar (2)
Spreading rod and alcohol

CULTURE

E. coli amp^s/kan^s

TECHNIQUES REQUIRED

Inoculating loop technique (Exercise 4)
Spreading rod (Exercise 27)
Pipetting (Appendix A)
Electrophoresis (Appendix G)

PROCEDURE

A. Plasmid Digestion (Figure 31.3a)

Keep all reagents ice-cold.

1. Label two sterile microcentrifuge tubes "1" and "2."
2. Aseptically add reagents as shown below.

Tube	pAMP	pKAN	Restriction Buffer	*Bam*HI/*Hind*III
1	5.5 µl	—	7.5 µl	2 µl
2	—	5.5 µl	7.5 µl	2 µl

3. Close the caps and mix the tubes for 1 to 2 seconds by centrifuging.

  Be sure the centrifuge is balanced by placing your tubes opposite each other.

4. Incubate the tubes at 35°C for 20 minutes to 2 hours, as instructed. Tubes can be frozen until the next lab period.

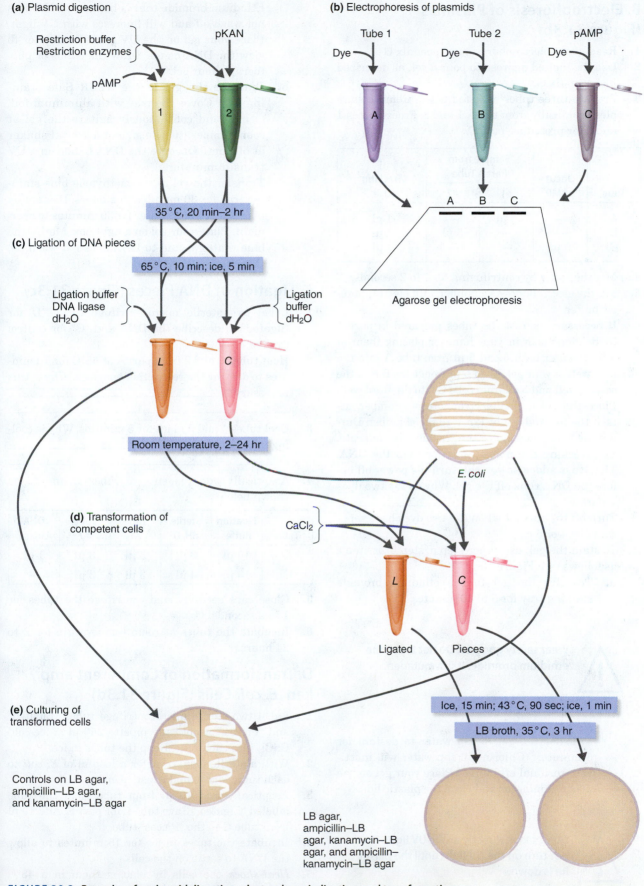

FIGURE 30.3 Procedure for plasmid digestion, electrophoresis, ligation, and transformation of *E. coli*.

B. Electrophoresis of Plasmids (Figure 31.3b)

1. Read about electrophoresis in Appendix G.
2. Use the melted agarose to pour a gel, as described in Appendix G.
3. Prepare three tubes as listed below. Remove samples aseptically from tubes 1 and 2. Freeze tubes 1 and 2 for procedure part C.

Tube	Uncut pAMP	Sample from Part A Tube #1	Sample from Part A Tube #2	Tracking Dye
A	—	5 µl	—	1 µl
B	—	—	5 µl	1 µl
C	5 µl	—	—	1 µl

4. Mix the tubes by centrifuging for 1 to 2 seconds.
5. Fill the electrophoresis chamber with electrophoresis buffer.
6. If necessary, defrost the tubes prepared in part A by holding them in your hand or placing them in a 37°C water bath. Load 5 µl from tube A into the first well of your gel. Load 5 µl from tube B into the second well and 5 ml from tube C into the third well.
7. Place the gel in the electrophoresis chamber, attach the lid, and apply 125 V to the chamber. During the run, you will see the tracking dye migrate. The tracking dye migrates faster than the DNA. The dye is added so you can turn the power off before the DNA runs off the gel. Why do the two dyes separate? _____

    Turn off the current when the two dyes have separated by 4 or 5 cm.
8. To stain the gel, use *either* step a, step b, *or* step c as follows (see Figure 30.1).
    a. Transfer the gel to the ethidium bromide staining tray for 5 to 10 minutes.

Wear safety goggles. Do not touch the ethidium bromide. It is a mutagen.

    Transfer the gel to tap water to destain for 5 minutes. (Chlorine in tap water will inactivate residual ethidium.) Place your gel on the transilluminator and close the plastic lid.

Do not look directly at the UV light. Do not turn on the UV light until the plastic lid is down.

Ethidium bromide that is bound to DNA does not wash off and will fluoresce with UV light. Place your gel on the UV transilluminator to view the DNA bands. Draw the bands or photograph your gel.

b. Transfer the gel to the SYBR Safe staining tray. Cover the tray with aluminum foil. Gently and continuously agitate the gel at room temperature (e.g., on an orbital shaker at 50 rpm). Observe the DNA bands on a UV transilluminator.

c. Transfer the gel to the methylene blue staining tray for 30 minutes to 2 hours. Destain by placing the gel in water for 30 minutes to overnight. Place your gel on a light box. Methylene blue that is bound to DNA does not wash off. Draw the bands.

C. Ligation of DNA Pieces (Figure 31.3c)

1. Label two sterile microcentrifuge tubes "L" for "ligated" to describe the DNA and "C" for control (unligated).
2. Heat tubes 1 and 2 from part A at 65°C for 10 minutes to destroy the restriction enzymes. Why is this necessary? _____

    _____
3. Cool tubes 1 and 2 in ice for 5 minutes. Why is cooling necessary? _____

    _____
4. Aseptically add reagents to tubes L and C, as shown below.

Tube	Ligation Buffer	Sterile dH₂O	Tube #1	Tube #2	DNA Ligase
L	10 µl	3 µl	3 µl	3 µl	1 µl
C	10 µl	4 µl	3 µl	3 µl	—

5. Close caps securely, and centrifuge the tubes for 1 to 2 seconds.
6. Incubate the tubes at room temperature for 2 to 24 hours.

D. Transformation of Competent amp^s/kan^s *E. coli* Cells (Figure 31.3d)

1. Label two sterile microcentrifuge tubes "Ligated" and "Pieces." Aseptically pipette 250 µl of ice-cold CaCl$_2$ into each tube. Keep the tubes in ice.
2. With a sterile loop, transfer a loopful of *E. coli* to each tube. Mix to make a suspension.
3. Aseptically add 10 µl from tube L to the tube labeled "Ligated." Save tube L for part E. Add 10 µl from tube C to the "Pieces" tube.
4. Incubate the tubes in ice for 15 minutes to allow the DNA to settle on the cells.
5. *Heat-shock* the cells by placing them in a 43°C water bath for 90 seconds so the DNA will enter the cells.

6. Return the tubes to the ice for 1 minute, and then add 700 μl sterile LB broth. Incubate the tubes in a 35°C water bath for at least 3 hours.

E. Culturing of Transformed Cells (Figure 31.3e)

1. Prepare the following controls. Inoculate one-half of an LB agar plate with a loopful of the amp^s/kan^s *E. coli* cells. Inoculate the other half with a loopful of the ligated plasmid DNA preparation (tube L). Similarly, inoculate an ampicillin–LB agar plate and a kanamycin–LB agar plate. How many plates do you have? _____

2. Add 100 μl from the "Ligated" tube to each of the following plates:

   LB agar
   Ampicillin–LB agar
   Kanamycin–LB agar
   Ampicillin–kanamycin–LB agar

3. Disinfect a spreading rod by dipping it in alcohol, quickly igniting the alcohol, and letting the alcohol burn off.

While the alcohol is burning off, hold the spreading rod pointed *down*. Keep the beaker of alcohol away from the flame.

Let the spreading rod cool.

4. Spread cells over one plate with the sterile spreading rod (see Figure 27.4 on page 216). Disinfect the spreading rod before spreading the cells over each of the remaining plates.

5. Repeat steps 2 through 4 to inoculate four plates from the "Pieces" tube.

6. Label the plates, and incubate them for 24 hours at 35°C.

7. Observe the plates, and record your results.

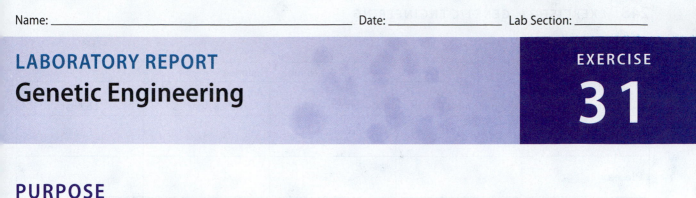

LABORATORY REPORT
Genetic Engineering

EXERCISE
31

PURPOSE _____

_____

_____

EXPECTED RESULTS

1. How many pieces will be produced by digesting the plasmids with the enzymes listed?

DNA	Enzyme	Number of Pieces	Sizes of RFLPs
pKAN	*Hind*III + *Bam*HI		
pAMP	*Hind*III + *Bam*HI		

2. *E. coli* will/will not grow on LB agar + amp + kan if recombination and transformation occur.

RESULTS

1. Sketch your electrophoresis results. Note the approximate sizes of the DNA fragments.

2. Controls:

Inoculum	Growth on		
	LB Agar	Ampicillin–LB Agar	Kanamycin–LB Agar
amp^s/kan^s *E. coli*			
DNA			

3. Transformation:

Inoculum	Number of Colonies			
	Ampicillin–LB Agar	Kanamycin–LB Agar	Ampicillin–Kanamycin–LB Agar	LB Agar
Ligated				
Pieces				

CONCLUSIONS

1. Did you cut the plasmids? How do you know? _____

_____

2. What do the results of your transformation experiment indicate? _____

_____

_____

QUESTIONS

1. What is the purpose of each control? _____

_____

_____

_____

2. If there was no growth on the DNA + *E. coli* plates, what went wrong? _____

_____

_____

3. Why were the *E. coli* bacteria treated with $CaCl_2$ before the plasmids were added? _____

_____

_____

4. Do the enzymes used in this experiment produce blunt or sticky ends? Why was this type of enzyme used?

_____

_____

CRITICAL THINKING

1. If you got growth on the DNA + *E. coli* plates, how could you rule out contamination? Mutation?

2. How could you prove that ligation of the two plasmids occurred, and not two separate transformation events?

CLINICAL APPLICATION

Researchers inserted the gene for green fluorescent protein (*gfp*) after the promoters for *ssaG* and *rpsM* in *Salmonella* bacteria. The *ssaG* gene encodes a protein secreted from bacterial cells into a host, and the *rpsM* gene encodes a ribosomal subunit. Macrophages and the genetically modified *Salmonella* were mixed in tubes. Intensity of bacterial fluorescence was measured 6 hours later. The results are shown below.

	Average of Bacterial Fluorescence after 6 hr	
gfp Inserted at	Intracellular (after being phagocytized)	Extracellular Bacteria
ssaG	35.8%	1.9%
rpsM	34.8%	75.7%

a. What can you tell about the interaction of pathogen and host from this experiment?

b. What is the value of this experiment?

Ames Test for Detecting Possible Chemical Carcinogens

The requirements of health can be stated simply. Those fortunate enough to be born free of significant congenital disease or disability will remain well if three BASIC NEEDS *are met: they must be adequately fed; they must be protected from a wide range of hazards in the environment; and they must not* DEPART RADICALLY *from the pattern of personal behavior under which man evolved, for example, by smoking, overeating, or sedentary living.* –THOMAS MCKEOWN

OBJECTIVES

After completing this exercise, you should be able to:

1. Differentiate the term *mutagenic* from *carcinogenic*.
2. Explain the rationale for the Ames test.
3. Perform the Ames test.
4. Differentiate a revertant from a mutant.

BACKGROUND

Every day we are exposed to a variety of chemicals, some of which are **genotoxic**—that is, they can cause mutations, which can result in cancer. Historically, animal models have been used to evaluate the carcinogenic potential of a chemical, but the procedures are costly and time-consuming and result in the inadequate testing of some chemicals. Many chemical carcinogens induce cancer because they are **mutagens** that alter the nucleotide base sequence of DNA.

Bruce Ames and his coworkers at the University of California at Berkeley developed a fast, inexpensive assay for mutagenesis using a *Salmonella* auxotroph. Most chemicals that have been shown to cause cancer in animals have proved mutagenic in the **Ames test.** Considering that not all mutagens induce cancer, the Ames test is a screening technique that can be used to identify high-risk compounds that must then be tested for carcinogenic potential. The Ames test uses auxotrophic strains of *Salmonella enterica* Typhimurium, which cannot synthesize the amino acid histidine (his⁻). The strains are also defective in *dark excision* repair of mutations (***uvrB***), and an **rfa** mutation eliminates a portion of the lipopolysaccharide that coats the bacterial surface. The *rfa* mutation increases the cell wall permeability; consequently, more mutagens enter the cell. The *uvrB* mutation minimizes repair of mutations; as a result, the

bacteria are much more sensitive to mutations. To grow the auxotroph, histidine and biotin (because of the *uvrB*) must be added to the culture media.

In the Ames spot test, a small sample of the test chemical (a suspected mutagen) is placed on the surface of glucose–minimal salts agar (GMSA, Table 28.1 on page 221) seeded with a lawn of the *Salmonella* auxotroph. (A small amount of histidine allows all the cells to go through a few divisions.) If certain mutations occur, the bacteria may *revert* to a wild type, or prototroph, and grow to form a colony. Only bacteria that have mutated (reverted) to his⁺ (able to synthesize histidine) will grow into colonies. In theory, the number of colonies is proportional to the mutagenicity of the chemical (see **FIGURE 32.1**).

In nature many chemicals are neither carcinogenic nor mutagenic, but they are metabolically converted to mutagens by liver enzymes. The original Ames test

Salmonella colonies

FIGURE 32.1   Ames test. The chemical (ethidium bromide) diffused from the disk on the left and caused the his⁻ bacteria to revert to his⁺. The his⁻ cells on the control plate (right) are unable to grow on glucose–minimal salts agar.

could not detect the mutagenic potential of these in vivo (in-the-body) conversions. In 1973, Ames and his colleagues modified the spot test to add mammalian liver enzymes and the *Salmonella* auxotroph to melted soft agar (containing histidine) that is overlaid on GMSA. The mutagens to be tested are then placed on the agar overlay. *Salmonella* is exposed to the "activated" chemicals to test for mutagenicity. The treat-and-plate method is another modification in which the test chemical and liver enzymes are incubated in an aerobic environment. Then a sample of the activated chemicals is added to soft agar containing the *Salmonella* auxotroph and histidine and is poured onto GMSA.

In this exercise, you will use the spot test with liver enzymes. The liver enzymes are called S9. A slurry of liver cells that have been broken apart is centrifuged at 9000 $\times g$.[*] The supernatant contains the cytoplasm, including enzymes.

Salmonella enterica Typhimurium is a potential pathogen.

MATERIALS

FIRST PERIOD

Petri plates containing glucose–minimal salts agar (2)
Tubes containing 4 ml of soft agar (glucose–minimal
    salts + 0.05 mM histidine and 0.05 mM biotin) (2)
Tube containing rat liver enzymes
Sterile filter paper disks
Sterile 1-ml pipettes (3)
Forceps and alcohol
Paper disks soaked with suspected mutagens:
    Benzo(α)pyrene
    Ethidium bromide
    Nitrosamine
    2-Aminofluorene
    Cigarette smoke condensates
    Food coloring
    Hair dye
    Household compound (your choice)
    Maraschino cherries
    Hot dog
    Cosmetic or personal care item (your choice)

CULTURES

Salmonella enterica Typhimurium his⁻, *uvrB, rfa* **BSL-2**

---

[*]Precise centrifugation conditions are specified in terms of units of gravity ($\times$g) because centrifuges with rotating heads of different sizes produce different centrifugal force, even at the same speed.

TECHNIQUES REQUIRED

Pipetting (Appendix A)
Aseptic technique (Exercise 4)

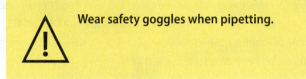

PROCEDURE First Period

1. Label one glucose–minimal salts agar plate "Liver Enzymes" and the other "No Enzymes."

Wear safety goggles when pipetting.

2. Aseptically pipette 0.1 ml *Salmonella* into one of the soft agar tubes, add 0.5 ml of liver enzymes, and quickly pour it over the surface of the "Liver Enzyme" plate (**FIGURE 32.2a**). Tilt the plate back and forth gently to spread the agar evenly. Let it harden.

3. Obtain another soft agar tube. Aseptically pipette 0.1 ml of *Salmonella* into the soft agar, and quickly pour it over the surface of the "No Enzymes" plate. Why does the soft agar contain histidine? _____

_____

Dip the forceps in alcohol, and with the tip pointed *down,* burn off the alcohol. Keep the beaker of alcohol away from the flame.

4. Aseptically place three to five disks saturated with the various suspected mutagens on the surface of the soft agar overlay of each plate (**FIGURE 32.2b**). If the chemical is crystalline, place a few crystals directly on the agar. Label the bottom of the Petri plates with the name of each chemical. Place one more sterile disk on each plate for a control. Why is this a control? _____

Be very careful. These are potentially dangerous compounds.

5. Incubate both plates, right-side up, at 35°C for 48 hours.

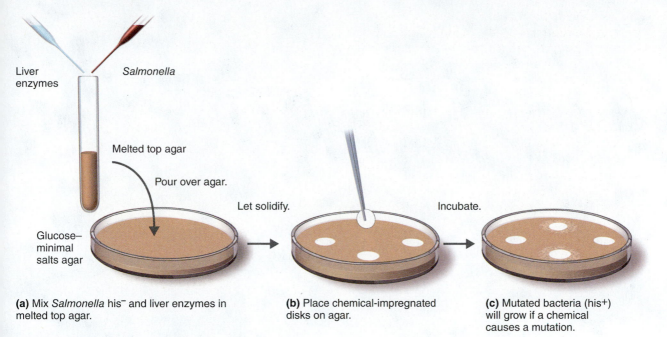

(a) Mix *Salmonella* his⁻ and liver enzymes in melted top agar.

(b) Place chemical-impregnated disks on agar.

(c) Mutated bacteria (his+) will grow if a chemical causes a mutation.

FIGURE 32.2   Procedure for the Ames test. (a) *Salmonella* his⁻ and liver enzymes are mixed in melted top agar, which is poured over glucose–minimal salts agar. **(b)** Disks impregnated with suspected mutagens are placed on the agar. **(c)** Only bacteria that have mutated (reverted) to his⁺ (able to synthesize histidine) will grow into colonies.

PROCEDURE   Second Period

Describe your results. As the chemical diffuses into the agar, a concentration gradient is formed. A mutagenic chemical will give rise to a ring of revertant colonies surrounding a disk (**FIGURE 32.2c**). If a compound is toxic, a zone of inhibition will also be observed around the disk.

LABORATORY REPORT
Ames Test for Detecting Possible Chemical Carcinogens

PURPOSE _____

_____

_____

HYPOTHESIS

The compounds tested will/will not be mutagenic.

RESULTS

1. Show the location of paper disks and bacterial growth on the plates. Number the disks to correspond to the table shown on the next page.

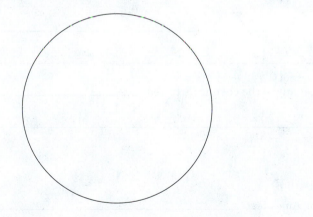

With liver enzymes   Without liver enzymes

2. Complete the following table.

Compound	With Liver Enzymes		Without Liver Enzymes		Mutagenic?
	Growth (large colonies)	Toxicity (zone of inhibition, mm)	Growth (large colonies)	Toxicity (zone of inhibition, mm)	
1 (Control)					
2					
3					
4					
5					
6					

CONCLUSIONS

1. Did you accept or reject your hypothesis based on your results? _____

_____

2. What compounds were mutagens? _____

3. Did the liver enzymes affect the mutagenicity of any of the chemicals tested? _____

_____

QUESTIONS

1. Why are mutants used as test organisms in the Ames test? _____

_____

_____

2. What is the advantage of this test over animal tests? Disadvantages? _____

_____

_____

CRITICAL THINKING

1. Does this technique give a minimum or a maximum mutagenic potential? Briefly explain.

2. Is it possible for a chemical to be negative in the Ames test and yet be a carcinogen? Briefly explain.

CLINICAL APPLICATION

In the following case, the Ames test was used to determine the mutagenic capability of 2-aminofluorene. What can you conclude from these data?

Test Substance	Relative Amounts of Growth in the Ames Test
2-Aminofluorene	+
2-Aminofluorene activated by:	
Liver enzymes	++++
Intestinal enzymes	++
Bacteroides fragilis enzymes	++
Intestinal enzymes + *B. fragilis* enzymes	+++

PART

9

The Microbial World

EXERCISES

Organisms with eukaryotic cells include algae, protozoa, fungi, and plants and animals. The eukaryotic cell is typically larger and structurally more complex than the prokaryotic cell. The DNA of a eukaryotic cell is enclosed within a membrane-bounded **nucleus.** In addition, eukaryotic cells contain membrane-bounded **organelles,** which are specialized structures that perform specific functions (see the illustration on the next page).

Mushrooms, molds, and yeasts are fungi (Exercise 33). Yeasts are possibly the best-known microorganisms. They are widely used in commercial processes and can be purchased in the supermarket for baking. Yeasts are unicellular fungi.

Van Leeuwenhoek was the first to observe the yeast responsible for fermentation in beer:

> *I have made divers observations of the yeast from which beer is made and I have generally seen that it is composed of globules floating in a clear medium (which I judged to be the beer itself).* *

Many algae are visible only through the microscope, whereas others can be a few meters long. Algae are important producers of oxygen and food for protozoa and other organisms. A few unicellular algae, such as the agents of "red tides" (*Alexandrium catanella* and related species), are toxic to animals, including humans, when ingested in large numbers.

Algae and cyanobacteria (Exercise 34) are photoautotrophs, using light as a source of energy and CO_2 for their carbon. Fungi, protozoa, and helminths are all chemoheterotrophs, which use organic molecules as a source of carbon and energy. Protozoa, originally called "infusoria," were of interest to early investigators. In 1778, Friedrich von Gleichen studied food vacuoles by feeding red dye to his infusoria. A refinement of

---

*Quoted in H. A. Lechevalier and M. Solotorovsky. *Three Centuries of Microbiology*. New York: Dover Publications, 1974, p. 502.

this experiment will be done in Exercise 35. Parasitic worms, or helminths, are animals (Exercise 36). Like all animals, they use other organisms for food. However, the parasitic worms live on and in a *living* host, their "prey." The compound light microscope is invaluable for identifying molds, algae, cyanobacteria, protozoa, and parasitic helminths; however, many yeasts and bacteria look alike through the microscope. These organisms are identified by biochemical and genetic testing.

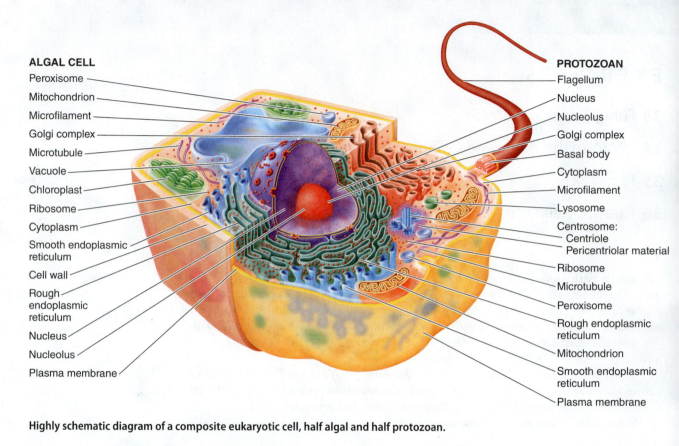

ALGAL CELL
- Peroxisome
- Mitochondrion
- Microfilament
- Golgi complex
- Microtubule
- Vacuole
- Chloroplast
- Ribosome
- Cytoplasm
- Smooth endoplasmic reticulum
- Cell wall
- Rough endoplasmic reticulum
- Nucleus
- Nucleolus
- Plasma membrane

PROTOZOAN
- Flagellum
- Nucleus
- Nucleolus
- Golgi complex
- Basal body
- Cytoplasm
- Microfilament
- Lysosome
- Centrosome:
  - Centriole
  - Pericentriolar material
- Ribosome
- Microtubule
- Peroxisome
- Rough endoplasmic reticulum
- Mitochondrion
- Smooth endoplasmic reticulum
- Plasma membrane

Highly schematic diagram of a composite eukaryotic cell, half algal and half protozoan.

CASE STUDY: Is It Just Dust?

Mary Ann is doing a training rotation in ophthalmology. She is learning to use the slit lamp to view a patient's cornea. Her current patient, Joaquin, wears contact lenses. Joaquin is an attorney who works indoors. However, the recent winds and dry weather are blowing dust into the air. He came in complaining of blurred vision, redness, tearing, and pain in his left eye. Mary Ann observed a large white mass and smaller gray masses in Joaquin's cornea. Mary Ann made a tentative diagnosis of keratitis and scraped Joaquin's cornea for a sample to send to the hospital laboratory. The lab immediately performed a wet mount and reported finding two filaments composed of three and four cells 10 μm long. The lab also inoculated nutrient agar and Sabouraud agar. Three weeks later, Mary Ann received the lab report describing mycelial colonies with black conidiospores on Sabouraud agar.

Questions

1. What is the cause of Joaquin's corneal infection?
2. Why did the lab work take three weeks?
3. How are weather conditions related to keratitis?

Fungi: Yeasts and Molds

And what is a weed? A plant whose VIRTUES *have not been discovered.* - RALPH WALDO EMERSON

OBJECTIVES

After completing this exercise, you should be able to:

1. Characterize and classify fungi.
2. Compare and contrast fungi and bacteria.
3. Identify common saprophytic molds.

BACKGROUND

Fungi possess eukaryotic cells and can exist as unicellular or multicellular organisms. They obtain nutrients by absorbing dissolved organic material through their cell walls and plasma membranes. Fungi (with the exception of yeasts) are aerobic. Unicellular yeasts, multicellular molds, and macroscopic species such as mushrooms are included in the Kingdom Fungi. Compared to bacteria, fungi generally grow better in more acidic conditions and tolerate higher osmotic pressure and lower moisture. They are larger than bacteria, with more cellular and morphologic detail.

Yeasts

Yeasts are nonfilamentous, unicellular fungi that are typically spherical or oval in shape. Yeasts are widely distributed in nature and frequently found on fruits and leaves as a white, powdery coating. When budding yeasts reproduce asexually, cell division is uneven. A new cell forms as a protuberance (bud) from the parent cell (**FIGURE 33.1a**). In some instances, when buds fail to detach themselves, a short chain of cells called a **pseudohypha** forms (**FIGURE 33.1b**).

Metabolic activities are used to identify genera of yeasts. Yeasts are facultative anaerobes. Their metabolic activities are used in many industrial fermentation processes. Yeasts are used to prepare many foods, including bread, and beverages, such as wine and beer.

Molds

The multicellular filamentous fungi are called **molds.** Because of the wide diversity in mold morphology, morphology is very useful in identifying these fungi.

A macroscopic mold colony is called a **thallus** and is composed of a mass of strands called the **mycelium.**

(a) Budding   **(b)** Pseudohypha

FIGURE 33.1 **Budding yeasts. (a)** A bud forming from a parent cell. **(b)** Pseudohyphae are short chains of cells formed by some yeasts.

Each strand is a **hypha.** The **vegetative hyphae** grow in or on the surface of the growth medium. Aerial hyphae, called **reproductive hyphae,** originate from the vegetative hyphae and produce a variety of asexual reproductive **spores** (**FIGURE 33.2**). The hyphal strand of most molds is composed of individual cells separated by a cross wall, or **septum.** These hyphae are called **septate hyphae.** A few fungi, including *Rhizopus,* shown in **FIGURE 33.2a,** have hyphae that lack septa and are a continuous mass of cytoplasm with multiple nuclei. These are called **coenocytic hyphae.**

In a laboratory, molds are identified by the appearance of their colony (color, size, and so on), hyphal organization (septate or coenocytic), and the structure and organization of reproductive spores. Because of the importance of colony appearance and organization, culture techniques and microscopic examination of molds are very important.

Identification

Until recently, fungi were classified on the basis of their sexual spores. Fungi that do not produce sexual spores were put in a "holding" phylum called *deuteromycota*. Now fungi are classified into phyla based on similarities in their rRNA. This eliminated the need

261

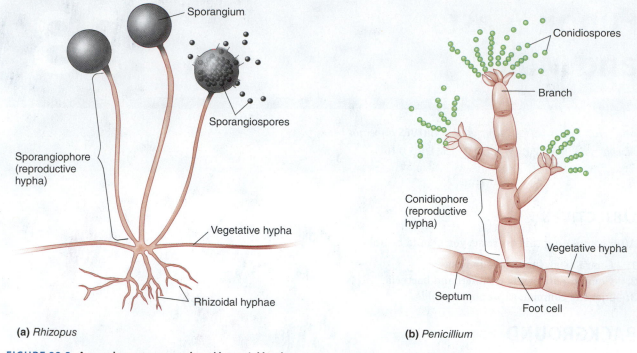

(a) *Rhizopus*

(b) *Penicillium*

FIGURE 33.2 **Asexual spores are produced by aerial hyphae.** **(a)** Sporangiospores are formed within a sporangium. **(b)** Conidiospores are formed in chains. One possible arrangement is shown here.

for deuteromycota as a taxonomic classification and revealed that water molds (Oomycota) are actually more closely related to algae such as diatoms and kelp than to fungi.

The fungus-like algae in the phylum **Oomycota** are included here because they look like fungi to the unaided eye and their growth requirements are similar to fungi. The oomycota are usually found in aquatic habitats and form sexual spores called oospores, formed by the fusion of two cells, and motile asexual spores called **zoospores.** One oomycote, *Phytophthora infestans*, causes late potato blight, which resulted in the Irish famine in the mid-nineteenth century.

Members of the phylum **Zygomycota** are saprophytic molds that have coenocytic hyphae. A **saprophyte** obtains its nutrients from dead organic matter;

in healthy animals and plants, they do not usually cause disease. A common example is *Rhizopus* (bread mold; **FIGURE 33.3a**). The asexual spores are formed inside a **sporangium,** or spore sac, and are called **sporangiospores** (see Figure 33.2a). Sexual spores called **zygospores** are formed by the fusion of two cells.

The **Ascomycota** include molds with septate hyphae and some yeasts. They are called sac fungi because their sexual spores, called **ascospores,** are produced in a sac, or **ascus.** The saprophytic molds usually produce **conidiospores** asexually (**FIGURE 33.2b**). The arrangements of the conidiospores are used to identify these fungi. Common examples are *Penicillium* (**FIGURE 33.3b**) and *Aspergillus* (**FIGURE 33.3c**).

The **Basidiomycota** include the fleshy fungi, or mushrooms, and have sexual spores called

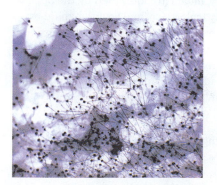

(a) *Rhizopus stolonifer*

1 mm

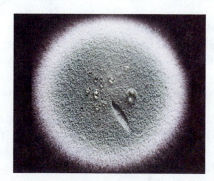

(b) *Penicillium chrysogenum*

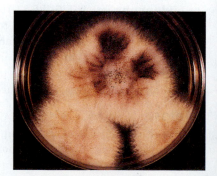

(c) *Aspergillus niger*

FIGURE 33.3 **Common molds.**

TABLE 33.1   CHARACTERISTICS OF COMMON SAPROPHYTIC FUNGI

Phylum	Growth Characteristics	Asexual Reproduction	Sexual Reproduction
Fungi			
Zygomycota ("conjugation fungi")	Coenocytic hyphae	Sporangiospores	Zygospores
Ascomycota ("sac fungi")	Septate hyphae; yeastlike	Conidiospores, budding	Ascospores
Basidiomycota ("club fungi")	Septate hyphae; includes fleshy fungi (mushrooms)	Fragmentation	Basidiospores
Fungus-like algae			
Oomycota ("water molds")	Coenocytic hyphae; cellulose cell walls	Zoospores	Oospores

basidiospores produced by the club-shaped stalk called a **basidium.**

A classification scheme for saprophytic fungi is shown in **TABLE 33.1.** A majority of the fungi classified are saprophytes, but each phylum contains a few genera that are **pathogens,** causing disease in plants and animals.

Laboratory Culture

Fungi, especially molds, are important clinically, and spores in the air are also the most common source of contamination in the laboratory.

In the laboratory, **Sabouraud agar,** a selective medium, is commonly used to culture fungi. Sabouraud agar has very simple nutrients (glucose and peptone) and a low pH, which inhibits the growth of most other organisms. Many of the techniques useful in working with bacteria can be applied to fungi.

MATERIALS

YEAST CULTURE

Petri plates containing Sabouraud agar (2)
Bottle containing glucose–yeast extract broth
Sterile cotton swab
Coverslip
Test tube
Balloon
Fruit or leaves
Methylene blue (second period)

MOLD CULTURE

Petri plates containing Sabouraud agar (2)
Scalpel and forceps
Beaker with alcohol
Sterile Petri dish
Coverslips

CULTURES

YEAST CULTURE

Saccharomyces cerevisiae (baker's yeast)

MOLD CULTURES

Rhizopus stolonifer
Aspergillus niger
Penicillium chrysogenum

PREPARED SLIDES

Zygospores
Ascospores

TECHNIQUES REQUIRED

Compound light microscopy (Exercise 1)
Wet-mount technique (Exercise 2)
Colony morphology (Exercise 3)
Plate streaking (Exercise 11)
Dissecting microscope (Appendix E)

PROCEDURE   Yeast Culture

First Period

1. Cut the fruit or leaves into small pieces. Place them in the bottle of glucose–yeast extract broth. Cover the mouth of the bottle with a balloon. Incubate the bottle at room temperature until growth has occurred.

2. Divide a Sabouraud agar plate in half. Inoculate one-half with baker's yeast. Inoculate the other half of the medium, following *either* procedure a *or* procedure b.

    a. Swab the surface of your tongue with a sterile swab. Inoculate one-half of the agar surface with the swab. Discard the swab in the disinfectant. Why will few bacteria grow on this medium? _____

    b. Using a sterile inoculating loop, streak one-half of the agar surface with a loopful of broth from the bottle just prepared in step 1. Replace the balloon.

3. Incubate the plate, inverted, at room temperature until growth has occurred.

Second Period

1. Record the appearance of the plant-infusion broth after incubation. Was gas produced? _____
2. Describe the different-appearing colonies.
   BSL-2 Prepare wet mounts with methylene blue and coverslips from different-appearing colonies. Examine the wet mounts with a microscope. Record your observations.

(a) Aseptically place an agar square on the slide.

PROCEDURE Mold Culture

First Period

1. Contaminate one Sabouraud agar plate in any manner you desire. Expose it to the air (outside, hall, lab, or wherever) for 15 to 30 minutes, or touch it. Incubate the plate, inverted, at room temperature for 5 to 7 days.
2. Inoculate the other Sabouraud agar plate with the mold culture assigned to you. Make one line in the center of the plate. Why don't you streak it? _____

   _____

   Incubate the plate, inverted, at room temperature for 5 to 7 days.
3. To set up a slide culture of your assigned mold, first clean a slide and dry it. Then disinfect a scalpel and forceps by putting them in alcohol and burning off the alcohol.

(b) Inoculate the mold onto one edge.

(c) Place a coverslip on the agar.

 Keep the flame away from the beaker of alcohol.

4. Using **FIGURE 33.4** as a reference, follow the steps below to prepare a slide culture.
   a. Using the scalpel and forceps, aseptically cut a square of agar from a Sabouraud agar plate, and place the square on the slide.
   b. Gently shake the mold culture to resuspend it, and inoculate one edge of the agar with a loopful of mold.
   c. Place a coverslip on the inoculated agar.
   d. Put a piece of wet paper towel in the bottom of a Petri dish. Why? _____

   _____

   Place the slide culture on the towel, and close the Petri dish. Incubate it at room temperature for 2 to 5 days. Do *not* invert the dish.
5. Observe the prepared slides showing sexual spores. Carefully diagram each of the spore formations in the spaces provided in the Laboratory Report.

Second Period

1. Examine plate cultures of *each* mold (without a microscope), and describe their color and appearance.

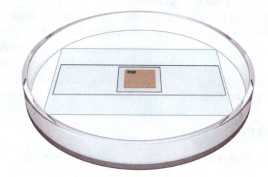

(d) Place the slide on a moist paper towel in a Petri dish.

FIGURE 33.4 Preparing a slide culture.

Then examine them with a dissecting microscope. Look at both the top and the underside.

 Do not open any of the plates because mold spores can easily be spread in the air.

2. Examine your contaminated Sabouraud plate, and describe the results.
3. Examine slide cultures of each mold using a dissecting microscope. Record your observations.

LABORATORY REPORT
Fungi: Yeasts and Molds

PURPOSE _____

_____

_____

EXPECTED RESULTS

1. What will be observed in the glucose–yeast broth after incubation? _____

2. Will mold be found on the Sabouraud agar plates that were exposed to the air? _____

RESULTS

Yeast Culture

Plants used: _____

Describe the appearance of the glucose–yeast extract broth after _____ days' incubation. _____

_____

Was gas produced? _____

Sabouraud agar:

Inoculum	Colony Appearance	Color	Size	Wet Mount
S. cerevisiae				
Area Sampled:				
Tongue:				
Bottle:				

Any bacteria seen? _____

Mold Culture

	Organism			
Colony Appearance	*Rhizopus stolonifer*	*Aspergillus niger*	*Penicillium chrysogenum*	**Unknown from Contaminated Plate**
Macroscopic				
Hyphae color				
Spore color				
Underside color				
Microscopic				
Diagram _____ ×				

Slide Cultures

	Organism		
Colony Appearance	*Rhizopus stolonifer*	*Aspergillus niger*	*Penicillium chrysogenum*
Macroscopic			
Hyphae color			
Spore color			
Underside color			
Microsopic			
Diagram _____ ×			

Prepared Slides

Sexual spores

Phylum:	Zygomycetes	Ascomycetes
Genus:	_____	_____
Total magnification:	____ ×	____ ×

CONCLUSIONS

1. Did your results agree with your predicted results? _____

2. Did you culture yeast from your mouth? _____ From the plants? _____ How do you know? _____

_____

3. Compare and contrast yeast and bacteria regarding their appearance both on solid media and under the

microscope. _____

_____

_____

4. Could you identify the phylum of mold from your contaminated plate? _____

_____

QUESTIONS

1. What was the purpose of the balloon on the glucose–yeast extract broth bottle? _____

_____

2. Define the term *yeast*. _____

_____

_____

3. Why are yeast colonies larger than bacterial colonies? _____

_____

_____

4. Why do media that are used to culture fungi contain sugars? _____

_____

_____

CRITICAL THINKING

1. Why are antibiotics frequently added to Sabouraud agar for isolating fungi from clinical samples?

2. How can fungi cause respiratory tract infections?

3. Why weren't pathogenic fungi used in this exercise?

CLINICAL APPLICATIONS

1. A 45-year-old HIV-positive woman using continuous intravenous antibiotic infusion to manage a kidney infection developed a 39°C (102.2°F) fever. Cultures of blood, the needle tip, and the insertion site revealed large ovoid cells that reproduced by budding.
   a. What is your preliminary identification of the infecting organism?

b. How might the antibiotic treatment have affected this patient?

c. What do the findings suggest about the portal of entry for the infection?

2. An 82-year-old man was hospitalized for pneumonia. He was treated with penicillin and discharged after 7 days when his symptoms subsided. Two weeks later, the man was readmitted to the hospital for pneumonia. No bacteria or viruses were cultured from a biopsy specimen. Broad-spectrum antibiotic therapy was administered. The patient developed pneumonia again after returning home from an 8-day hospital stay. A bronchoscopy showed septate hyphae in his lung tissue. An investigation of his home showed that he did not keep pets, but some sparrows had nested outside the living room window; he was not a gardener or outdoorsman. *Aspergillus* was cultured from his humidifier.

a. What is your preliminary identification of the pathogen?

b. Why was the diagnosis delayed?

c. Why did the man's infection recur?

Phototrophs: Algae and Cyanobacteria

OBJECTIVES

After completing this exercise, you should be able to:

1. List criteria that are used to classify algae.
2. List general requirements for the growth of phototrophic organisms.
3. Compare and contrast algae with fungi and bacteria.

BACKGROUND

Algae is the common name for photosynthetic eukaryotic organisms that lack true roots, stems, and leaves. Algae may be found in the ocean and in freshwater and on moist tree bark and soil (**FIGURE 34.1**). Algae may be unicellular, colonial, filamentous, or multicellular. They exhibit a wide range of shapes: from the giant brown algae or kelp and delicate marine red algae to spherical green-algal colonies. Algae are identified according to pigments, storage products, the chemical composition of their cell walls, flagella, and analysis of the rRNA sequence. Identification of algae requires microscopic examination.

Most freshwater algae belong to the groups listed in **TABLE 34.1**. Of primary interest to microbiologists are the **cyanobacteria.** Although cyanobacteria have prokaryotic cells and belong to the Domain Bacteria, they are often observed in water samples along with algae. Cyanobacteria have chlorophyll *a,* which is found in all plants and algae. Like animal cells, however, cyanobacteria store carbohydrates as glycogen. Cyanobacteria can fix nitrogen in specialized cells called heterocysts, and they often contain gas vacuoles, which enable cyanobacteria to float, increasing exposure to the sun for photosynthesis.

The growth of phototrophs is essential in providing oxygen and food for other organisms; however, some filamentous algae, such as *Spirogyra,* are a nuisance to humans because they clog filters in water systems (**FIGURE 34.2**). And others, such as *Alexandrium,* produce toxins that are harmful to vertebrates. Phototrophs can be used to determine the quality of water. Polluted waters containing excessive nutrients from sewage or other sources have more cyanobacteria and fewer diatoms than clean waters do. Additionally, the *number* of algal cells indicates water quality. Counts of more than 1000 algal cells per milliliter indicate that excessive nutrients are present.

In this exercise, we will evaluate the effect of sunlight and inorganic chemicals on the growth of algae.

TABLE 34.1   SOME CHARACTERISTICS OF MAJOR GROUPS OF PHOTOTROPHS FOUND IN FRESHWATER

	Bacteria	Algae			
Characteristics	Cyanobacteria	Euglenoids	Diatoms	Green Algae	Oomycotes
Color	Blue-green	Green	Yellow-brown	Green	White
Cell wall	Bacteria-like	Lacking	Visible with regular markings	Visible	Visible
Cell type	Prokaryote	Eukaryote	Eukaryote	Eukaryote	Eukaryote
Flagella	Absent	Present	Absent	Present in some	On zoospores
Cell arrangement	Unicellular or filamentous	Unicellular	Unicellular or colonial	Unicellular, colonial, or filamentous	Filamentous
Nutrition	Autotrophic	Facultatively heterotrophic	Autotrophic	Autotrophic	Heterotrophic
Produce O_2	Yes	Yes	Yes	Yes	No

FIGURE 34.1 *Chlamydomonas nivalis.* Red snow is due to the green alga *C. nivalis*. The green color of chlorophyll is masked by red carotenoid pigments produced in the presence of intense light.

50 µm

FIGURE 34.2 *Spirogyra.* The helical chloroplasts are visible in the green alga *Spirogyra*.

MATERIALS

Pond water samples:
A. Incubated in the light for 4 weeks.
B. Incubated in the dark for 4 weeks.
C. With nitrates and phosphates added; incubated in the light for 4 weeks.
D. With copper sulfate added; incubated in the light for 4 weeks.

PREPARED SLIDES

Spirogyra
Diatom
Cyanobacteria

TECHNIQUES REQUIRED

Compound light microscopy (Exercise 1)
Hanging-drop procedure (Exercise 2)

PROCEDURE

1. Prepare a hanging-drop slide from a sample of pond water A. Take your drop from the bottom of the container. Why? _____
2. Examine the slide using the low and high-dry objectives. Identify the algae present in the pond water. Refer to **FIGURE 34.3** for identification. Draw those algae that you cannot identify. Record the relative amounts of each type of alga from 4+ (most abundant) to + (one representative seen).
3. Repeat the observation and data collection for the remaining pond water samples.
4. Examine the prepared slides, and then record your observations using Figure 34.3 as a reference.

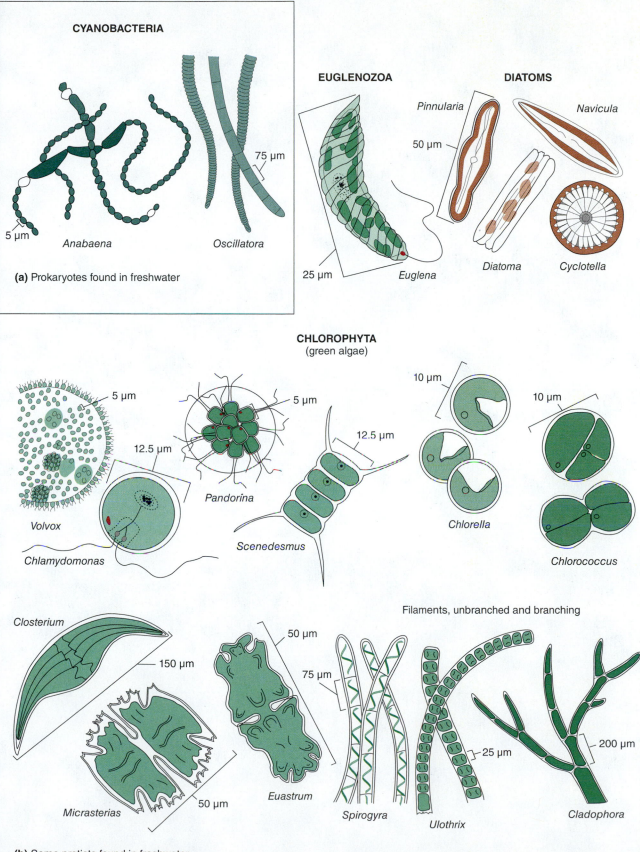

(a) Prokaryotes found in freshwater

(b) Some protists found in freshwater

FIGURE 34.3 Phototrophic microorganisms.

LABORATORY REPORT
Phototrophs: Algae and Cyanobacteria

PURPOSE _____

_____

_____

EXPECTED RESULTS

1. Will more phototrophs be found in the dark sample or in the light sample? _____

2. Which sample will have more algae: the water sample with the high phosphate and nitrate, or the sample with

   no additional nitrate or phosphate? _____

3. Do you expect to see more algae in the water sample with copper or in the sample with no copper? _____

   _____

RESULTS

Name or Sketch Alga or Cyanobacterium	Relative Abundance in Pond Water			
	Incubated in Light	Incubated in Dark	Incubated with NO_3^- and PO_4^{3-}	Incubated with $CuSO_4$
Total number of species				
Total number of organisms				

Which sample is the control? _____

PREPARED SLIDES

Sketch each phototroph.

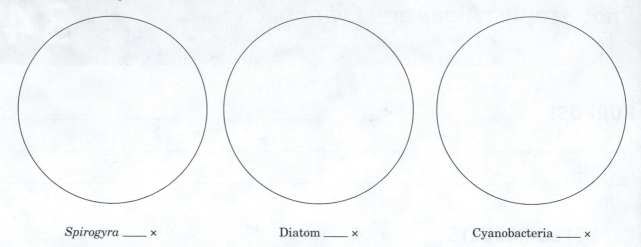

Spirogyra ____ ×   Diatom ____ ×   Cyanobacteria ____ ×

CONCLUSIONS

Compare the number and various groups of algae observed in each environment with the control.

Dark (Sample B) versus Light (Sample A)

1. Did your findings agree with your expected results? If not, briefly explain why. _____

_____

_____

Nitrate and Phosphate (Sample C) versus No Nitrate and Phosphate (Sample A)

2. Did your findings agree with your expected results? If not, briefly explain why. _____

_____

_____

Copper (Sample D) versus No Copper (Sample A)

3. Did your findings agree with your expected results? If not, briefly explain why. _____

_____

_____

QUESTIONS

1. Why can algae and cyanobacteria be considered indicators of productivity as well as of pollution? _____

_____

_____

2. How can algae be responsible for producing more oxygen than land plants produce? _____

_____

_____

3. Why aren't all algae included in the Kingdom Plantae? _____

_____

_____

_____

4. Describe one way in which algae and fungi differ. _____

_____

_____

How are they similar? _____

_____

_____

_____

CRITICAL THINKING

1. Cyanobacteria were once called "blue-green algae." What characteristics would lead to this name? What caused biologists to reclassify them as bacteria?

2. Oomycotes, including the *Phytophthora* parasite that destroys crop plants and oak trees, until recently were classified as fungi. What property could have led to the conclusion that they were fungi?

CLINICAL APPLICATIONS

1. Under certain environmental conditions, the microscopic alga *Karenia brevis* grows quickly, creating blooms that can make the ocean appear red or brown. *K. brevis* produces powerful toxins called brevetoxins, which have killed millions of fish. In addition to killing fish, brevetoxins can become concentrated in the tissues of mollusks that feed on *K. brevis*. People who eat these bivalves (e.g., clams, mussels, and oysters) may suffer from paralytic shellfish poisoning, a food poisoning that can cause severe gastrointestinal and neurological symptoms, such as tingling in the fingers or toes. Use your results to explain why these so-called "red tides" occur in coastal water.

2. Outbreaks of cyanobacterial intoxication associated with lakes and ponds are reported annually. What would cause an increased number of cases in summer months? Why aren't cyanobacterial intoxications associated with swimming pools?

Protozoa

OBJECTIVES

After completing this exercise, you should be able to:

1. List three characteristics of protozoa.
2. Explain how protozoa are classified.

BACKGROUND

Protozoa are unicellular eukaryotic organisms. Many protozoa live in soil and water, and some are normal microbiota in animals. A few species of protozoa are parasites.

Protozoa are heterotrophs, and most are aerobic. They feed on other microorganisms and on small particulate matter. Protozoa lack cell walls; in some, the outer covering is a thick, elastic membrane called a **pellicle.** Cells with a pellicle require specialized structures to take in food. The **contractile vacuole** may be visible in some specimens (**FIGURE 35.1**). This organelle fills with freshwater and then contracts to eliminate excess water from the cell, allowing the organism to live in low-solute environments. Would you expect more contractile vacuoles in freshwater protozoa or in marine protozoa? _____

In this exercise, we will examine live, free-living protozoa, as well as prepared slides of three parasitic

protozoa. The **amebas** (**FIGURE 35.1a**) move by extending lobelike projections of cytoplasm called **pseudopods.** As pseudopods flow from one end of the cell, the rest of the cell flows toward the pseudopods.

The **Euglenozoa** (**FIGURE 35.1b**) have one or more flagella. Although many euglenozoa are heterotrophs, the organism used in this exercise is a facultative heterotroph. It grows photosynthetically in the presence of light and heterotrophically in the dark.

The **Ciliates** of the phylum **Ciliophora** (**FIGURE 35.1c**) have many cilia extending from the cell. In some ciliates, the cilia occur in rows over the entire surface of the cell. In ciliates that live attached to solid surfaces, the cilia occur only around the oral groove. Why only around the oral groove? _____

Food is taken into the **oral groove** through the cytostome (mouth) and into the cytopharynx, where a **food vacuole** forms.

Diplomonads and **Parabasalids** generally have multiple flagella and lack mitochondria. Diplomonads have two nuclei and live in the digestive tracts of animals. The absence of mitochondria is probably not a disadvantage in this anoxic (without oxygen) environment. Parabasalids are symbionts in many animals. **Apicomplexa** are nonmotile, obligate intracellular parasites. These protozoa have a unique organelle, the

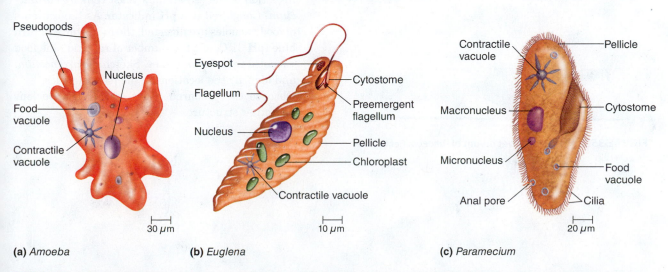

(a) *Amoeba*   (b) *Euglena*   (c) *Paramecium*

FIGURE 35.1 **Protozoa. (a)** An ameba moves by extending pseudopods. **(b)** *Euglena* has a whiplike flagellum.
(c) *Paramecium* has cilia over its surface.

apicoplast, that contains its own DNA. Apicomplexa have complex life cycles that ensure their transmission to new hosts.

In this exercise, we will examine food uptake as well as motility in various protozoa.

MATERIALS

Methylcellulose, 1.5%
Acetic acid, 5%
Pasteur pipettes
Coverslips

CULTURES

Amoeba
Paramecium
Paramecium feeding on Congo red–yeast suspension
Euglena

PREPARED SLIDES

Giardia
Plasmodium
Trypanosoma

TECHNIQUES REQUIRED

Compound light microscopy (Exercise 1)
Wet-mount technique (Exercise 2)

PROCEDURE

1. Prepare a wet mount of *Amoeba*. Place a drop from the bottom of the *Amoeba* culture on a slide. Place one edge of the coverslip into the drop, and let the fluid run along the coverslip (**FIGURE 35.2**). Gently

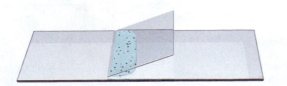

FIGURE 35.2 Preparing a wet mount of *Amoeba*. Gently lower the coverslip.

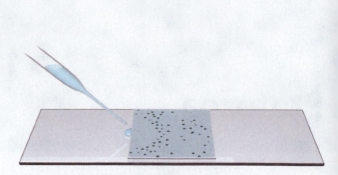

FIGURE 35.3 Preparing a wet mount of *Euglena*. Add a drop of acetic acid at one edge of the coverslip. Allow it to diffuse into the wet mount.

lay the coverslip over the drop. Observe the ameboid movement, and diagram it.

2. Prepare a wet mount of *Euglena*. Follow one individual, and diagram its movement. Can you see *Euglena*'s red "eyespot"? _____

   Why do you suppose it is present in photosynthetic strains and not in nonphotosynthetic strains? ____

   _____

   Allow a drop of acetic acid to seep under the coverslip (**FIGURE 35.3**). How does *Euglena* respond? ___

   _____

3. Prepare a wet mount of *Paramecium,* and observe its movement (see Figure 35.2). Why do you suppose it rolls and *Amoeba* does not? _____

4. Place a drop of methylcellulose on a slide. Make a wet mount of the *Paramecium* culture that has been feeding on a Congo red–yeast suspension in the methylcellulose. The *Paramecium* will move more slowly in the viscous methylcellulose. Observe the ingestion of the red-stained yeast cells by *Paramecium*. Congo red is a pH indicator. As the contents of food vacuoles are digested, the indicator will turn blue (pH 3). Count the number of red and blue food vacuoles in a *Paramecium*. Sketch a *Paramecium,* and identify the locations of the food vacuoles.

5. Observe the prepared slides, noting the nucleus and other structures.

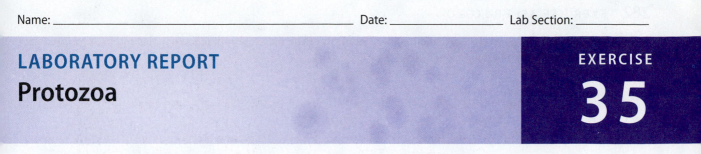

LABORATORY REPORT
Protozoa

EXERCISE
35

PURPOSE _____

_____

_____

EXPECTED RESULTS

1. How will the movement of a protozoan with pseudopods differ from the movement of a protozoan with flagella? __

    _____

2. From the movement of a protozoan with cilia? _____

RESULTS

Use a series of diagrams to illustrate:

1. The movement of *Amoeba* across the field of vision. ____ ×

2. The movement of *Euglena* and its flagellum. ____ ×

3. The movement of *Paramecium* and its cilia. ____ ×

4. The ingestion of food, the formation of a food vacuole, and the movement of food vacuoles in *Paramecium*. Note any color changes in the food vacuoles. ____ ×

PREPARED SLIDES

Sketch a field with each parasite.

Trypanosoma	*Giardia*	*Plasmodium*
____ ×	____ ×	____ ×
Label: Flagellum	Flagella	Nucleus and cytoplasm

Disease
caused: _____   _____   _____

Phylum: _____   _____   _____

CONCLUSIONS

1. Did your results agree with your expected results? _____

2. How did *Euglena* respond to the acetic acid? _____

_____

3. Which of the live organisms observed in this exercise would you bet on in a race? _____

4. Describe the arrangement of cilia on *Paramecium*. _____

_____

QUESTIONS

1. *Trypanosoma* and *Plasmodium* are both found in blood. How do they differ in their locations relative to red

    blood cells? _____

    _____

2. What advantage does *Trypanosoma*'s shape provide? _____

    _____

    _____

CRITICAL THINKING

1. Why is *Euglena* often used to study algae *and* protozoa?

2. Why are *Giardia* and *Trypanosoma* not classified into the same phylum? Which two genera in this exercise are most closely related?

3. *Pneumocystis* had been classified as a protozoan since its discovery in 1908. However, rRNA sequencing now shows that it is a fungus. Why might it be important to have an accurate classification of this organism?

CLINICAL APPLICATION

Laboratory eyewashes should be flushed once a month to remove *Acanthamoeba* accumulations from the pipes. Why is this removal necessary for eyewashes but not for other water outlets?

Parasitic Helminths

OBJECTIVES

After completing this exercise, you should be able to:

1. Differentiate nematode, cestode, and trematode.
2. Explain how helminth infections are diagnosed.

BACKGROUND

Helminths are multicellular eukaryotic animals that generally possess digestive, circulatory, nervous, excretory, and reproductive systems. Some are free-living in soil and water, whereas others are parasites of humans and other animals and plants. Parasitic helminths are studied in microbiology because they cause infectious diseases and because most of these diseases are diagnosed by microscopic examination of eggs or larvae. Eggs may have striations (lines), a spine, or an operculum (a hatch by which the larva leaves).

Helminths infect more than one-third of the world population. Helminth infections differ from bacterial or protozoan infections because the worms do not usually increase in number in the host. Symptoms are usually due to mechanical damage, eating of host tissues, or competing for vitamins. In this exercise, we will examine prepared slides of parasitic helminths.

Life Cycle

Parasitic helminths are highly modified compared to free-living helminths. They often lack sense organs such as eyes, and they may even lack a digestive system. Their reproductive system, however, is often complex, with characteristics that promote infection of new hosts. Some flukes, for example, can produce 25,000 eggs per day.

Adult helminths may be **dioecious;** that is, male reproductive organs are in one individual, and female reproductive organs are in another. In those species, reproduction occurs only when two adults opposite in sex are present in the same host. Adult helminths may also be **monoecious,** or **hermaphroditic**—one animal has both male and female reproductive organs. Two hermaphrodites may copulate and simultaneously fertilize each other.

1. *Intermediate host*. Some parasites have a different host for each larval stage. These are called intermediate hosts. Humans can serve as the intermediate host for the dog tapeworm. The larva encysts as a hydatid cyst in a variety of tissues, including the lungs and liver.
2. *Definitive host*. The adult (reproductively mature) stage of a parasite lives in a definitive host. Humans can serve as the definitive host for beef, pork, and fish tapeworms.
3. *Eggs infective*. The eggs of some parasitic roundworms are infective for humans. Adult pinworms are found in the large intestine. From there, the female pinworm migrates to the anus to deposit her eggs on the perianal skin. The eggs can be ingested by the host or by another person exposed through contaminated clothing or bedding.
4. *Larvae infective*. Some parasites are infective for mammals in the larval stage. Larvae of filarial worms, such as *Wucheria,* that infect humans and the dog heartworm are transmitted to their mammalian host by mosquitoes. The larvae mature into adults in the mammal.

Flatworms

Platyhelminthes, or flatworms, are flattened from the dorsal to ventral surfaces. The classes of this phylum include trematodes and cestodes. Many trematodes, or **flukes,** have flat, leaf-shaped bodies with ventral and oral suckers (**FIGURE 36.1**). The suckers hold the animal in place. Flukes obtain food by absorbing it through their outer covering, called the **cuticle.** Flukes are given common names according to the tissue of the definitive host in which the adults live (examples include lung fluke, liver fluke, and blood fluke).

Cestodes, or **tapeworms,** are intestinal parasites. Their structure is shown in **FIGURE 36.2**. The head, or **scolex** (plural: *scoleces*), has suckers for attaching to the intestinal mucosa of the definitive host; some species also have small hooks for attachment. Tapeworms do not ingest the tissues of their hosts; in fact, they completely lack a digestive system. To obtain nutrients from the small intestine, they absorb food through their cuticle. The body consists of segments called **proglottids.** Proglottids are continually produced by the neck region of the scolex, as long as the scolex is attached and alive. Each proglottid matures as it is pushed away from the neck by new proglottids. Each

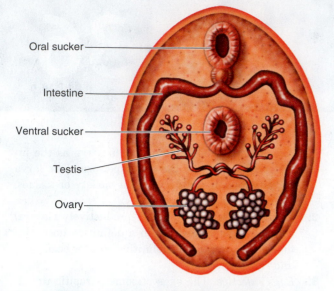

FIGURE 36.1 Generalized anatomy of a roundworm.

Oral sucker

Intestine

Ventral sucker

Testis

Ovary

Scolex

Neck

Testis

Genital pore

Ovary

Mature proglottid will disintegrate and release eggs

FIGURE 36.2 Generalized anatomy of a tapeworm.

proglottid contains both male and female reproductive systems, and eggs are fertilized as the proglottid reaches the middle of the worm. The proglottids farthest away from the scolex are basically bags of fertilized eggs that will be shed in feces.

Roundworms

Nematodes, or roundworms, are cylindrical and tapered at each end. Roundworms have a *complete* digestive system, consisting of a mouth, an intestine, and an anus (**FIGURE 36.3**). Most species are dioecious. The reproductive system consists of long tubules that serve as ovaries or testes. In females, the reproductive tubule (ovary) is usually double. Males are smaller than females and have one or two hardened **spicules** on their posterior ends that guide sperm to the female's genital pore. Species identification is often based on spicule structure.

MATERIALS

PREPARED SLIDES

Enterobius vermicularis (pinworm)
Wucheria bancrofti (filarial worm)
Trichinella spiralis cyst
Taenia sp. (tapeworm)
Echinococcus (hydatid cyst)
Clonorchis sinensis (liver fluke)
Fecal smears of eggs

TECHNIQUES REQUIRED

Compound light microscope (Exercise 1)
Dissecting microscope (Appendix E)

PROCEDURE

1. Using a compound light microscope, examine the male and female pinworms. Sketch the parts of the worms that show anatomic differences. Using a dissecting microscope, measure the length of the male and female pinworms.
2. Observe the blood smear containing *Wucheria*. Estimate the length and width of the worm by com-

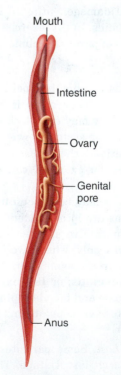

Mouth

Intestine

Ovary

Genital pore

Anus

FIGURE 36.3 Generalized anatomy of a fluke.

paring it to blood cells (white blood cells = 15 μm; red blood cells = 7 μm). Note that these filarial worm larvae are filled with a column of nuclei. Do the nuclei extend all the way to the tail? _____

3. Examine and sketch a *Trichinella* cyst with a compound light microscope. Using a dissecting microscope, measure the size of *Trichinella* cysts. In what tissue are the nematodes encysted? _____

4. Using a dissecting microscope, examine the tapeworm slide. Measure the length and width of immature and mature proglottids.

   a. Examine the neck region. Measure the width of a proglottid. Count the number of proglottids that fit into 1 millimeter. Estimate the length of one proglottid (divide 1 mm by the number of proglottids).

   b. Using a compound light microscope, examine and sketch a hydatid cyst. What structures are visible in the cyst? _____

   _____

5. Examine the liver fluke. Measure the length of the ovaries and the total body.

6. Examine the fecal smears containing worm eggs. Sketch one egg of each species observed. Be aware that the slides may have debris, so you will need to search for the regularly shaped eggs.

LABORATORY REPORT
Parasitic Helminths

PURPOSE _____

_____

_____

EXPECTED RESULT

What will distinguish tapeworms and roundworms? _____

_____

RESULTS

Roundworms

1. Pinworm. Sketch to show differences between the sexes.

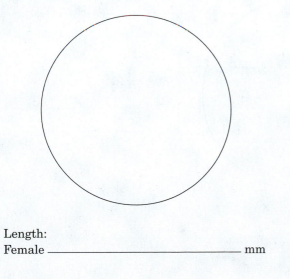 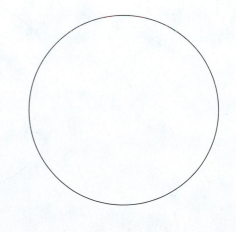

Length:
Female ——————————— mm   Male ——————————— mm

2. *Wucheria.*

   Length ——————————— μm   Width ——————————— μm

   Sketch, labeling the nuclei.

   ————×

3. *Trichinella* cyst.

   Length: ——————————— mm

   Tissue: ———————————

   Sketch the cyst.

   ————×

Flatworms

1. What organism did you observe? ———————————

   Record your measurements.

Tapeworm	Mature Proglottid	Immature Proglottid	Neck-Region Proglottid
Width			
Length			

2. Hydatid cyst. Sketch and label the visible structures.

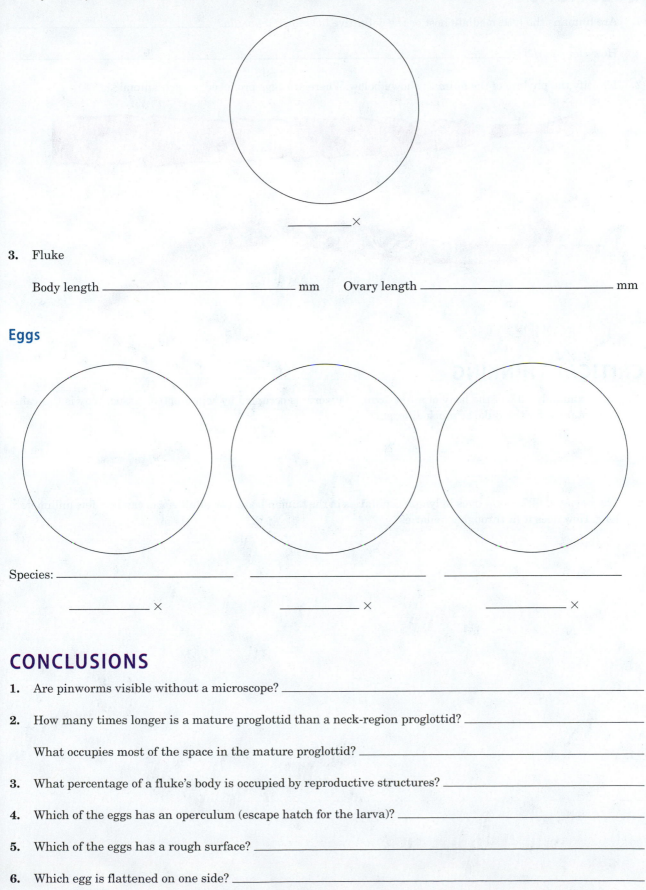

_____×

3. Fluke

Body length _____ mm   Ovary length _____ mm

Eggs

Species: _____   _____   _____

_____×   _____×   _____×

CONCLUSIONS

1. Are pinworms visible without a microscope? _____

2. How many times longer is a mature proglottid than a neck-region proglottid? _____

   What occupies most of the space in the mature proglottid? _____

3. What percentage of a fluke's body is occupied by reproductive structures? _____

4. Which of the eggs has an operculum (escape hatch for the larva)? _____

5. Which of the eggs has a rough surface? _____

6. Which egg is flattened on one side? _____

QUESTIONS

1. Are humans the intermediate host or the definitive host for *Trichinella*? _____

    How can you tell? _____

2. Identify the phylum of the animals shown below. Where are eggs produced in each animal?

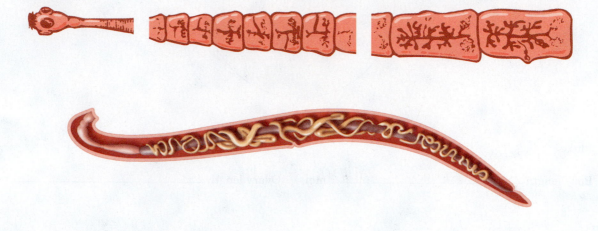

CRITICAL THINKING

1. Approximately 10% of the body of a free-living flatworm is occupied by reproductive tissue. Why is the value you obtained for the fluke so much different?

2. The nematode *Wucheria* lives in lymph capillaries in the human body. The adult worm can be a few millimeters long. How does it fit through capillaries?

CLINICAL APPLICATIONS

1. A woman found a worm in her laundry basket. The worm measured approximately 15 centimeters in length (**FIGURE a**). Eggs were removed from the worm and examined at 400× magnification (**FIGURE b**). What is the worm? On what criteria do you base your identification?

(a)   1 cm

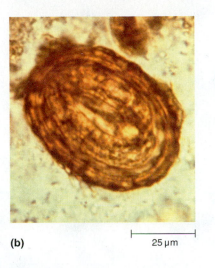

(b)   25 µm

2. A 25-year-old previously healthy woman had a physical examination, including a chest X ray, as one of the requirements for her employment. The X-ray film showed a cystlike lesion in her left lung. The cyst measured 3 to 4 cm in diameter (see the figure on the next page). What is your diagnosis? On what criteria do you base that diagnosis?

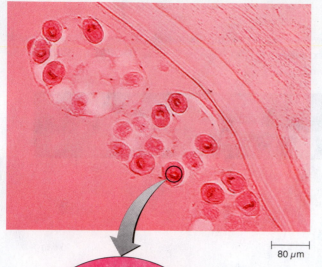

80 µm

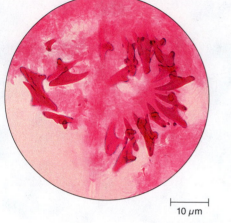

10 µm

Viruses

EXERCISES

Viruses are fundamentally different from the bacteria studied in previous exercises, as well as from all other organisms. Wendell M. Stanley, who was awarded the Nobel Prize in 1946 for his work on tobacco mosaic virus, described the virus as

> one of the great riddles of biology. We do not know whether it is alive or dead, because it seems to occupy a place midway between the inert chemical molecule and the living organism.*

Viruses are submicroscopic, filterable, infectious agents. Viruses are too small to be seen with a light microscope and can be seen only with an electron microscope. Filtration is frequently used to separate viruses from other microorganisms. A suspension containing viruses and bacteria is filtered through a membrane filter with a small pore size (0.45 μm) that retains the bacteria but allows viruses to pass through.

Viruses are *obligate intracellular parasites* with a simple structure and organization. Viruses contain DNA or RNA enclosed in a protein coat. Viruses exhibit quite strict specificity for host cells. Once inside the host cell, the virus uses the host cell's anabolic machinery and material to cause synthesis of virus particles that can infect new cells.

Viruses are widely distributed in nature and have been isolated from virtually every eukaryotic and prokaryotic organism. Based on their host specificity, viruses are divided into general categories, such as **bacterial viruses (bacteriophages), plant viruses,** and **animal viruses.** A virus's host cells must be grown in order to culture viruses in a laboratory.

In Exercise 37, we will isolate bacteriophages and estimate the number of phage particles. In Exercise 38, we will isolate plant viruses in plant leaves.

*Quoted in H. A. Lechevalier and M. Solotorovsky. *Three Centuries of Microbiology*. New York: Dover Publications, 1974, p. 25.

CASE STUDY: Where There's Smoke There Are Plaques!

The Green Thumb greenhouse was getting geared up for eager gardeners after a long, hard winter. Tomato plants were being subdivided and planted in individual containers. Carrie, the owner of the greenhouse, was concerned because many of the transplants were developing brown spots and some had wilted. Carrie had majored in microbiology in college and was determined to find the cause of the disease in the transplants. Late in the day, Carrie stopped in unexpectedly and discovered several of the temporary workers smoking while they worked. Carrie set up an experiment to determine the etiology (cause) of the disease. Carrie collected damaged leaves from the tomato plants and made a slurry of the leaves using a blender. She also made a slurry of some of the tobacco the workers had been smoking. She filtered the suspensions through a 0.45-μm filter she obtained from the local college. She carefully rubbed sand on the leaves of young tomato and tobacco plants. She applied the filtrate to the leaves, set up controls, and put the plants in an isolated part of the greenhouse. After several days, plaques (spots) appeared on the leaves of both the tomato plants and the tobacco plants. The control plants were disease-free.

Questions

1. Carrie's experiment was modeled after Koch's postulates. Explain.
2. Carrie's experiment really recreated the experiments of Adolf Mayer and Dimitri Iwanowski. What did their historical experiments indicate about viruses?
3. Why did Carrie rub the tobacco plants and the tomato plants with sand before exposing the leaves to the slurries?
4. In the greenhouse, how could the plants have become infected?
5. What does Carrie's experiment indicate about the host range of the virus?

Isolation and Titration of Bacteriophages

OBJECTIVES

After completing this exercise, you should be able to:

1. Isolate a bacteriophage from a natural environment.
2. Describe the cultivation of bacteriophages.
3. Determine the titer of a bacteriophage sample using the broth-clearing and plaque-forming methods.

BACKGROUND

Bacteriophages (the term was coined around 1917 by Félix d'Hérelle and means "bacteria eater") parasitize most, if not all, bacteria in a very specific manner. Some bacteriophages, or **phages,** such as T-even bacteriophages, have a **complex structure** (**FIGURE 37.1a**). The protein coat consists of a polyhedral head and a helical tail, to which other structures are attached. The head contains the nucleic acid. To initiate an infection, the bacteriophage **adsorbs** onto the surface of a bacterial cell by means of its tail fibers and base plate. The bacteriophage injects its nucleic acid into the bacterium during **penetration** (**FIGURE 37.1b**). The tail sheath contracts, driving the tail core through the cell wall and injecting the nucleic acid into the bacterium.

Isolation of Bacteriophages

Bacteriophages can be grown in liquid or solid cultures of bacteria. When solid media are used, the **plaque-forming method** allows the bacteriophage to be located. Host bacteria and bacteriophages are mixed together in melted agar, which is then poured into a Petri plate containing hardened nutrient agar. Each bacteriophage that infects a bacterium multiplies, releasing several hundred new viruses. The new viruses infect other bacteria, and more new viruses are produced. All the bacteria in the area surrounding the original virus are destroyed, leaving a clear area, or **plaque,** against a confluent "lawn" of bacteria. The lawn of bacteria is produced by the growth of uninfected bacterial cells.

In this exercise, we will isolate a bacteriophage from host cells in a natural environment (i.e., in sewage). Because the numbers of phages in a natural source are low, the desired host bacteria and additional nutrients are added as an **enrichment procedure.**

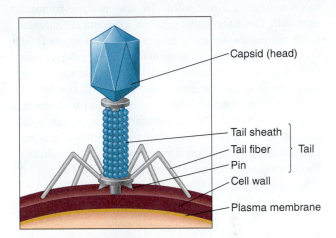

(a) Diagram of a bacteriophage

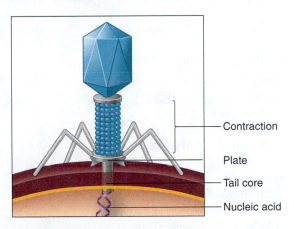

(b) Penetration of a host cell by the bacteriophage

FIGURE 37.1 T-even bacteriophage. (a) Diagram of a bacteriophage, showing its component parts and adsorption onto a host cell. **(b)** Penetration of a host cell by the bacteriophage.

After incubation, the bacteriophage can be isolated by centrifugation of the enrichment media and membrane filtration. **Filtration** has been used for removing microbes from liquids for purposes of sterilization since 1884, when Pasteur's associate Charles Chamberland made the first filter out of porcelain. Today most filtration of viruses is done using nitrocellulose or polyvinyl membrane filters with pore sizes (usually 0.45 μm) that physically exclude bacteria from the filtrate.

Assays

You will measure the viral activity in your sample by performing sequential dilutions of the viral preparation and assaying for the presence of viruses. In the **broth-clearing assay,** the **endpoint** is the highest dilution (smallest amount of virus) producing lysis of bacteria and clearing of the broth. The **titer,** or concentration, that results in a recognizable effect is the reciprocal of the endpoint. In the **plaque-forming method,** the titer is determined by counting plaques. Each plaque theoretically corresponds to a single infective virus in the initial suspension. Some plaques may arise from more than one virus particle, and some virus particles may not be infectious. Therefore, the titer is determined by counting the number of **plaque-forming units (PFU).** The titer, plaque-forming units per milliliter, is determined by counting the number of plaques and dividing by the amount plated times the dilution. For example, 32 plaques from 0.1 ml plated of a $1:10^3$ dilution is equal to

$$\frac{32}{0.1 \times 10^{-3}} = 3.2 \times 10^5 \text{ PFU/ml}$$

In this exercise, we will determine the viral activity by a plaque-forming assay and a broth-clearing assay.

MATERIALS

ISOLATION OF BACTERIOPHAGE

Use sewage or bacteriophages.

    Sewage as source: **BSL-2**
        Raw sewage (45 ml)
        $10\times$ nutrient broth (5 ml)
        50-ml graduated cylinder
        Funnel
    Bacteriophage as source:
        Bacteriophage
        Trypticase soy broth (20 ml)
Sterile 125-ml Erlenmeyer flask
Sterile 5-ml pipette (1)
Centrifuge tubes
Screw-capped tube
Sterile membrane filter (0.45 μm)
Sterile membrane filter apparatus
Safety goggles
Gloves

TITRATION OF BACTERIOPHAGE

Tubes containing 9 ml of trypticase soy broth (6)
Petri plates containing nutrient agar (6)
Tubes containing 3 ml of melted soft trypticase soy
    agar (0.7% agar) (6)
Sterile 1-ml pipettes (7)
Safety goggles

CULTURES

Escherichia coli broth
Bacteriophage culture *or* bacteriophage from isolation
    procedure

TECHNIQUES REQUIRED

Pour plate technique (Exercise 11)
Pipetting (Appendix A)
Serial dilution technique (Appendix B)
Membrane filtration (Appendix F)

PROCEDURE First Period

Isolation of Bacteriophage

    Wear gloves and goggles when working with raw sewage.

1.   Follow enrichment procedure a *or* b.
    a.   **BSL-2** **Isolation from sewage.** Wear gloves and safety glasses. Using the graduated cylinder and funnel, add 45 ml of sewage to 5 ml of $10\times$ nutrient broth in a sterile flask. Add 5 ml of *E. coli* broth. Mix gently. Place gloves in the biohazard container. Why is the broth 10 times more concentrated than normal? _____

    _____

    b.   **Bacteriophage.** Add 0.5 ml of bacteriophage to 20 ml of trypticase soy broth in a sterile flask. Add 2 ml of *E. coli* broth. Mix gently.
2.   Incubate the enrichment for 24 hours at 35°C.

PROCEDURE Second Period

    Wear gloves and goggles when working with liquid cultures.

Isolation of Bacteriophage

1.   Decant 10 ml of the enrichment into a centrifuge tube. Place the tube in a centrifuge.

    Balance the centrifuge with a similar tube containing 10 ml of water.

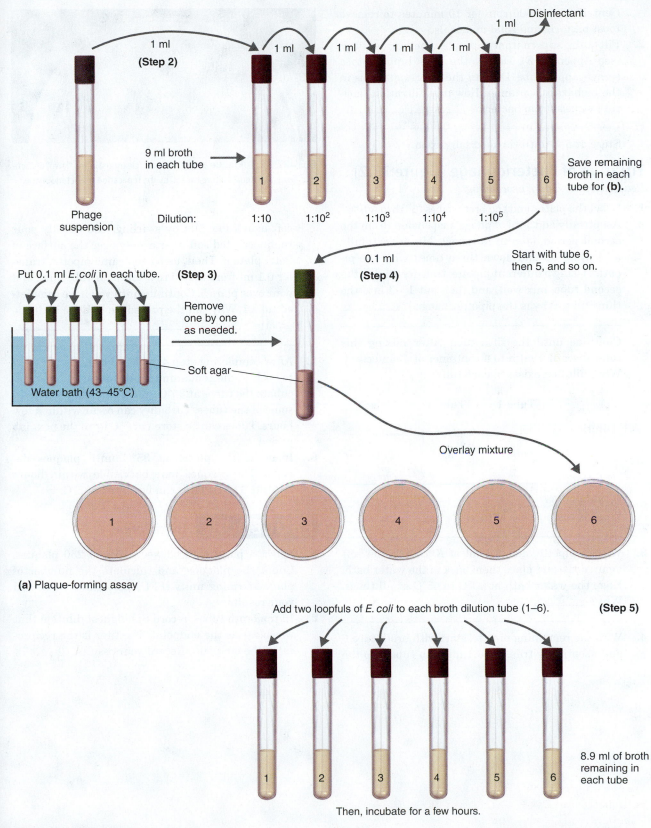

(a) Plaque-forming assay

(b) Broth-clearing assay

FIGURE 37.2 Procedure for titration of bacteriophages.

Centrifuge at 2500 rpm for 10 minutes to remove most bacteria and solid materials.

2. Filter the supernatant through a membrane filter (see Appendix F). Decant the clear liquid into a screw-capped tube. Discard the filter apparatus in the biohazard container. How does filtration separate viruses from bacteria? _____
_____

Store at 5°C until the next lab period.

Titration of Bacteriophage (Figure 37.2)
Read carefully before proceeding.

1. Label the plates and the broth tubes "1" through "6."
2. Aseptically add 1 ml of phage suspension (from the second period, step 2) to tube 1. Mix by carefully aspirating up and down three times with the pipette. Using a different pipette, transfer 1 ml to the second tube, mix well, and then put 1 ml into the third tube. Why is the pipette changed? _____
_____

Continue until the fifth tube. After mixing this tube, discard 1 ml into a container of disinfectant. What dilution exists in each tube?

	Tube 1	Tube 2	Tube 3
Dilution:	_____	_____	_____

	Tube 4	Tube 5	Tube 6
Dilution:	_____	_____	_____

What is the purpose of tube 6? _____
_____

3. With a pipette, add 0.1 ml of *E. coli* to the soft agar tubes, and place them back in the water bath. Keep the water bath at 43°C to 45°C at all times. Why? _____
_____

4. With the remaining pipette, start with broth tube 6 and aseptically transfer 0.1 ml from tube 6 to the

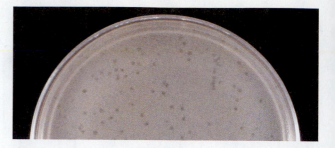

FIGURE 37.3 Plaque assay. Plaques (clearings) in the "lawn" of *Escherichia coli* bacterial culture caused by the replication of bacteriophage.

soft agar tube. Mix by swirling, and quickly pour the inoculated soft agar evenly over the surface of Petri plate 6. Then, using the same pipette, transfer 0.1 ml from tube 5 to a soft agar tube, mix, and pour over plate 5. Continue until you have completed tube 1. Why can the procedure be done with one pipette? _____
_____

5. After completing step 4, add two loopfuls of *E. coli* to each of the remaining broth tubes, and mix. Incubate the tubes at 35°C until turbidity develops in some of the tubes. Turbidity can occur within a few hours. Tubes can be stored at 5°C until the next lab period.
6. Incubate the plates at 35°C until plaques develop. The plaques may be visible within hours (**FIGURE 37.3**). Plates can be stored at 5°C.

PROCEDURE Third Period

1. Select a plate with between 25 and 250 plaques. Count the plaques, and calculate the number of plaque-forming units (PFU) per milliliter. Record your results.
2. In the broth tubes, record the highest dilution that was clear as the endpoint. The titer is the reciprocal of the endpoint. Record your results.

LABORATORY REPORT
Isolation and Titration of Bacteriophages

PURPOSE _____

_____

_____

HYPOTHESES

1. What evidence would indicate that you successfully isolated bacteriophage? _____

_____

2. Filtration is used to separate bacteria from the bacteriophage. What results would indicate that you have been

   successful? _____

_____

RESULTS

Plaque-Forming Assay

Choose one plate with 25 to 250 plaques.

Draw what you observed.

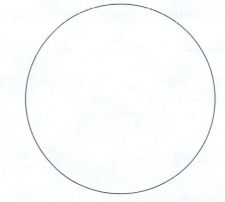

   Number of plaques = _____

   Dilution used for that plate = _____

Broth-Clearing Assay

Indicate whether each tube was turbid or clear. Incubated _____ hours.

Tube	Bacterial Growth	Dilution
1		
2		
3		
4		
5		
6		

CONCLUSIONS

Plaque-Forming Assay

1. PFU/ml = _____
   Show your calculations.

2. Are any bacteria present within the plaques? What do they indicate? _____

   _____

3. Are all the plaques the same size? Briefly explain why or why not. _____

   _____

Broth-Clearing Assay

1. What was the endpoint? _____

2. What was the titer? _____

3. What does the endpoint represent in a broth-clearing assay? _____

   _____

   _____

   _____

4. Which of these assays is more accurate? Briefly explain. _____

_____

_____

QUESTIONS

1. Why did you add *Escherichia coli* to sewage, which is full of bacteria? _____

_____

_____

2. If there were no plaques on your plates, offer an explanation. How would you explain turbidity in all of the

tubes in a broth-clearing assay? _____

_____

CRITICAL THINKING

1. How would you develop a pure culture of a phage?

2. How would you isolate a bacteriophage for a species of *Bacillus*?

CLINICAL APPLICATIONS

1. Bacteriophages are being explored as a treatment for antibiotic-resistant bacteria. Among the advantages of bacteriophages is that they are self-replicating and self-limiting. Explain how this is different from antibiotic therapy.

2. Rotavirus grows in cells at the tips of intestinal villi and causes gastroenteritis. Abrupt onset of vomiting and diarrhea occurs 1 to 3 days after the virus is contracted and lasts 3 days. Use the lytic cycle to explain the disease pattern.

*Smoking was **DEFINITELY DANGEROUS** to your health in seventeenth-century Russia. Czar Michael Federovitch executed anyone on whom tobacco was found. But Czar Alexei Mikhailovitch was **EASIER** on smokers; he merely tortured them until they told who their suppliers were.* – **A N O N Y M O U S**

OBJECTIVES

After completing this exercise, you should be able to:

1. Isolate a plant virus.
2. Describe the cultivation of a plant virus.
3. Determine the host range of a plant virus.

BACKGROUND

Plant viruses are important economically in that they are the second leading cause of plant disease. (Fungi are the leading cause of plant disease.) For most plant viruses to infect a plant, the virus must enter through an abrasion in the leaf or stem. Insects called **vectors** carry viruses from one plant to another. Most viruses cause either a **localized infection,** in which the leaf will have necrotic lesions (brown plaques), or a **systemic infection,** in which the infection will run throughout the entire plant. One of the most studied viruses is tobacco mosaic virus (TMV), which was the first virus to be purified and crystallized.

In this experiment, we will attempt to isolate TMV from various tobacco products and determine its host range.

MATERIALS

FIRST PERIOD

Tobacco plant
Tomato plant
Bean plant (*Chenopodium*)
Tobacco products of various types (bring some, if possible)
Mortar and pestle
Fine sand or carborundum
Wash bottle of water
Paper labels or labeling tape
Cotton
Sterile water

TECHNIQUE REQUIRED

None

PROCEDURE First Period

1. Select a plant, and label the pot with your name and lab section.
2. Place labels around the petioles (see **FIGURE 38.1**) of several leaves, with names corresponding to the various tobacco products available. Label one petiole "C" for control.
3. *Wash your hands carefully before and after each inoculation.*

FIGURE 38.1 Labeling a plant. Place labels close to the leaves that will be treated, as shown.

305

4. Spray the control leaf with a small amount of water, and dust the leaf with carborundum or sand.

 Be careful not to inhale carborundum dust. It is a lung irritant.

Gently rub the leaf with a cotton ball dampened in sterile water. Wash the leaf with a wash bottle of water. What is the purpose of this leaf?_____

_____

5. Select a tobacco product, place a "pinch" of it into a mortar, and add sterile water. Grind the mixture into a slurry with a pestle. Spray one leaf with water, dust it with carborundum or sand, and gently rub it with a cotton ball soaked with the tobacco slurry. Put the cotton ball in the biohazard container. Wash the leaf with water. Record the name of the tobacco source.

 Put the cotton ball in the To Be Autoclaved container.

6. Using the same tobacco product, repeat steps 1 through 5 on a different species of plant.
7. Repeat steps 5 and 6 with another tobacco product, cleaning the mortar and pestle with soap and water between products.
8. Replace the plant in the rack.

FIGURE 38.2 **Plant virus assay.** Chlorosis (loss of green color) and plaques (spots) in this tomato leaf are due to infection by tobacco mosaic virus.

PROCEDURE Next Periods

Check the plant at each laboratory period for up to 3 to 4 weeks for brown plaques (localized infection) or wilting (systemic infection). See **FIGURE 38.2**. Water your plants as needed. Discard plants in the To Be Autoclaved area.

LABORATORY REPORT
Plant Viruses

PURPOSE _____

_____

_____

HYPOTHESIS

A localized viral infection of the plant will cause _____ .

RESULTS

Obtain data from your classmates for the other plant species and tobacco products.

Identify Plant Species and Tobacco Product Used		Appearance of Leaves							
		Lab Period							
		1	2	3	4	5	6	7	8
1.	Control								
	Test								
2.	Control								
	Test								
3.	Control								
	Test								
4.	Control								
	Test								
5.	Control								
	Test								

Draw a plant showing locations of the labeled leaves and any signs of infection.

CONCLUSIONS

1. Do you accept/reject your hypothesis? _____ What evidence do you have? _____

    _____

2. Did systemic disease occur? How can you tell? _____

    _____

3. What can you conclude about the host range of TMV? _____

    _____

QUESTIONS

1. If the control leaf was damaged, what happened? _____

    _____

2. Why is the leaf rubbed with sand or carborundum? _____

    _____

CRITICAL THINKING

1. A reservoir of infection is a continual source of that infection. What is a likely reservoir for TMV?

2. How could TMV be used to make inexpensive therapeutic proteins, such as insulin, or to make vaccines?

CLINICAL APPLICATION

An important disease of citrus crops worldwide causes wilting, leaf yellowing and curling, and death due to phloem blockage. Design an experiment to prove this disease is caused by a virus.

PART 11

Interaction of Microbe and Host

EXERCISES

39 Epidemiology

40 Koch's Postulates

In 1883, after detecting microorganisms in the environment, Robert Koch asked,

> *What significance do these findings have? Can it be stated that air, water, and soil contain a certain number of microorganisms and yet are without significance to health?**

Koch felt that the causes, or **etiologic agents,** of infectious diseases could be found in the study of microorganisms. He proved that anthrax and tuberculosis were caused by specific bacteria, and his work provided the framework for the study of the etiology of any infectious disease.

Today we refer to Koch's protocols for identifying etiologic agents as **Koch's postulates** (see the illustration on the next page; also see Exercise 40). The study of how and when diseases occur is called **epidemiology** (Exercise 39).

CASE STUDY: Travelers Beware!

The semester-abroad program "Term in South America" was in its eighth week. The program was off to a great start with an energetic, hardworking group of students. A group dinner was held at a favorite restaurant, El Terro, in Arequipa, Peru. Students enjoyed a wide selection of food served buffet style, including chicken, pork, pasta, guinea pig, salad, and soup. After the buffet the students were treated to a performance of traditional folk dances and songs. Everyone was in an upbeat mood. The next afternoon, however, a student named Sam came to the group leader Ted Olson, complaining of persistent diarrhea and abdominal cramps. His temperature was elevated at 37.8°C (100°F), and he was very uncomfortable. Within a few hours several other students had similar symptoms, and Dr. Olson knew he had a problem. All the students took Pepto-Bismol and went to bed. During the night, Sam had significant prolonged diarrhea, so Dr. Olson gave him a bottle of water with oral rehydration salts. The next morning Sam appeared better and, against the advice of Dr. Olson, started eating food. Within a few hours he was back in bed with a temperature of 38.3°C (101°F), diarrhea, and abdominal cramps.

---

*Quoted in R. N. Coetsch, ed. *Microbiology: Historical Contributions from 1776 to 1908.* New Brunswick, NJ: Rutgers University Press, 1960, p. 130.

Dr. Olsen took Sam to the International Urgent Care Clinic, which took a stool sample, drew blood, and did a physical exam. Because of his diarrhea, Sam was given intravenous fluids and anti-nausea medications. A stool sample was examined because the physician suspected Sam had a parasite. To his surprise, the stool contained a large number of gram-negative vibrio-shaped bacteria, but no parasites. The microbiology lab identified the bacteria as *Campylobacter jejuni*. The physician asked Sam about his recent meals, and Sam told him about the buffet and the other ill students. Sam began to feel better with the IV fluids and medication. He was sent back to his residence, and he recovered in 24 hours.

Questions

1. What food in the buffet was probably the source of the infection?
2. What complication, even though it is unusual, is a concern for *Campylobacter* infection?
3. Is this disease spread from human to human?
4. Why is culturing *Campylobacter jejuni* difficult?
5. The focus of the public health professional would be epidemiology. What would be the goal?

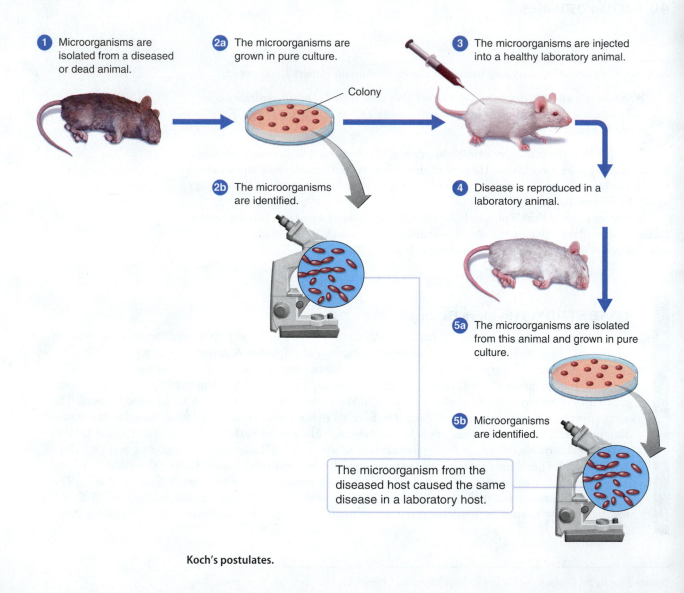

Koch's postulates.

Epidemiology

OBJECTIVES

After completing this exercise, you should be able to:

1. Define the following terms: *epidemiology, epidemic, reservoir,* and *carrier.*
2. Describe three methods of disease transmission.
3. Define the terms *index case* and *case definition.*
4. Determine the source of a simulated epidemic.

BACKGROUND

In every infectious disease, the disease-producing microorganism, the **pathogen,** must come in contact with the **host,** the organism that harbors the pathogen. **Communicable diseases** can be spread either directly or indirectly from one host to another. Some microorganisms cause disease only if the body is weakened or if a predisposing event, such as a wound, allows them to enter the body. Such diseases are called **noncommunicable diseases**—that is, they cannot be transmitted from one host to another. The science that deals with when and where diseases occur and how they are transmitted in the human population is called **epidemiology. Endemic diseases** such as pneumonia are constantly present in the population. When many people in a given area acquire the disease in a relatively short period of time, it is referred to as an **epidemic disease.** The first reported patient in a disease outbreak is the **index case.** One of the first steps in analyzing a disease outbreak is to make a **case definition,** which should include the typical symptoms of patients identified as cases in an outbreak investigation.

Disease Transmission

Diseases can be transmitted by **direct contact** between hosts. **Droplet infection,** which occurs when microorganisms are carried on liquid drops from a cough or sneeze, is an example of direct contact. Diseases can also be transmitted by contact with contaminated inanimate objects, or **fomites.** Drinking glasses, bedding, and towels are examples of fomites that can be contaminated with pathogens from feces, sputum, or pus.

Some diseases are transmitted from one host to another by vectors. **Vectors** are insects and other arthropods that carry pathogens. In **mechanical transmission,** insects carry a pathogen on their feet and may transfer the pathogen to a person's food. For example, houseflies may transmit typhoid bacteria from the feces of an infected person to food. Transmission of a disease by an arthropod's bite is called **biological transmission.** An arthropod ingests a pathogen while biting an infected host. The pathogen can multiply or mature in the arthropod and then be transferred to a healthy person in the arthropod's feces or saliva.

The continual source of an infection is called the **reservoir.** Humans who harbor pathogens but who do not exhibit any signs of disease are called **carriers.**

Tracking a Disease

An **epidemiologist** compiles data on the incidence of a disease and its method of transmission and tries to locate the source of infection to decrease the incidence. The time course of an epidemic is shown by graphing the number of cases and their date of onset. This **epidemic curve** gives a visual display of the outbreak's magnitude and time trend. An epidemic curve provides a great deal of information. First, you will usually be able to tell where you are in the course of the epidemic and may even be able to project its future course. Second, if you have identified the disease and know its usual incubation period, you may be able to estimate a probable time period of exposure and can then develop a questionnaire focusing on that time period.

By graphing the cases, an epidemiologist may be able to locate the index case. With more information from patients, you may be able to determine whether the outbreak has resulted from a common-source exposure, from person-to-person spread, or both. The latter occurs, for example, when someone gets an infection from a food and then transmits the infection to family members who did not eat that food. The infected family members are called **secondary cases.**

In this exercise, an epidemic will be simulated. Although you will be in the "epidemic," you will be the epidemiologist who, by deductive reasoning and with luck, determines the source of the epidemic.

FIGURE 39.1 **Inoculating the plate.** After touching the palm of one classmate with your gloved fingers, immediately touch sector 1 of the plate. Record the person's name and swab number. After the finger–palm touch with a second classmate, touch sector 2. Continue until you have inoculated all five sectors.

MATERIALS

Petri plate containing nutrient agar
Latex or vinyl glove, or small plastic sandwich bag
One unknown swab per student (one swab has *Serratia marcescens, Rhodotorula rubra*, or *Kocuria roseus* on it. Your instructor will tell you which organism.)

TECHNIQUE REQUIRED

Colony morphology (Exercise 3)

PROCEDURE First Period

1. Divide the Petri plate into five sectors labeled "1" through "5" (**FIGURE 39.1**).
2. Record the number of your swab in your Laboratory Report.

Carefully read steps 3 through 9 before proceeding.

3. Put a glove on your left hand. Carefully remove your swab without touching the cotton. Holding the swab with your right hand, rub it on the palm of your left hand (i.e., on the glove).
4. Discard the swab in a container of disinfectant.
5. Using your gloved left hand, touch the gloved palm of a classmate when the instructor gives the signal. Your gloved fingers touch the other's gloved palm, and vice versa.
6. After the finger–palm touch, touch the fingers of your gloved left hand to the first sector of the nutrient agar. Record the person's name and swab number.
7. Repeat the finger–palm touch with a different classmate. Then touch your fingers to the second sector of the nutrient agar. Record the person's name and swab number.
8. Repeat steps 5 and 6, doing the finger–palm touch with another classmate and then touching the third sector of the nutrient agar.
9. Repeat steps 5 and 6, doing the finger–palm touch with two other classmates. Remember to touch your fingers to the fourth and fifth sector of the nutrient agar after each finger–palm touch. Keep good records. Do not touch a classmate more than once. Touch a total of five individuals.
10. Discard the glove in the To Be Autoclaved basket.
11. Incubate the plate, inverted, at room temperature.

PROCEDURE Second Period

Record your results. Data can be keyed into a computer database. Using your class's data, deductively try to determine who had the contaminated fomite.

LABORATORY REPORT
Epidemiology

PURPOSE _____

_____

_____

HYPOTHESIS

You will/will not be able to identify the index case.

RESULTS

1. Your swab # _____ Which organism was on the contaminated swab? _____

2. What is the case definition for this classroom "epidemic"? _____

_____

Sector	Student's Name	Swab Number	Appearance of Colonies on Nutrient Agar
1.			
2.			
3.			
4.			
5.			

Attach the class's data.

Index case? _____

CONCLUSIONS

1. Were you able to identify the index case? _____

2. What number was the contaminated swab? _____ Explain how you arrived at your conclusion.

   _____

3. Diagram the path of the epidemic from the index case to all "infected" class members. Identify any secondary contacts.

QUESTIONS

1. Could you be the "infected" individual and not have growth on your plate? Explain. _____

   _____

   _____

2. Do all people who contact an infected individual acquire the disease? _____

   _____

   _____

3. How can an epidemic stop without medical intervention (e.g., quarantine, chemotherapy, vaccines)? _____

   _____

   _____

4. Are any organisms other than the culture assigned for this experiment growing on the plates? How can you

   tell? _____

   _____

5. What was the method of transmission of the "disease" in this experiment? _____

   _____

   _____

CRITICAL THINKING

1. Assume you work in the Infectious Disease Branch of the Centers for Disease Control and Prevention. You are notified of the following two incidents. (1) On June 20, cruise ship X reported that 84 of 2318 passengers had reported to the infirmary with norovirus gastroenteritis during a 7-day vacation cruise. According to federal regulations, when the incidence of acute gastroenteritis among passengers and crew exceeds 3%, an outbreak is defined and requires a formal investigation. Is this an outbreak? _____
(2) A nursing home reported 125 cases of norovirus gastroenteritis among residents and staff during the week of June 23.

Data collected from the cruise ship and nursing home are shown in the following table. Use these data to answer the questions.

	Cruise Ship Data		Nursing Home Data	
	Date	Number of Cases	Date	Number of Cases
Cruise 1	6/9	2	6/23	1
	6/10	4	6/24	8
	6/11	5	6/25	12
	6/12	3	6/26	12
	6/13	3	6/27	50
	6/14	2	6/28	32
	6/15	1	6/29	10
Cruise 2	6/16	2	6/30	8
	6/17	10		
	6/18	13		
	6/19	13		
	6/20	84		
	6/21	46		
	6/22	20		
Cruise 3	6/23	10		
	6/24	23		
	6/25	39		
	6/26	41		
	6/27	18		
	6/28	17		
	6/29	9		

a. On a separate page, graph the epidemic curve for these data. Graph the number of cases versus the date.

b. How is norovirus transmitted? _____

c. What happened on cruise 2?

d. What would you do before cruise 4?

e. Three nursing home residents were passengers on cruise 2. What can you conclude?

2. A health department received a report from hospital A that 15 patients had been admitted on October 12 with unexplained pneumonia. On October 21, hospital B, located 15 miles from hospital A, reported a higher-than-normal pneumonia census for the first 2 weeks of October. *Legionella pneumophila* was eventually identified in 23 patients; 21 were hospitalized, and 2 died. To identify potential exposures associated with *L. pneumophila*, a questionnaire was developed, and a case-control study was initiated on November 2 to identify the source of infection. Three healthy controls were selected for each confirmed case; controls were matched by age, gender, and underlying medical conditions. Of the 15 case patients for whom a history was available, 14 had visited a large home improvement center 2 weeks before onset of illness. Results of the questionnaire are shown below.

	Case Patients	Healthy Controls
Number of patients	15	45
Visited home improvement center	14	12
Average time at center (min)	79	29
Looked at whirlpool spa X	13	9
Looked at whirlpool spa Y	13	1
Visited greenhouse sprinkler system display	10	10
Visited decorative fish pond	14	12
Used drinking fountain	13	10
Used urinals	10	4
Used restroom hot water	6	4
Used restroom cold water	8	8

a. What is the most likely source of this outbreak of legionellosis? _____

b. How would you prove this was the source?

c. How was this disease transmitted? _____

d. Provide an explanation of the infected patient who did not go to the home improvement center.

CLINICAL APPLICATION

During a 3-month period, acute hepatitis B virus (HBV) infection was diagnosed in 9 residents of a nursing home. Serological testing of all residents revealed that 9 people had acute HBV infection, 2 had chronic infection, 5 were immune, and 58 were susceptible. Medical charts of residents were reviewed for history of medications and use of ancillary medical services. Infection control practices at the nursing home were assessed through interviews with personnel and direct observations of nursing procedures. A summary of the medical charts is shown below.

	Case Patients*	Susceptible Residents[†]
Received insulin injections	11	58
Patients having capillary blood taken by fingersticks	11	6
≥60 fingersticks/month	7	0
<60 fingersticks/month	4	6
Average number of venous blood draws/month	23	6
Patients having both capillary and venous blood drawn	11	39
Visited by dentist	1	5
Visited by podiatrist	10	52
Received blood transfusion	0	0

*A case patient has hepatitis B.

[†]Susceptible residents live in the nursing home but do not have hepatitis B.

a. What is the usual method of transmission for hepatitis B? _____

b. What is the probable source of infection in hospitals? _____

c. How was *this* infection transmitted? _____

Koch's Postulates

OBJECTIVES

After completing this exercise, you should be able to:

1. Define the following terms: *etiologic agent, pathogenicity*, and *virulence*.
2. List and explain Koch's postulates.
3. Perform Koch's postulates.

BACKGROUND

The **etiologic agent** is the cause of an infectious disease. Microorganisms are the etiologic agents of a wide variety of infectious diseases. Microbes that cause diseases are called *pathogens*, and the process of disease initiation and progress is called **pathogenesis.** The interaction between the microbe and host is complex. Whether a disease occurs depends on the host's vulnerability, or **susceptibility,** to the pathogen and on the virulence of the pathogen. **Virulence** is the degree of pathogenicity. Factors influencing virulence include extracellular enzymes, capsules, and toxin production.

The actual cause of many diseases is hard to determine. Although many microorganisms can be isolated from a diseased tissue, their presence does not prove that any or all of them caused the disease. A microbe may be a secondary invader or part of the normal microbiota or transient microbiota of that area. While working with anthrax and tuberculosis, Robert Koch established four criteria, now called **Koch's postulates,** to help identify a particular organism as the causative agent for a particular disease. (See the figure on page 310.) Koch's postulates are the following:

1. The same organism must be present in every case of the disease.
2. The organism must be isolated from the diseased tissue and grown in pure culture in the laboratory.
3. The organism from the pure culture must cause the disease when inoculated into a healthy, susceptible laboratory organism.
4. The organism must again be isolated—this time from the inoculated organism—and must be shown to be the same pathogen as the original organism.

These criteria are used by most investigators, but they cannot be applied to all infectious diseases. For example, viruses cannot be cultured on artificial media, and they are not readily observable in a host. Moreover, many viruses cause diseases only in humans and not in laboratory animals.

In this exercise, we will demonstrate Koch's postulates with one of two different bacteria: *Bacillus thuringiensis* isolated from milky spore disease of tomato hornworm larvae, or *Pectobacterium carotovorum* isolated from soft rot of carrots.

MATERIALS

BACILLUS: MILKY SPORE DISEASE

Petri plate containing nutrient agar
Petri plate containing starch agar
Tube containing sterile nutrient broth
Tomato hornworm larvae (2)
Sterile Pasteur pipettes (2)
Scalpel and scissors (second period)
Beaker of alcohol (second period)
Gram-staining reagents (third period)

DEMONSTRATION (SECOND PERIOD)
Dissected control larvae

PECTOBACTERIUM: CARROT SOFT ROT

Petri plate containing nutrient agar
Carrot
Scalpel or paring knife
Potato peeler
Forceps
Alcohol
Disinfectant (bleach)
Sterile water
Sterile Petri dish with filter paper
Gram-staining reagents (second and third period)

CULTURES

BACILLUS: MILKY SPORE DISEASE
Bacillus thuringiensis

PECTOBACTERIUM: CARROT SOFT ROT
Pectobacterium carotovorum

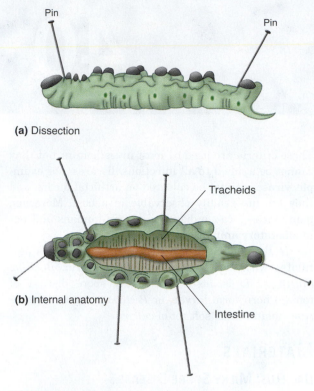

(a) Dissection

Tracheids

(b) Internal anatomy

Intestine

FIGURE 40.1 **Hornworm larva necropsy.**

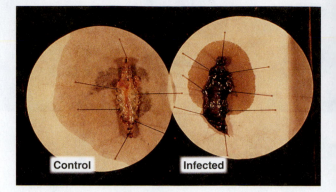

Control   Infected

FIGURE 40.2 **Milky spore disease.** Control larva on left and infected larva on right, which ingested the toxin-producing *Bacillus thuringiensis.*

TECHNIQUES REQUIRED

Compound light microscopy (Exercise 1)
Gram staining (Exercise 7)
Plate streaking (Exercise 11)
Starch hydrolysis (Exercise 13)

PROCEDURE *Bacillus:* Milky Spore Disease

First Period

1. Divide the starch and nutrient agar Petri plates into three sectors; label them "A," "B," and "C." Streak a loopful of nutrient broth onto sector A and a loopful of *Bacillus* onto sector B of each plate. Incubate the plates, inverted, at room temperature until growth occurs in sector B, and then store them in the refrigerator.
2. Make a smear of the *Bacillus* culture and heat-fix it; store it in your drawer.
3. Carefully examine your larvae. Include records of their size, weight, and color.
4. Infect one larva by adding a few drops of *B. thuringiensis* culture to its food supply.
5. Feed the remaining larva in the same manner with nutrient broth. What is the purpose of this step? _____
_____

6. Observe the larvae over the next week. As the infected larva begins to show symptoms of the disease, and as the disease progresses, record the symptoms.
7. The inoculated larva should die within 1 week. Once it is dead, record the size, weight, and external appearance of both larvae; place the dead larva in a small piece of wet paper towel in a Petri dish; and keep it in the refrigerator until you have time to necropsy the larva. *If your larva does not die, provide some reasons for its survival.*

Second Period

1. Observe the intestine of the dissected control larva.
2. Work with the infected larva on a disinfectant-saturated towel. Disinfect your instruments by putting them in alcohol and burning off the alcohol immediately before and after use.

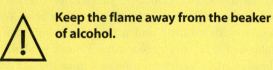

> ⚠️ **Keep the flame away from the beaker of alcohol.**

Slit the ventral surface open from anus to mouth; then make horizontal cuts to expose the intestine (**FIGURE 40.1** and **FIGURE 40.2**). Examine and record your findings. Transfer a loopful of the hemolymph (blood) to sector C on your starch and nutrient agar plates stored previously in the refrigerator (step 1 during the first period), and incubate them at room temperature for 48 hours. Make a smear of hemolymph, and heat-fix it on the same slide you used in step 2 during the first period. Discard the larva as instructed.

> ⚠️ **These insects are agricultural pests and cannot be released.**

Third Period

Prepare a Gram stain of the smears, and examine the growth on the plates. Growth on the plates can also be Gram stained. Flood the starch plate with Gram's iodine. Record your results.

PROCEDURE *Pectobacterium:* Carrot

Soft Rot*

First Period

1. Wash the carrot well, and peel it to eliminate the outer surface; then dry it, wash it with disinfectant, and rinse it with sterile water. Dip your scalpel in alcohol, burn off the alcohol, and cut the carrot into four cross-sectional slices 5 to 8 mm thick.

Keep the flame away from the beaker of alcohol.

2. Put the four slices on filter paper in the bottom of a Petri plate. Saturate the filter paper with sterile water (**FIGURE 40.3**).

FIGURE 40.4 Carrot soft rot caused by *Pectobacterium carotovorum.*

3. Place the four carrot slices on the filter paper. Inoculate the center of three slices with a loopful of the *Pectobacterium* culture. Why not all four? _____

_____

Incubate the plate right-side up at room temperature until soft rot appears (**FIGURE 40.4**). You may add more sterile water if the disease process is slow.

4. Divide the nutrient agar plate in half. Inoculate one-half of the nutrient agar with the *Pectobacterium* broth; incubate the plate, inverted, for 48 hours at room temperature, and then refrigerate it. Make a smear of the *Pectobacterium.* Heat-fix the smear and store it in your drawer.

Second Period

1. Streak an inoculum from the diseased carrot on the remaining half of the Petri plate. Incubate the plate, inverted, at room temperature for 48 hours.

2. Make a smear from the diseased carrot. Perform a Gram stain on both smears, and record your observations.

Third Period

Observe the nutrient agar plate, and record your results. Prepare a Gram stain from the nutrient agar cultures if time permits.

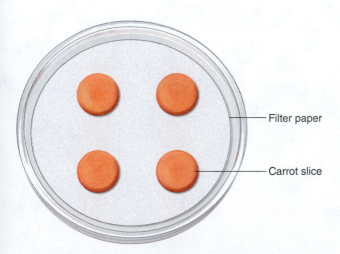

FIGURE 40.3 Place the four carrot slices on the filter paper, and saturate the filter paper with sterile water.

- Filter paper
- Carrot slice

---

*Adapted from R. S. Hogue. "Demonstration of Koch's Postulates." *American Biology Teacher* 33: 174–175, 1971.

LABORATORY REPORT
Koch's Postulates

PURPOSE _____

_____

_____

HYPOTHESIS

You will/will not be able to demonstrate Koch's postulates.

RESULTS

Bacillus: Milky Spore Disease

GRAM STAINS

Specimen: *Bacillus thuringiensis*

Morphology: _____

Gram reaction: _____

Hemolymph from dead larva

Morphology: _____

Gram reaction: _____

CULTURES

Describe the growth on your plates.

Inoculum	Nutrient Agar	Gram Stain	Starch Agar	Starch Hydrolysis
Broth (control)				
Bacillus culture				
Hemolymph				

Describe the larvae by filling in the following table.

	Infected Larva	Control Larva
Initial appearance Date, time		
Length		
Weight		
Color		
Activity		
After death of infected larva Date, time		
Length		
Weight		
Color		
Appearance of intestine		
Appearance of hemolymph		

CONCLUSIONS

1. What evidence would indicate that the microbe obtained from the dead larva was *B. thuringiensis*? _____

_____

_____

2. What parts of Koch's postulates did this experiment fulfill? _____

_____

QUESTIONS

1. How would you determine the ID_{50} for *B. thuringiensis*? _____

_____

_____

2. What is the normal method of transmission of *Bacillus thuringiensis*? _____

3. *Bacillus thuringiensis* endospores in an inert matrix are sold in garden supply stores. Describe the properties

   of this insecticide that make it better than chemical insecticides. _____

_____

_____

RESULTS

Pectobacterium: Carrot Soft Rot

GRAM STAINS

Specimen: *Pectobacterium carotovorum*

Morphology: _____

Gram reaction: _____

Soft rot

Morphology: _____

Gram reaction: _____

CULTURES

Describe the growth on your plates.

Inoculum	Colony Description	Gram Stain
Pectobacterium culture		
Soft rot		

Describe the appearance of the carrots. _____

_____

_____

CONCLUSIONS

1. What evidence would indicate that the soft rot is from *Pectobacterium carotovorum*? _____

_____

_____

2. What parts of Koch's postulates did this experiment fulfill? _____

_____

QUESTIONS

1. Why is the disease called *soft rot*? _____

_____

2. Why will the addition of water speed the disease process? _____

_____

_____

3. What is the normal method of transmission of *Pectobacterium carotovorum*? _____

_____

CRITICAL THINKING

1. In the procedure, did the Gram stain and culture results conclusively fulfill Koch's postulates?

2. Why are Koch's postulates *not* useful in determining etiology of a viral disease?

CLINICAL APPLICATIONS

1. Eighty-one patients became ill and 14 died in an outbreak at St. Elizabeth's Hospital, Washington, D.C., in 1965. Epidemiologic evidence suggested a link between infection and windblown dust from excavations on hospital grounds. In 1968, 144 cases of a self-limited illness occurred in employees at the health department building in Pontiac, Michigan. Investigations at that time demonstrated that the etiologic agent was present in the condenser of a malfunctioning air-conditioning system. In 1974, at least 20 persons attending a convention at the Bellevue Stratford Hotel in Philadelphia developed pneumonia, and 2 died. In 1976, 182 people became ill and 29 died at another convention at the Bellevue Stratford Hotel. What would you need to do to identify the causative agent of these outbreaks?

2. Near Four Corners, New Mexico, on May 14, two people living in the same household died within 5 days of each other. Their illnesses were characterized by flulike symptoms with rapid respiratory failure. The third case was the death of an 8-year-old girl in Mississippi. Over the next 120 days, 53 more people became ill near Four Corners, and 26 died. One man became ill 2 weeks after returning home from a visit to New Mexico. On July 9, a previously healthy woman living in eastern Texas died following acute respiratory distress. The woman had not traveled outside Texas. What evidence do you have that this is an infectious disease?

PART 12 Immunology

EXERCISES

In this part, we will examine our defenses against invasion by microorganisms. In 1883, Elie Metchnikoff began his investigations into the body's defenses with this entry in his diary:

> These wandering cells in the body of the larva of a starfish, these cells eat food, they gobble up carmine granules—but they must eat up microbes too! Of course—the wandering cells are what protect the starfish from microbes! Our wandering cells, the white cells of blood—they must be what protects us from invading germs.*

These white blood cells and certain chemicals are considered agents of **innate** or **nonspecific immunity** because they will combat any microorganism that invades the body. Factors involved in innate immunity are the topic of Exercise 41.

The role of specific defenses—or **adaptive immunity**—was demonstrated when Emil von Behring developed what he called diphtheria "antitoxin" and, in 1894, successfully treated humans with it. Adaptive immunity involves the production by the body of specific proteins called **antibodies,** which are directed against distinct microorganisms. If microorganism A invades the body, antibodies are produced against A; if microorganism B invades the body, antibodies are produced against B. The antibodies produced against microorganism B will not react with microorganism A. The chemical substance that induces antibody production is called an **antigen.** In the examples just mentioned, microorganisms A and B possess antigens on their surfaces.

---

*Quoted in P. DeKruif. *Microbe Hunters*. New York: Harcourt, Brace, & World, 1953, p. 195.

Adaptive and innate immunity involve blood. **Blood** consists of a fluid called **plasma** and formed elements: cells and cell fragments. The cells that are of immunologic importance are the white blood cells. **Serum** is the fluid portion that remains after blood has clotted. We will study the techniques and mechanisms of **serology,** or antigen–antibody reactions in vitro, in Exercises 42 through 44.

CASE STUDY: A Cross Match in the Frontier!

Dr. Yale and his summer research student Ben traveled from Nome, Alaska, by seaplane to a remote island off the coast. The research site had a small community of scientists studying the effects of global warming on the increase of vector-borne diseases in the animal population. After settling into the camp routine for a week on the island, Ben was scheduled to take his regular turn preparing the evening meal. But he was exhausted, and in his haste to prepare the meal, he cut himself badly. He severed an artery and lost a significant amount of blood before his mentor applied a tourniquet.

One of the members of the research group, Michelle, had EMT training and was concerned; Ben was pale and very lethargic. To complicate the crisis, a fog had blanketed the bay and ruled out any air flight to the mainland. Michelle phoned medical personnel at Nome, and they suggested that as a last resort, an emergency transfusion might be needed. Michelle had a limited range of clinical supplies and equipment, with no blood-typing serum. She gathered the group together and explained the situation. They all agreed to help, and Michelle quickly outlined her plan.

As soon as everyone understood her approach, Michelle obtained blood in tubes with an anticoagulant from Ben and from each of the individuals who had agreed to help. She centrifuged the tubes and separated the plasma from the red blood cells. She was very careful and used universal precautions in working with the blood samples. For each potential donor, she set up a major cross match by putting on a slide two drops of Ben's plasma with two drops of the donor's red blood cells. Then, on another slide, she added two drops of each potential donor's plasma to Ben's red blood cells as a minor cross match. She mixed each combination with a toothpick and examined each one for agglutination. She found several potential donors who exhibited no agglutination on the slides. Just as she finished the cross matches the fog began to dissipate, and to her relief, Ben was airlifted to a local hospital.

Questions

1. What causes agglutination on the slide?
2. What makes one of these procedures a major cross match and the other a minor cross match?
3. If the cross match indicated compatibility, could a transfusion reaction still occur if the blood were given to Ben?
4. Why are universal precautions needed? What are they?
5. How does plasma differ from serum?

Innate Immunity

OBJECTIVES

After completing this exercise, you should be able to:

1. Differentiate between adaptive immunity and innate immunity.
2. List and discuss the functions of at least two chemicals involved in innate immunity.
3. Use a spectrophotometer.
4. Determine the effects of normal serum bactericidins.

BACKGROUND

Our ability to ward off diseases through our body's defenses is called **immunity.** Immunity can be divided into two kinds: adaptive (specific) and innate (nonspecific). **Adaptive immunity** is the defense against a specific microorganism. **Innate immunity** refers to all our defenses that protect us from invasion by any microorganism. Physical barriers, such as the skin and mucous membranes, are part of innate immunity. Inflammation and phagocytosis play major roles in innate immunity, as do chemicals produced by the body. As early as 1888, Metchnikoff noted that bacteria "show degenerative changes" when inoculated into samples of mammalian blood.

Lysozyme and complement are examples of chemicals involved in innate immunity. These proteins are collectively called **bactericidins. Lysozyme** is an enzyme found in body fluids that is capable of breaking down the cell walls of gram-positive bacteria and a few gram-negative bacteria. **Complement** is a group of proteins found in serum that are involved in enhancing phagocytosis and lysis of bacteria. Complement can be activated by bacterial cell wall polysaccharides and antigen–antibody reactions.

In this exercise, we will examine chemicals involved in innate immunity.

MATERIALS

LYSOZYME ACTIVITY

Lysozyme buffer, 4.5 ml
Egg-white lysozyme, $1:10^5$ dilution, 2.5 ml

Spectrophotometer tubes (2)
1-ml pipette
5-ml pipette (5)
Petri dish
Spectrophotometer

NORMAL SERUM BACTERICIDINS

Nutrient agar, melted and cooled to 45°C
Sterile Petri dishes (9)
Sterile 1-ml pipettes (10)
Sterile 99-ml dilution blanks (3)
Sterile serological tubes (9)
Serological test-tube rack
Sterile 0.85% saline solution
Normal animal serum
Animal serum heated to 56°C for 30 minutes
Safety goggles

CULTURES (AS ASSIGNED)

Micrococcus luteus suspension, 5 ml
Escherichia coli
Staphylococcus epidermidis
Staphylococcus aureus BSL-2

TECHNIQUES REQUIRED

Pour plate procedure (Exercise 11)
Pipetting (Appendix A)
Serial dilution technique (Appendix B)
Spectrophotometry (Appendix C)
Graphing (Appendix D)

PROCEDURE First Period

Work only with your own tears or saliva.

Lysozyme Activity

1. Collect tears or saliva in a Petri dish, as explained by your instructor.

Wear safety goggles when pipetting.

2. Prepare a 1:10 dilution of tears or saliva by adding 0.5 ml of tears or saliva to a 4.5-ml lysozyme buffer. What is the purpose of the buffer? _____

    _____

3. Mix 2.5 ml of the tear or saliva preparation with 2.5 ml of *Micrococcus luteus* suspension in a spectrophotometer tube. Carefully pipette the contents of the tube up and down three times to mix.

4. Record the absorbance at 540 nm at 30 seconds, 60 seconds, 120 seconds, 180 seconds, 240 seconds, and 5 minutes, and at 5-minute intervals for the next 15 minutes (Appendix C).

5. Repeat steps 3 and 4, using egg-white lysozyme.

6. Plot your data in the Laboratory Report.

Place the Petri plate, dilution tube, and spectrophotometer tube in disinfectant.

Normal Serum Bactericidins (Figure 41.1)

Work in groups as assigned.

1. Aseptically prepare $1:10^2$, $1:10^4$, and $1:10^6$ dilutions of the culture assigned to you (**FIGURE 41.1a**). Wear safety goggles and gloves.

2. Label three sterile serological tubes "6A," "6B," and "6C." Aseptically transfer 0.1 ml of culture from the highest dilution (1: ____) to each tube (**FIGURE 41.1b**).

3. Repeat step 2 with the other dilutions, using the same pipette. Label each tube appropriately. How many tubes do you have? _____

4. To each A tube, add 0.4 ml of unheated normal serum; to each B tube, add 0.4 ml of serum heated to 56°C for 30 minutes; to each C tube, add 0.4 ml of saline (**FIGURE 41.1c, d,** and **e**). What is the purpose of the C tubes? _____

5. Mix all tubes and incubate them at 35°C for 1 hour (**FIGURE 41.1f**).

6. Label nine sterile Petri dishes in the same manner as the tubes: "$1:10^2$ A," "$1:10^2$ B," "$1:10^2$ C," "$1:10^4$ A," "$1:10^4$ B," and so on.

7. Remove 0.1 ml from the A tube of the highest dilution, and place it in the appropriate Petri dish. Using the same pipette, transfer 0.1 ml from the $1:10^4$ A tube to the corresponding Petri dish. Repeat this procedure with the $1:10^2$ A tube (**FIGURE 41.1g**). Why are the samples transferred from the highest dilution first? _____

8. Using another pipette, repeat step 7 with the B tubes. With your remaining pipette, repeat step 7 with the C tubes.

9. Pour melted, cooled nutrient agar into each dish to a depth of approximately 5 mm. Gently swirl to mix the contents, and allow the agar to solidify (**FIGURE 41.1h**).

10. Incubate the plates for 24 to 48 hours at 35°C.

PROCEDURE   Second Period

Normal Serum Bactericidins

1. Record the number of colonies on each plate. Plates with fewer than 25 colonies should be reported as TFTC (too few to count) and those with more than 250 colonies as TNTC (too numerous to count).

2. Calculate the number of colony-forming units (CFU) per milliliter for the control. Choose one plate with between 25 and 250 colonies:

$$\frac{\text{Number of colonies}}{0.1 \text{ ml} \times \text{dilution}} = \text{CFU/ml}$$

    Calculate the number of colony-forming units per milliliter for the unheated and heated sera.

3. Calculate the percent of increase or decrease in bacterial numbers caused by exposure to heated and unheated serum as compared to the control.

$$\text{Percent change} = \frac{|\text{Control} - \text{Experiment}|}{\text{Control}} \times 100$$

4. Compare your data with those of a group using the other bacterial species.

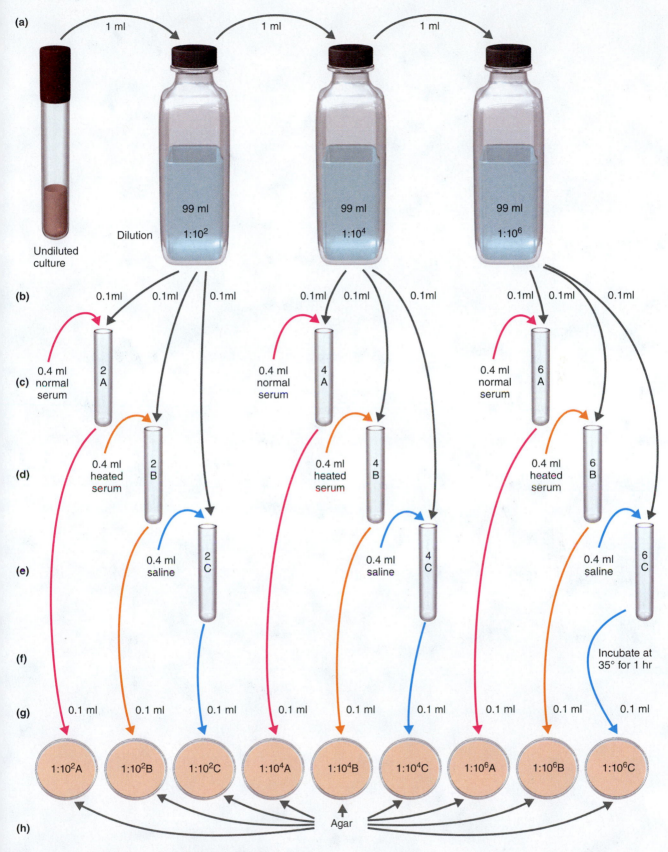

FIGURE 41.1 Determining the bactericidal activity of serum.

LABORATORY REPORT
Innate Immunity

PURPOSE _____
_____
_____

HYPOTHESES

1. Lysozyme activity will be found in saliva and tears. Agree/disagree

2. Bactericidins in serum are destroyed by heat. Agree/disagree

RESULTS

Lysozyme Activity

Time	Absorbance		
	Saliva	Tears	Egg-White Lysozyme
30 sec			
60 sec			
120 sec			
180 sec			
240 sec			
5 min			
10 min			
15 min			
20 min			

Plot your data for lysozyme activity using a computer application or on the graph paper on the next page. Absorbance is marked on the Y-axis and time on the X-axis. Make one line for tears or saliva and another line for egg-white lysozyme.

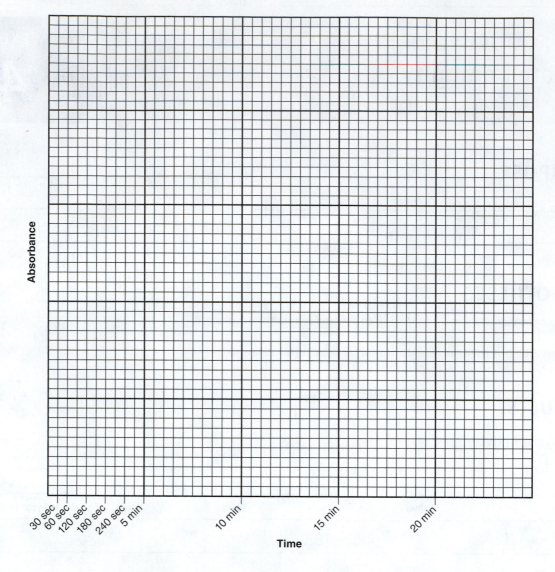

Absorbance / Time

30 sec   60 sec   120 sec   180 sec   240 sec   5 min   10 min   15 min   20 min

Normal Serum Bactericidins

	Number of Colonies			
	1:10²	1:10⁴	1:10⁶	CFU/ml
Unheated serum				
Heated serum				
Control				

Your calculations:

CONCLUSIONS

1. Did your results confirm your hypotheses?

   a. Lysozyme activity _____.

   b. Bactericidins _____.

2. What effect did lysozyme have on the bacterial suspension? _____

   _____

3. Did you test tears or saliva? _____ Compare your data with someone who tested the other.

   Which secretion has more lysozyme activity? _____ How can you tell? _____

   _____

   _____

   _____

	% Increase or Decrease	
	Staphylococcus sp.	*E. coli*
Unheated serum		
Heated serum		

   Your calculations:

4. Does normal serum have bactericidal properties? Explain. _____

   _____

   _____

QUESTIONS

1. Why are tears and saliva potential biohazards? _____

   _____

2. What can you conclude about the effect of heat on the bactericidal properties of normal serum?

   _____

CRITICAL THINKING

1. What other factors contribute to the innate immunity of your eyes? Of your mouth?

2. Design an experiment to determine whether plant fluids have bactericidal properties.

3. How would you experimentally show that the bactericidal properties of the normal serum were due to innate immunity and not to adaptive immunity?

CLINICAL APPLICATION

A very sick, lethargic child is brought to the emergency room with a high temperature and a history of bacterial infections. Laboratory tests reveal elevated white blood cell counts with fully functioning phagocytic cells. Antibody levels are in the normal range. What defect might be a factor in the child's increased susceptibility to the bacterial infections?

Agglutination Reactions: Slide Agglutination

OBJECTIVES

After completing this exercise, you should be able to:

1. Compare and contrast the terms *agglutination* and *hemagglutination*.
2. Use agglutination to identify a pathogenic bacterium.
3. Determine ABO and Rh blood types.
4. Determine possible compatible transfusions.

BACKGROUND

The surfaces of bacterial cells contain antigens that can be used in agglutination reactions. **Agglutination reactions** occur between *particulate* antigens—such as cell walls, flagella, or capsules *bound* to cells—and antibodies. **Agglutination,** or the clumping of bacteria by antibodies, is a useful laboratory diagnostic technique. When the cells involved are red blood cells, the reaction is called **hemagglutination.**

Agglutination of Bacteria

In an agglutination test, an unknown bacterium is suspended in a saline solution on a slide and mixed with a drop of known antiserum. This test is done with different antisera on separate slides. The bacteria will agglutinate when mixed with antibodies produced against the same species and strain of bacterium. A positive test can identify the bacterium. In some diagnostic tests, the antiserum is coupled to latex particles to enhance the visibility of the positive agglutination test.

Hemagglutination

Hemagglutination reactions are used in the typing of blood. The presence or absence of two very similar carbohydrate antigens (designated A and B) located on the surface of red blood cells is determined using specific antisera (**FIGURE 42.1**). Hemagglutination occurs when anti-A antiserum is mixed with type A red blood cells. When anti-A antiserum is mixed with type B red blood cells, no hemagglutination occurs. People with type AB blood possess both A and B antigens on their red blood cells, and those with type O blood lack A and B antigens (see Figure 42.1).

Many other blood antigen series exist on human red blood cells. One of the surface protein antigens on red blood cells is designated the Rh factor. The **Rh factor** is a complex of many antigens. The Rh

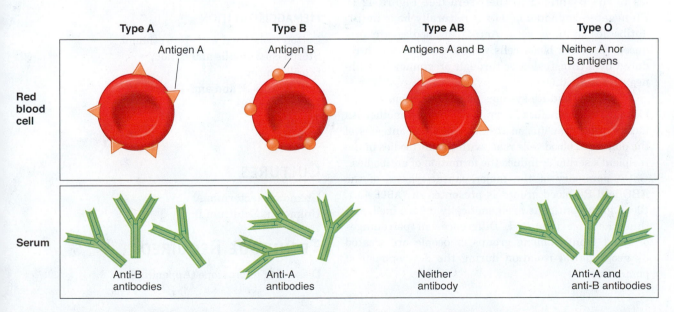

Type A	Type B	Type AB	Type O
Antigen A	Antigen B	Antigens A and B	Neither A nor B antigens

Red blood cell

Serum

| Anti-B antibodies | Anti-A antibodies | Neither antibody | Anti-A and anti-B antibodies |

FIGURE 42.1 Relationship of antigens and antibodies involved in the ABO blood group system.

TABLE 42.1	THE ABO AND RH BLOOD GROUP SYSTEMS					
	Blood Group					
Characteristic	A	B	AB	O	Rh$^+$	Rh$^-$
Antigen present on the red blood cells	A	B	Both A and B	Neither A nor B	D	No D
Antibody normally present in the serum	Anti-B	Anti-A	Neither anti-A nor anti-B	Both anti-A and anti-B	No anti-D	No anti-D*
Serum causes agglutination of red blood cells of these types	B, AB	A, AB	None	A, B, AB	Neither Rh$^+$ nor Rh$^-$	Neither Rh$^+$ nor Rh$^-$*
Percent occurrence in a mixed white population	41	10	4	45	85	15
Percent occurrence in a mixed black population	27	20	7	46	90	10
Percent occurrence in a mixed Asian population	28	27	5	40	98	2

*Anti-D antibodies are not naturally present in the serum of Rh$^-$ people. Anti-D antibodies can be produced upon exposure to the D antigen through blood transfusions or pregnancy.

Source: Adapted from G. J. Tortora, B. R. Funke, and C. L. Case. *Microbiology: An Introduction*, 12th ed. San Francisco, CA: Pearson Education, 2015.

factor that is used routinely in blood typing is the **Rh$_0$ antigen**, or **D antigen.** Individuals are Rh-positive when D antigen is present. The presence of the Rh factor is determined by a hemagglutination reaction between anti-D antiserum and red blood cells with D antigen on their surfaces.

A person possesses antibodies to the alternate A-B antigen. Thus, people of blood type A will have antibodies to the B antigen in their sera (see Figure 42.1). Rh-negative individuals do not naturally have anti-D antibodies in their sera. Anti-D antibodies are produced when red blood cells with D antigen are introduced by a transfusion or through pregnancy into Rh-negative individuals.

The ABO and Rh systems place restrictions on how blood may be transfused from one person to another. An incompatible transfusion results when the antigens of the donor red blood cells react with the antibodies in the recipient's serum or induce the formation of antibodies.

A summary of the major characteristics of the ABO and Rh blood groups is presented in **TABLE 42.1**. Blood group antigens are genetically determined, and thus they are inherited. Differences in percentages of blood groups among groups of people are created by geographical isolation during the development of populations.

MATERIALS

BACTERIAL AGGLUTINATION

Glass slide
Staphylococcus antiserum
Pasteur pipette (2) and bulbs
Toothpicks
Wax pencil

HEMAGGLUTINATION

Cotton moistened with 70% ethanol
Sterile cotton balls and bandage
Sterile lancet
Anti-A, anti-B, and anti-D antisera
Glass slides (2)
Toothpicks
Wax pencil

CULTURES

Unknown bacterium A
Unknown bacterium B

TECHNIQUE REQUIRED

Dissecting microscope (Appendix E)

PROCEDURE

Bacterial Agglutination

1. With a wax pencil, draw two circles on a clean glass slide, and label one "A" and the other "B."
2. With a Pasteur pipette, add one drop of unknown bacterium A to circle A on the slide. Discard the pipette in the disinfectant. Add one drop of unknown bacterium B to circle B. Discard the pipette in the disinfectant.
3. Add one drop of *Staphylococcus* antiserum to each circle.
4. Mix each suspension with a toothpick. Discard each toothpick in the disinfectant.
5. Hold the slide by its ends, rock gently for 30 to 40 seconds, and observe for small to large clumps, which indicate agglutination. A dissecting microscope may help determine which bacterium exhibits agglutination. Discard the slide in the disinfectant. Record your results.

Hemagglutination

Carefully read the safety precautions for working with blood in the laboratory in the box below:

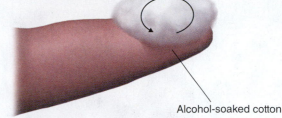
Wipe in a circular pattern

Alcohol-soaked cotton

FIGURE 42.2 Disinfecting the skin. Rub the skin with 70% ethanol.

1. With a wax pencil, draw two circles on a clean glass slide, and label one "A" and the other "B." Draw a circle on the second slide, and label it "D."
2. Disinfect a finger with cotton saturated in ethanol (**FIGURE 42.2**). Pierce the disinfected finger with a sterile lancet.
3. Let a drop of blood fall into each circle. Stop the bleeding with a sterile cotton ball, and apply an adhesive bandage.
4. To the A circle, add one drop of anti-A antiserum. Add one drop of anti-B antiserum to the B circle and one drop of anti-D antiserum to the D circle.
5. Mix each suspension with toothpicks. Use a different toothpick for each antiserum. Why? _____
   _____

> **Discard the slides, toothpicks, cotton balls, and lancet in the disinfectant.**

6. Observe for agglutination, and determine your blood type (**FIGURE 42.3**). A dissecting microscope may help you see hemagglutination.

> ### Safety Precautions for Working with Blood in the Laboratory
>
> - Do not perform this lab experiment if you are sick.
> - Work only with **your own** blood[*] or wear gloves.
> - **Disinfect** your finger with 70% alcohol. Use any finger except your thumb. Unwrap a **new, sterile** lancet, and pierce the disinfected finger with the sterile lancet.
> - Place the used lancet **in disinfectant** in a container designated by the instructor or a biohazard "sharps" container.
> - Stop the bleeding with a sterile cotton ball, and **apply an adhesive bandage.**
> - **Dispose of** the used cotton **in disinfectant.**
> - **Discard** used slides and toothpicks **in disinfectant.**
> - **Wash** any spilled blood from your work area **with disinfectant.**
>
> *Blood obtained from blood banks is tested for hepatitis B virus and HIV. However, no test method can offer complete assurance that laboratory specimens do not contain these viruses.

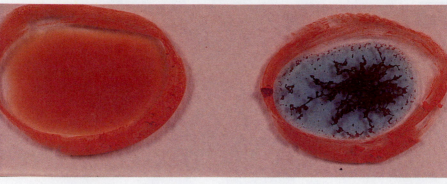

Anti-B   **Anti-A**

FIGURE 42.3 Slide agglutination test for the ABO blood group. A negative reaction with anti-B and a positive reaction with anti-A (indicated by agglutination). Results indicate blood group A.

LABORATORY REPORT
Agglutination Reactions:
Slide Agglutination

EXERCISE
42

PURPOSE _____

_____

_____

EXPECTED RESULT

A positive agglutination with bacteria or red blood cells will give what result? _____

_____

RESULTS

Bacterial Agglutination

Bacterium	Description of the Results	Agglutination?
Unknown A		
Unknown B		

Hemagglutination

Antiserum	Hemagglutination
Anti-A	
Anti-B	
Anti-D	

CONCLUSIONS

1. Which unknown bacterial sample was *Staphylococcus*? _____

2. What is your blood type? ABO _____ Rh _____

3. On the blackboard, tabulate the blood types for the class or laboratory section. Calculate the percentage of each blood type, and compare your results with the percent distribution given in Table 42.1.

Blood Type	Number of Students	Percent
A		
B		
AB		
O		
Rh$^+$		

QUESTIONS

1. How do agglutination tests detect microbes in a patient's blood or fluids differ from techniques to detect antibodies to microbes in a patient's serum? _____

   _____

2. What is the blood type of the person on your left? _____

   Is your blood potentially compatible? _____ Explain briefly. _____

   _____

3. What is the blood type of the person on your right? _____

   Is your blood potentially compatible? _____ Explain briefly. _____

   _____

CRITICAL THINKING

1. How would agglutination reactions be used to locate the source of an epidemic?

2. Individuals may have antibodies to the A and B blood antigens not found on their red blood cells. What is the origin of these antibodies in someone who has never had a transfusion?

CLINICAL APPLICATIONS

1. What is hemolytic disease of the newborn? How does RhoGAM prevent it?

2. A woman with type A blood can have a healthy baby with type B blood. Why doesn't the baby develop hemolytic disease of the newborn?

3. A woman with type A, Rh-negative blood can have a healthy baby with type B, Rh-positive blood and usually will not develop antibodies to the Rh antigen. Why?

Agglutination Reactions: Microtiter Agglutination

OBJECTIVES

After completing this exercise, you should be able to:

1. Define the terms *agglutination* and *titer*.
2. Determine the titer of antibodies by the agglutination method.
3. Provide two applications for agglutination reactions.

BACKGROUND

Agglutination reactions are used to determine whether a particular bacterium is present in a specimen, as shown in Exercise 42. Agglutination reactions are also used to determine whether antibodies have formed in a patient's serum because of exposure to a microbe.

An **agglutination titration** can estimate the concentration (*titer*) of antibody in serum. This test is used to determine whether a particular organism may be causing the patient's symptoms. If an increase in titer is shown in successive daily tests, the patient probably has an infection caused by the organism used in the test.

In the titration, bacteria are mixed with dilutions of serum (e.g., 1:10, 1:20, 1:40, and so on). The *endpoint* is the greatest dilution of serum showing an agglutination reaction. The *reciprocal* of this dilution is the *titer*. For example, in **FIGURE 43.1a** the titer is 20.

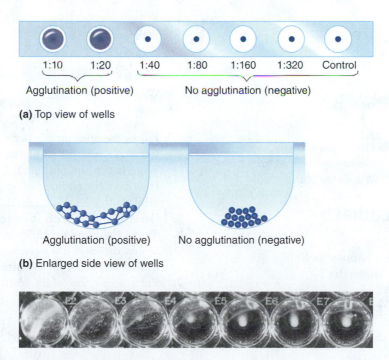

(a) Top view of wells

Agglutination (positive)   No agglutination (negative)

(b) Enlarged side view of wells

(c) Appearance of wells after refrigeration for 24 hours

FIGURE 43.1 Agglutination titration. (a) Each well in this microtiter plate contains, from left to right, only half the concentration of serum in the preceding well. Each well contains the same concentration of bacterial cells. **(b)** In a positive reaction, sufficient antibodies are present in the serum to link the antigens together, forming an antibody–antigen mat that sinks to the bottom of the well. In a negative reaction, insufficient antibodies are present to cause the linking of antigens. **(c)** Appearance of wells after refrigeration for 24 hours. The first two wells in the row show agglutination. The endpoint is 1:20.

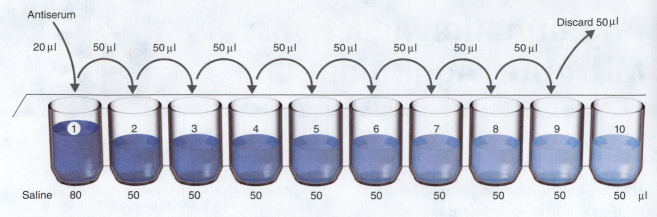

FIGURE 43.2 **Dilution of antiserum.** First, 20 μl of patient's serum is mixed with 80 μl of saline in well 1. Serial dilutions are made by transferring 50 μl from one well to the next, through well 9. Discard 50 μl from well 9.

In this exercise, we will perform a test to diagnose typhoid fever, the **Widal test.** Typhoid fever is caused by *Salmonella enterica* Typhi. However, the Widal test also detects nontyphoidal *Salmonella*, which enables us to use *S. enterica* Typhimurium, which is less virulent, in this exercise.

MATERIALS

0.85% saline solution
"Patient's" serum
Microtiter plate and lid
Micropipettes, 10 to 100 μl
Micropipette tips (12)

CULTURES

Commercial *Salmonella* O antigen *or*
Salmonella enterica Typhimurium (heated for
   30 minutes at 56°C in a water bath)

TECHNIQUES REQUIRED

Pipetting (Appendix A)
Serial dilution technique (Appendix B)
Dissecting microscope (Appendix E)

PROCEDURE First Period

1. Label 10 wells in the microtiter plate "1" through "10" (**FIGURE 43.2**).
2. Add 80 μl of saline to the first well and 50 μl of saline to each of the remaining wells.
3. Add 20 μl of patient's serum to the first well. Mix up and down three times, and, with a new pipette tip, transfer 50 μl to the second well. Change the pipette tip, mix, and transfer 50 μl to the third well, and so on. Continue until you have reached

the ninth well. Discard 50 μl from that well, as shown in Figure 43.2. The patient's serum has now been diluted. The dilutions are as follows:

Well 1	1:5
Well 2	1:10
Well 3	1:20
Well 4	1: _____
Well 5	1: _____
Well 6	1: _____
Well 7	1: _____
Well 8	1: _____
Well 9	1: _____
Well 10	no serum

Fill in the missing dilutions.

4. Carefully add 50 μl of the antigen to each well. What is the antigen? _____ How does the addition of 50 μl of antigen affect the dilutions? _____
5. Cover the plate with the microtiter plate lid.
6. Shake the plate carefully in a horizontal direction to mix the contents in each well. Place the plate in a 35°C incubator for 60 minutes.
7. Refrigerate the plate until the next laboratory period.

PROCEDURE Second Period

1. Observe the bottom of each well for agglutination. A dissecting microscope may help you to see agglutination. Which well serves as the control? _____

What should occur in this well? _____

Determine the endpoint and the titer. Discard the plate in the biohazard container.

LABORATORY REPORT

Agglutination Reactions: Microtiter Agglutination

PURPOSE _____

_____

_____

HYPOTHESIS

The endpoint will be the highest dilution of antiserum that results in clumping. Agree/disagree

RESULTS

Well #	Final Dilution	Agglutination
1		
2		
3		
4		
5		
6		
7		
8		
9		
10		

Diagram the appearance of the bottom of a positive well and of a negative well:

Positive   Negative (Well 10)

CONCLUSIONS

1. Did your result confirm your hypothesis? _____

2. How did you determine the presence of agglutination? _____

3. What was the endpoint? _____

4. What was the antibody titer? _____

QUESTIONS

1. What was the antiserum used in this experiment? _____

2. What are acute and convalescent sera? If the titer of the convalescent serum is much greater than that of the

   acute serum, what does this indicate? _____

   _____

CRITICAL THINKING

1. Rubella virus can cause birth defects when the virus is acquired during a pregnancy. When the baby is born, the baby's blood is tested for IgM and IgG antibodies to the rubella virus. Results for one baby indicate no IgM antibodies but high IgG antibodies in the baby's blood. What are the implications of this result?

2. An agglutination test was performed on a patient's blood. The titer was 512, and upon further testing, the antibody involved was identified as IgM. Can any conclusions be drawn?

CLINICAL APPLICATION

The following antibody titers were obtained for three patients:

Patient	Antibody Titer		
	Day 1	Day 5	Day 12
A	128	128	128
B	128	256	512
C	0	0	0

What can you conclude about each of these patients?

ELISA Technique

OBJECTIVES

After completing this exercise, you should be able to:

1. Explain how ELISA tests can be used clinically to detect antibodies or antigens.
2. Differentiate direct from indirect ELISA tests.
3. Determine the titer of antibodies by the ELISA method.

BACKGROUND

Diseases can be diagnosed by using known antibodies to detect the presence of an antigen or by detecting specific antibodies in a patient. These types of tests are called **immunoassays.** Immunoassays are based on detectable interactions between antigens and antibodies such as precipitation, agglutination, or complement fixation. Increased sensitivity in detecting antigens can be achieved by labeling antibodies with substances such as radioactive chemicals (e.g., iodine-125), fluorescent compounds, magnetic beads, or enzymes. **ELISA (enzyme-linked immunosorbent assay)** or **EIA (enzyme immunoassay)** is the most widely used immunoassay in labs today. ELISAs use enzyme labels because of their stability, reproducibility, safety, ease of detection, and relatively low cost. The enzymes horseradish peroxidase and alkaline phosphatase are most often used. When the appropriate substrate is added, the enzyme reacts with the substrate to make a colored product, which can be detected visually or by a spectrophotometer. The amount of product produced is directly proportional to the amount of enzyme—and, therefore, antibody—present, allowing the technique to be quantitative as well as qualitative.

Direct ELISA Technique

In the *direct* ELISA technique, enzyme-labeled antibodies are used to identify an antigen. In the direct ELISA technique, a known antibody is adsorbed to the wells in a microtiter plate. The unknown microorganism from a patient is added to the wells. If the antibody in the well is specific for the microorganism, the microbe will be bound to the antibody. A second antibody specific for the antigen is then added. This antibody is linked to an enzyme. The reaction between the antigen and antibody is made visible by addition of the substrate for the enzyme (**FIGURE 44.1a**).

Indirect ELISA Technique

Indirect ELISAs are used to detect a specific antibody in a patient's serum. In the indirect technique, the known antigen is attached to the wells of a microtiter plate. Dilutions of the suspected antibody are added, and after washing, enzyme-labeled anti-antibody is added. A color change after the substrate is added indicates that the antibody was present (**FIGURE 44.1b**). The ELISA test for HIV antibodies is one example of the indirect technique.

In this exercise, we will perform an ELISA test to determine whether the patient's serum has antibodies to *Salmonella*.

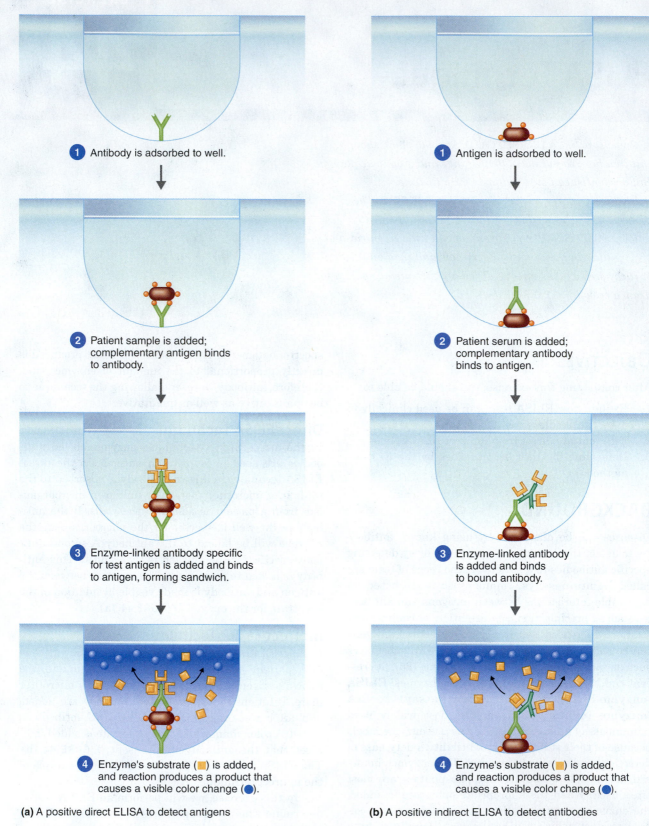

① Antibody is adsorbed to well.

② Patient sample is added; complementary antigen binds to antibody.

③ Enzyme-linked antibody specific for test antigen is added and binds to antigen, forming sandwich.

④ Enzyme's substrate (▥) is added, and reaction produces a product that causes a visible color change (●).

(a) A positive direct ELISA to detect antigens

① Antigen is adsorbed to well.

② Patient serum is added; complementary antibody binds to antigen.

③ Enzyme-linked antibody is added and binds to bound antibody.

④ Enzyme's substrate (▥) is added, and reaction produces a product that causes a visible color change (●).

(b) A positive indirect ELISA to detect antibodies

FIGURE 44.1 The ELISA method. The components are usually contained in small wells of a microtiter plate.

FIGURE 44.2 ELISA titration method. The endpoint is 1:16.

MATERIALS

FIRST PERIOD

Coating buffer
Flat-bottomed microtiter plate
Clear plastic tape
Micropipettes, 50 to 100 µl
Micropipette tips

SECOND PERIOD

Washing buffer
Blocking buffer
Patient's serum
Alkaline phosphatase–labeled anti-antibodies
BCIP/NBT substrate
Micropipettes, 50 to 100 µl
Micropipette tips

CULTURES

Commercial *Salmonella* O antigen *or*
Salmonella enterica Typhimurium (heated for
   30 minutes at 56°C in a water bath)

TECHNIQUES REQUIRED

Pipetting (Appendix A)
Serial dilution technique (Appendix B)
Microtiter dilutions (Exercise 43)

PROCEDURE First Period

1. Add 100 µl of coating buffer to each well of one row
   (wells 1–12) of the microtiter plate.
2. Add 100 µl of *S. enterica* Typhimurium to each
   well.
3. Seal the wells with a strip of plastic tape, and
   refrigerate the plate at 5°C for 1 to 7 days.

PROCEDURE Second Period

1. Remove your plate from the refrigerator, and carefully remove the tape. Shake the inverted plate
   with a quick shake to remove the liquid into disinfectant.

2. Fill the wells with washing buffer, and shake to
   remove. Wash two more times.
3. Add 100 µl of blocking buffer. Leave for 30 to 90 minutes as directed by your instructor.
4. Perform dilutions of the patient's serum by placing 100 µl in the first well. Mix up and down three
   times and, with a new pipette tip, transfer 100 µl
   to the second well. Mix up and down three times,
   change pipette tips, and transfer 100 µl to the third
   well, and so on. Continue until you have reached the
   11th well. Discard 100 µl from that well (see Figure 44.2). The patient's serum has now been diluted
   from 1:2 to 1: __ in well 11. What is in well 12? ____
   What is the purpose of well 12? _____

5. Incubate the plate at 35°C for 60 minutes.
6. Shake the inverted plate with a quick shake to
   remove the contents. Wash three times with washing buffer as described in step 2.
7. Add 100 µl of alkaline phosphatase–labeled anti-
   antibody to each well (1–12). Seal the wells with
   tape and incubate the plate at 35°C for 45 minutes.
   Plates can be sealed and stored at 5°C until the
   next lab period. What is the antigen? _____

8. Remove the tape carefully, shake out the contents,
   and wash the wells three times with washing
   buffer.
9. Add 100 µl of the alkaline phosphatase substrate
   (BCIP/NBT) to each well in the row.
10. Leave at room temperature for 10 to 30 minutes
    until color develops; well 12 will be colorless.
11. Record the results. The highest dilution with a blue
    color is the endpoint (**FIGURE 44.2**). The titer is the
    reciprocal of the dilution of the endpoint.

LABORATORY REPORT
ELISA Technique

PURPOSE _____

_____

_____

HYPOTHESIS

The ELISA technique will/will not result in a blue color in the wells if antibodies to *Salmonella* bacteria are present.

RESULTS

Well #	Final Dilution	Color
1		
2		
3		
4		
5		
6		
7		
8		
9		
10		
11		
12 (control)		

CONCLUSIONS

1. Does your hypothesis agree with your results? _____

2. What was the endpoint? _____

3. What was the antibody titer? _____

QUESTIONS

1. Was the test performed in this exercise a direct or an indirect ELISA test? _____

2. Why was the control well colorless? _____

   _____

   _____

3. What advantage does the ELISA test have over other immunological techniques, such as fluorescent antibody

   tests or the use of radioactive immunoassays? _____

   _____

CRITICAL THINKING

Compare and contrast direct and indirect ELISA techniques.

CLINICAL APPLICATIONS

1. Describe how you would determine whether a bacterium isolated from a patient was *Salmonella* using the
   ELISA technique with a known *Salmonella* antiserum.

2. Why would the ELISA technique be valuable for identifying viruses?

PART

13

Microorganisms and Disease

EXERCISES

Many microorganisms grow abundantly both inside and on the surface of the normal adult body. The microorganisms that establish more or less permanent residence without producing diseases are known as **normal microbiota.** Microorganisms that may be present for a few days or months are called **transient microbiota.** At one time, bacteria and fungi were thought to be plants, and thus the term *microflora* was used.

A microorganism that causes disease is called a **pathogen.** The pathogens that cause disease consist of many different organisms. Robert Koch speculated as follows in one of his early publications:

> *On this . . . I take my stand, and, till the cultivation of bacteria from spore to spore shows that I am wrong, I shall look on pathogenic bacteria as consisting of different species.**

In a clinical laboratory, samples from diseased tissue are cultured, and pathogens must be distinguished from normal and transient microbiota. Identifying pathogens is imperative for initiating proper treatment and tracing the source of the infection. Morphological characteristics (Part 3 of this lab manual), differential stains (Part 3), and biochemical testing (Part 5) provide information used to identify microorganisms in a clinical laboratory.

---

*Quoted in T. D. Brock, ed. *Milestones in Microbiology.* Washington, DC: American Society of Microbiology, 1961, p. 99.

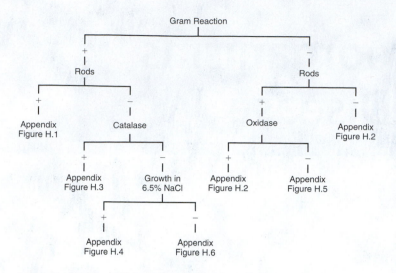

A dichotomous key to Appendix H. There is no single dichotomous key. This is one example of a key to Appendix H. In this key, the first test is a Gram stain. If the bacterium is gram-positive, move to the left side of the diagram, ignoring all of the gram-negative bacteria. If the bacterium is a gram-positive coccus, you need to do a catalase test. If the bacterium is catalase-positive, go to Appendix H, Figure H.3 to complete the identification.

Dichotomous keys are essential tools for sorting through the information obtained from these laboratory tests in an organized manner (see the figure). In a dichotomous key, identification is based on successive questions, and each question has two possible answers. (*Dicho-* means "cut in two.") After answering one question, the microbiologist is directed to another question until an organism is identified. For example, a dichotomous key for bacteria would begin with an easily determined characteristic, such as Gram reaction, and move on to the ability to ferment a sugar. There is no single key for all bacteria, and there is no one correct way to make a dichotomous key. Keys are made for specific groups of bacteria as suits the needs of the microbiologist. Several dichotomous keys to help you identify bacteria in Part 13 appear in Appendix H.

In Exercises 45 through 49, we will examine the variety of bacteria associated with the human body. In Exercise 50, we will identify bacteria in a simulated clinical sample. In Exercise 51, unknown bacteria will be identified on the basis of metabolic characteristics by using commercial rapid identification methods.

CASE STUDY: Grandpa's Ill

Seventy-eight-year-old Ernie had been sick for several days and was struggling to breathe. He finally agreed to go to the emergency room to find out what was causing his distress. The nurse practitioner carefully listened to Ernie describe his concern about not getting enough oxygen and his discomfort. Suspecting pneumonia, the nurse practitioner ordered a chest X ray, and a cannula was placed in Ernie's nose to deliver supplementary oxygen. After a review of the chest X ray, Ernie was admitted to the hospital. His (rust-colored) sputum was cultured on blood agar with an optochin disk placed next to the subculture. In 24 hours, alpha-hemolysis appeared with the subculture sensitive to optochin. Gram-positive diplococci were seen in a Gram stain. An antibiotic sensitivity test was set up, and the microbe was found to be resistant to penicillin but sensitive to erythromycin and several other chemotherapeutic drugs. A serum sample was obtained for serological testing for epidemiological purposes. Ernie's pneumonia responded to erythromycin, and he soon regained his ability to breathe normally. He was discharged and reminded to complete the chemotherapy regimen even though he felt great.

Questions

1. Based on the lab results, what bacterium is the cause of Ernie's pneumonia?
2. How is serological testing by agglutination done?
3. Why do a catalase test? Can it be done on blood agar?
4. Why was the sputum rust-colored?
5. Is this typical or atypical pneumonia? Explain.
6. If beta-hemolysis had occurred, a bacitracin sensitivity test would have been useful in identifying the bacterium. Why?

Bacteria of the Skin

OBJECTIVES

After completing this exercise, you should be able to:

1. Isolate and identify bacteria from the human skin.
2. Provide an example of normal skin microbiota.
3. List characteristics used to identify the staphylococci.
4. Explain why many bacteria are unable to grow on human skin.

BACKGROUND

The skin is generally an inhospitable environment for most microorganisms. The dry layers of keratin-containing cells that make up the epidermis (the outermost layer of the skin) are not easily colonized by most microbes. Sebum, secreted by oil glands, inhibits bacterial growth, and salts in perspiration create a hypertonic environment. Perspiration and sebum are nutritive for certain microorganisms, however, which establishes them as part of the normal microbiota of the skin.

Normal microbiota of the skin tend to be resistant to drying and to relatively high salt concentrations. More bacteria are found in moist areas, such as the axilla (armpit) and the sides of the nose, than on the dry surfaces of arms or legs. Transient microbiota are present on hands and arms in contact with the environment.

Skin Microbiota

Despite the adverse environment, the microbial population of the skin is diverse. Most information on skin microbiota is based on identifying bacteria that can be cultured. Recent analyses for ribosomal RNA have identified more than 40 bacterial species from swabbing the skin on the human forehead. Some of these, such as *Methylophilus*, don't show up on routine culture media, and some are potentially new species.

The most common genera on skin are *Propionibacterium* and *Staphylococcus*.

Propionibacterium live in hair follicles on sebum from oil glands. The propionic acid they produce maintains the pH of the skin between 3 and 5, which suppresses the growth of other bacteria. Most bacteria on the skin are gram-positive and salt-tolerant. Mannitol salt agar is selective for salt-tolerant organisms and is differential in that mannitol-fermenting organisms

FIGURE 45.1 Mannitol salt agar. *Staphyloccus aureus* (on the left) ferments mannitol, causing the phenol red to turn yellow, and *S. epidermidis* (on the right) tolerates the high salt concentration and grows on mannitol salt agar (MSA).

will produce acid, turning the indicator in the medium yellow (**FIGURE 45.1**).

Staphylococcus aureus, which forms golden yellow colonies, is part of the normal microbiota of the skin and is also considered a pathogen. *S. aureus*, which produces **coagulase,** an enzyme that coagulates (clots) the fibrin in blood, is pathogenic. A test for the presence of coagulase is used to distinguish *S. aureus* from other species of *Staphylococcus*.

Although many different bacterial genera live on human skin, in this exercise we will determine whether a potential *Staphylococcus* is present by observing mannitol fermentation and colony pigmentation. BSL-2 laboratories will attempt to isolate and identify a catalase-positive, gram-positive coccus.

MATERIALS

FIRST PERIOD

Petri plate containing mannitol salt agar
Sterile cotton swab
Sterile saline

SECOND PERIOD BSL-2

Petri plate containing mannitol salt agar
3% hydrogen peroxide

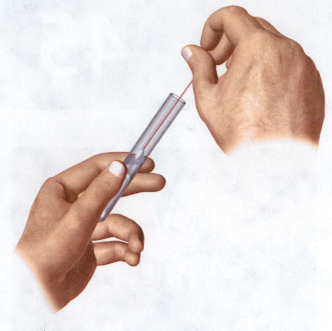

(a) Aseptically moisten a sterile cotton swab in saline.

(b) After swabbing half the plate, use a sterile loop to streak the inoculum over the agar.

FIGURE 45.2 **Taking a sample from the skin.**

Gram-staining reagents
Toothpick

THIRD PERIOD BSL-2

Fermentation tubes
Coagulase plasma (as needed)

TECHNIQUES REQUIRED

Gram staining (Exercise 7)
Plate streaking (Exercise 11)
Selective media (Exercise 12)
Fermentation tests (Exercise 14)
Catalase test (Exercise 17)

PROCEDURE First Period

1. Wet the swab with saline, and push the swab against the wall of the test tube to express excess saline (**FIGURE 45.2a**). Swab any surface of your skin. Possible areas include the sides of the nose, axilla, an elbow, or a pus-filled sore.
2. Swab one-third of the plate with the swab.

 Discard the swab in disinfectant.

Using a sterile loop, streak back and forth into the swabbed area a few times, and then streak away from the inoculum (**FIGURE 45.2b**), to cover

about one-third of the agar. Sterilize your loop and spread the bacteria over the rest of the agar. Seal the plates with tape.
3. Incubate the plate, inverted, at 35°C for 24 to 48 hours.

PROCEDURE Second Period

Be careful—these bacteria could be pathogens.

1. Examine the colonies. Record the appearance of the colonies. Look at the colonies for pigment production. Remember, a pigment is a color produced by the cells, not a change in the pH indicator. Record any mannitol fermentation (change in the indicator around colonies). (See Figure 45.1.)
2. BSL-2 Perform a Gram stain of the colonies with yellow halos, and test for catalase production. Perform the catalase test by making a suspension of the desired colony in a loopful of water on a slide with a toothpick and adding a drop of H_2O_2. Discard the toothpick in the biohazard container.
3. BSL-2 Subculture a catalase-positive, gram-positive coccus from a colony growing in an area with a yellow halo on another mannitol salt plate. Do not attempt to subculture a colony that has not been tested for catalase. Why? _____

PROCEDURE Third Period

1. **BSL-2** Using the key in Appendix H, Figure H.3, proceed to identify your isolate.

    a. To test for coagulase, place a loopful of rehydrated coagulase plasma on a clean slide. Add a loopful of water, and make a heavy suspension of the bacteria to be tested. Observe for clumping of the bacterial cells (clumping = coagulase-positive; no clumping = coagulase-negative; **FIGURE 45.3**).

    b. Inoculate the appropriate fermentation tubes. Incubate the fermentation tubes at 35°C for 24 to 48 hours.

(a) Negative test

Clumping

(b) Positive test

FIGURE 45.3 Coagulase test.

LABORATORY REPORT
Bacteria of the Skin

PURPOSE _____

_____

_____

HYPOTHESIS

You will culture _____ from your skin.

RESULTS

Source of inoculum: _____

First Mannitol Salt Plate

	Colony		
	1	2	3
Colony description			
Pigment			
Mannitol fermentation			

BSL-2 Which colony was isolated? _____

Catalase reaction: _____

Gram stain

Reaction: _____ Morphology: _____ Arrangement: _____

BSL-2 Second Mannitol Salt Plate

Coagulase test

Clumping: _____ Positive or negative: _____

Fermentation tests. Perform the necessary tests.

	Growth	Acid	Gas
Glucose			
Trehalose			
Xylose			

CONCLUSIONS

1. Do your results confirm your hypothesis? _____ What organism did you identify? _____

2. Is it a pathogen? _____

QUESTIONS

1. Why is mannitol salt agar used as a selective medium for normal skin microbiota? _____
    _____

2. List three identifying characteristics of *Staphylococcus aureus*. _____
    _____

3. *Staphylococcus aureus* typically forms colonies with a _____ pigment.

    What function might the pigment serve? _____

CRITICAL THINKING

1. What is coagulase? How is it related to pathogenicity?

2. Assume that you isolated *S. aureus* from your skin. How would you determine whether it is penicillin-resistant?

CLINICAL APPLICATION

An 8-year-old boy went to the pediatrician complaining of pain in his right hip. He had no previous injury to the hip or leg. His temperature was 38°C. A bone scan revealed damage to the head of the femur. Bacterial cultures of blood and fluid from the hip joint were positive for a gram-positive, catalase-positive, coagulase-positive coccus that ferments mannitol. Use Appendix H to identify the species of this bacterium. _____

Bacteria of the Respiratory Tract

OBJECTIVES

After completing this exercise, you should be able to:

1. List representative normal microbiota of the respiratory tract.
2. Differentiate among the streptococci on the basis of the pattern of hemolysis.

BACKGROUND

The respiratory tract can be divided into two systems: the upper and lower respiratory tracts. The **upper respiratory tract** consists of the nose and throat, and the **lower respiratory tract** consists of the larynx, trachea, bronchial tubes, and alveoli. The lower respiratory tract is normally sterile because of the efficient functioning of the ciliary escalator. The upper respiratory tract is in contact with the air we breathe—air contaminated with microorganisms.

The throat is a moist, warm environment, allowing many bacteria to establish residence. Species of many different genera—such as *Staphylococcus, Streptococcus, Neisseria,* and *Haemophilus*—can be found living as normal microbiota in the throat. Despite the presence of potentially pathogenic bacteria in the upper respiratory tract, the rate of infection is minimized by microbial **antagonism.** Certain microorganisms of the normal microbiota suppress the growth of other microorganisms through competition for nutrients and production of inhibitory substances.

Identifying Streptococcal Species

Streptococcal species are the predominant organisms in throat cultures, and some species are the major cause of bacterial sore throats (acute pharyngitis). Streptococci are gram-positive and catalase-negative. Streptococcal species are identified by biochemical characteristics, including hemolytic reactions, and antigenic characteristics (Lancefield system). Hemolytic reactions are based on hemolysins that are produced by streptococci while growing on blood-enriched agar. Blood agar is usually made with (5.0%) defibrinated sheep blood, (0.5%) sodium chloride to minimize spontaneous hemolysis, and nutrient agar. Three patterns of hemolysis can occur on blood agar:

1. **Alpha-hemolysis:** Green, cloudy zone around the colony. Partial destruction of red blood cells is due to bacteria-produced hydrogen peroxide (**FIGURE 46.1a**).
2. **Beta-hemolysis:** Complete hemolysis, giving a clear zone with a clean edge around the colony (**FIGURE 46.1b**).
3. **Gamma-hemolysis:** No hemolysis, and no change in the blood agar around the colony (**FIGURE 46.1c**).

(a) Alpha-hemolysis   **(b)** Beta-hemolysis   **(c)** Gamma-hemolysis

FIGURE 46.1 Hemolysis on blood agar. (a) Note the greenish color around the colonies in alpha-hemolysis.
(b) Beta-hemolysis results in complete clearing. **(c)** Absence of hemolysis is called gamma-hemolysis or ahemolytic.

Alpha-hemolytic and gamma-hemolytic streptococci are usually normal microbiota, whereas beta-hemolytic streptococci are frequently pathogens. *Streptococcus pneumoniae*, a causative agent of pneumonia, cannot be differentiated from other alpha-hemolytic streptococci on blood agar. **Optochin inhibition** with an optochin disk is used to identify this pathogen. A zone of inhibition equal to or greater than 15 mm indicates optochin sensitivity. **Bile solubility** is also used to distinguish *S. pneumoniae* from the other alpha-hemolytic streptococci. The addition of bile salts (sodium deoxycholate) activates an enzyme that destroys the cell wall, and the cells lyse—colonies will disappear after the addition of bile salts. The other alpha-hemolytic streptococci do not have this enzyme.

The streptococci can be antigenically classified into Lancefield groups A through O by antigens in their cell walls. Over 90% of streptococcal infections are caused by beta-hemolytic group A streptococci. These bacteria are assigned to the species *S. pyogenes*. *S. pyogenes* is sensitive to the antibiotic bacitracin; other streptococci are resistant to bacitracin (**FIGURE 46.2**).

MATERIALS THROAT CULTURE

FIRST PERIOD

Petri plate containing blood agar
Sterile cotton swab

SECOND PERIOD `BSL-2`

Toothpicks
Hydrogen peroxide, 3%

MATERIALS *STREPTOCOCCUS* `BSL-2`

FIRST PERIOD

Petri plate containing blood agar
Gram-staining reagents
Sterile cotton swabs (2)
Forceps and alcohol
Optochin disk
Bacitracin disk

SECOND PERIOD

10% bile salts
Tubes containing 2 ml of nutrient broth (2)
Gram-staining reagents

CULTURES

Streptococcus pyogenes `BSL-2`
Streptococcus pneumoniae `BSL-2`

TECHNIQUES REQUIRED

Gram staining (Exercise 7)
Plate streaking (Exercise 11)

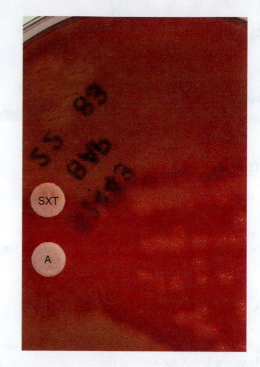

FIGURE 46.2 Identifying *Streptococcus pyogenes*. *S. pyogenes* is beta-hemolytic and inhibited by bacitracin (A disk). *S. pyogenes* is resistant to sulfamethoxazole/trimethoprim (SXT).

Catalase test (Exercise 17)
Kirby-Bauer technique (Exercise 25)

PROCEDURE Throat Culture

First Period

Work only with swabs collected from your own throat.

1. Swab your throat with a sterile cotton swab. The area to be swabbed is between the "golden arches" (glossopalatine arches), as shown in **FIGURE 46.3**. Do not hit the tongue.
2. After obtaining an inoculum from the throat, swab one-half of a blood agar plate. Streak the remainder of the plate with a sterile loop (Figure 45.2b on page 360).

Discard the swab in disinfectant.

3. Incubate the plate, inverted, at 35°C for 24 hours. Use tape to seal the plate.

Second Period

1. Observe the plate for hemolysis.
2. **BSL-2** The next step is a catalase test. Why a catalase test? _____

Transfer some colonies to a slide with a toothpick, and perform a catalase test. Discard the toothpick in the biohazard container. Why can't the catalase test be done on blood agar? _____
_____

Be careful—these bacteria may be pathogens.

PROCEDURE *Streptococcus* **BSL-2**

First Period

1. Inoculate each half of a blood agar plate, one half with *S. pyogenes* and the other half with *S. pneumoniae*. Use swabs to obtain confluent growth.

Be careful—these bacteria are pathogens.

2. Dip forceps in alcohol, and burn off the alcohol.

Keep the forceps tip pointed down while burning. Keep the beaker of alcohol away from the flame.

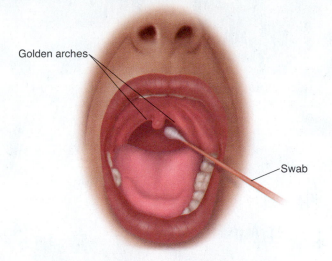

FIGURE 46.3 Swabbing the throat with a sterile swab.

Using the forceps, place and press a bacitracin disk and an optochin disk on each half. Space the disks so that zones of inhibition may be observed.
3. Incubate the plate, inverted, at 35°C for 24 hours.

Second Period

1. Observe for hemolysis and inhibition of growth by bacitracin and optochin.
2. Prepare Gram stains from smears of each organism. Do the organisms differ microscopically? _____
_____
3. Using a sterile loop, prepare a suspension of each organism in a tube of nutrient broth.
4. Add a few drops of 10% bile salts to each tube. Observe the tubes after 15 minutes for lysis of the cells. How will you detect bile solubility? _____

LABORATORY REPORT
Bacteria of the Respiratory Tract

PURPOSE _____

_____

_____

HYPOTHESIS

Your throat culture will grow _____-hemolytic bacteria.

RESULTS

Throat Culture

	Colony			
	1	2	3	4
Appearance of colonies on blood agar				
Hemolysis				
BSL-2 Catalase reaction				

BSL-2 Labs: *Streptococcus* BSL-2

	Organism	
	S. pyogenes	*S. pneumoniae*
Gram stain		
Gram reaction		
Morphology		
Arrangement		
Blood agar plate		
Appearance of colonies		
Hemolysis		
Inhibition by		
Optochin		
Bacitracin		
Bile solubility		

CONCLUSIONS

1. Do your results confirm your hypothesis? _____ Are any of the bacteria from your throat culture potential pathogens? _____

2. Subculturing to a blood agar plate and placing bacitracin and optochin disks on the area subcultured will help determine whether pathogenic streptococci are present. Explain. _____

_____

QUESTIONS

1. Is blood agar selective or differential? _____ Briefly explain. _____

2. Is the Gram stain of significant importance in identifying the organisms studied in this exercise?

Explain. _____

_____

3. From a throat culture, you have isolated gram-positive cocci that you cannot identify as staphylococci or streptococci. A test for one enzyme can be used to distinguish between these bacteria *quickly*. What is that enzyme?

_____

CRITICAL THINKING

1. In the last decade, a few bacitracin-resistant *Streptococcus pyogenes* bacteria have been isolated from patients with pharyngitis. How does this affect identification of *S. pyogenes*? What role might over-the-counter bacitracin-containing throat lozenges play in this resistance?

2. Assume that you isolated non–acid-fast, gram-positive, catalase-positive rods from your throat. Does this represent a disease state? Briefly explain.

CLINICAL APPLICATION

A 45-year-old man was admitted to the hospital with chest and back pain. On examination, his chest was dull to percussion. A chest X-ray film shows lower-left-lung infiltrates. A sputum culture reveals alpha-hemolytic, gram-positive, catalase-negative cocci that were inhibited by optochin. The bacteria produce acid from lactose. Use Appendix H to identify the species of this bacterium. _____

Bacteria of the Mouth

OBJECTIVES

After completing this exercise, you should be able to:

1. List characteristics of streptococci found in the mouth.
2. Describe the formation of dental caries.
3. Explain the relationship between sucrose and caries.

BACKGROUND

The mouth contains millions of bacteria in each milliliter of saliva. Some of these bacteria are transient microbiota carried on food. Some species of *Streptococcus* are part of the normal microbiota of the mouth. (An identification scheme for streptococci found in the mouth is shown in Appendix H, Figure H.4.)

Streptococcus mutans, *S. salivarius*, and *S. sanguinis* produce sticky polysaccharides specifically from sucrose. Bacterial exoenzymes hydrolyze sucrose into its component monosaccharides, glucose and fructose. The energy released in the hydrolysis is used by *S. mutans* and *S. sanguinis* for polymerization of the glucose to form a **dextran** capsule. Fructose released from the sucrose is fermented to produce lactic acid. Enzymes of *S. salivarius* have similar specificity for sucrose, but the fructose is polymerized into **levan,** and the liberated glucose is fermented.

The dextran capsule enables the bacteria to adhere to surfaces in the mouth. *S. salivarius* colonizes the surface of the tongue, and *S. sanguinis* and *S. mutans* the teeth. Masses of bacterial cells, dextran, and debris adhering to the teeth constitute dental **plaque.** Bacteria in the plaque produce lactic acid, which erodes tooth enamel, initiating dental caries. Streptococci and other bacteria, such as *Lactobacillus* species, are able to grow on the exposed dentin and tooth pulp.

Other carbohydrates, such as glucose or starch, may be fermented by bacteria, but they are not converted to dextran and hence do not promote plaque formation. "Sugarless" candies may contain mannitol or sorbitol, which cannot be converted to dextran, although they may be fermented by such normal microbiota as *S. mutans*. Acid production increases the extent of dental caries.

The number of *S. mutans* present in stimulated saliva has been correlated with the potential for the formation of caries. In this exercise, we will observe and determine the relative amount of *S. mutans* and other polysaccharide-producing streptococci in a stimulated saliva sample. The sucrose blood medium used contains sucrose to promote capsule formation, and bacitracin in mitis-salivarius-bacitracin (MSB) agar inhibits the growth of most oral bacteria, except *S. mutans*. Oral streptococci require an elevated CO_2 atmosphere. This can be achieved with a disposable plastic bag for one or two culture plates. In this technique, the bag is coated with antifogging chemicals, and a wet sodium bicarbonate tablet is added to generate CO_2 (**FIGURE 47.1a**). A candle jar can also be used to incubate several plates in an elevated-CO_2 atmosphere (**FIGURE 47.1b**).

MATERIALS

FIRST PERIOD

Rodac plate containing MSB agar
Petri plate containing sucrose blood agar
Chewing gum
Tongue blade
50-ml beaker
Sterile 1-µl calibrated loop
CO_2 jar or CO_2 envelope in a Brewer anaerobic jar

SECOND PERIOD BSL-2

3% hydrogen peroxide
Toothpicks

TECHNIQUES REQUIRED

Catalase test (Exercise 17)
Dissecting microscopy (Appendix E)

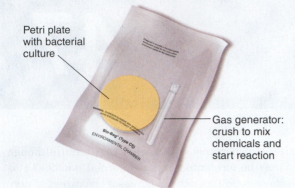

Petri plate with bacterial culture

Gas generator: crush to mix chemicals and start reaction

(a) CO₂ pouch. Your instructor will demonstrate how to crush the reagent packet to start the reaction. Place the pouch in a Brewer anaerobic jar.

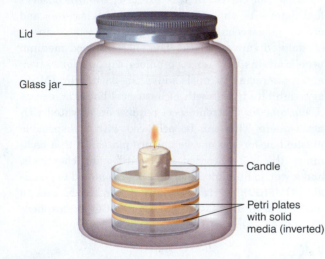

Lid

Glass jar

Candle

Petri plates with solid media (inverted)

(b) Candle jar. Light the candle and screw the lid onto the jar. Do not let the flame touch the Petri plates. The flame will produce CO_2 and will burn until the air in the jar has a lowered oxygen concentration.

FIGURE 47.1 CO_2-incubation techniques.

PROCEDURE First Period

Work only with your own saliva, and discard all contaminated materials in the To Be Autoclaved area.

Getting Ready

1. Label an MSB plate and a sucrose blood agar plate.
2. Chew a piece of chewing gum until the flavor is gone. Do not swallow your saliva.
3. Collect your saliva in a 50-ml beaker. Don't be embarrassed; everyone makes about 2 liters of saliva over the course of a day! Collect approximately 2 to 3 ml. Discard gum in the disinfectant.

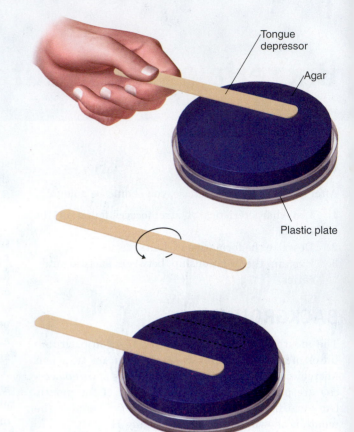

Tongue depressor

Agar

Plastic plate

FIGURE 47.2 Rodac plate inoculation. To inoculate the Rodac plate, press one side of the tongue blade on the agar. Turn the blade over, and press the other side on the remaining half of the agar.

MSB Agar

1. Rotate a sterile tongue blade in your mouth 10 times so both sides of the blade are thoroughly wetted with saliva. Remove the tongue blade through closed lips to remove excess saliva.
2. Press one side of the tongue blade onto one half of the surface of the MSB agar. Then press the other side of the blade onto the other side of the agar, as shown in **FIGURE 47.2**. Discard the tongue blade in disinfectant.
3. Use one of the CO_2-incubation techniques (shown in Figure 47.1). Seal the Petri plate with tape.
4. Incubate the plate, inverted, in the CO_2-incubation container for 72 to 96 hours at 35°C.

Sucrose Blood Agar

1. Using a 1-μl calibrated loop, streak one line of the saliva (from the 50-ml beaker) down the center of the sucrose blood agar. Then streak back and forth, perpendicular to that line (**FIGURE 47.3**). *Discard the loop in disinfectant.* What volume of saliva did you put on the agar? _____

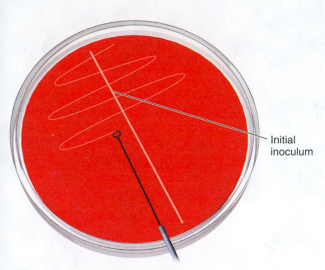

Initial inoculum

FIGURE 47.3 Sucrose blood agar inoculation. Using a calibrated loop, streak one line across the agar; then streak back and forth across that line.

2. Seal the plate with tape. Incubate the plate, inverted, in a CO_2-incubation chamber for 72 to 96 hours at 35°C (see Figure 47.1).

PROCEDURE Second Period

MSB Agar

Examine the MSB plate with a dissecting microscope (Appendix E) for the presence of *S. mutans*. *S. mutans* colonies are light blue to black, raised, and rough; their surface resembles etched glass or "burnt" sugar. Other bacteria occasionally grow on MSB; therefore, you must carefully observe colony morphology to accurately estimate the number of *S. mutans* present.

Sucrose Blood Agar

1. Examine the sucrose blood agar plate for production of large raised mucoid (gumdrop) colonies; these are likely to be *S. salivarius*. Look carefully for glistening colonies surrounded by an indentation of the agar surface that cannot be moved without tearing the agar; these are likely to be *S. sanguinis*. *S. mutans* colonies, if present, may be recognized by the presence of a drop of liquid polysaccharide on top of or surrounding the colony. Count the number of each colony type. Record the number of colony-forming units (CFUs) per milliliter of saliva.

$$CFU/ml = \frac{Number\ of\ colonies}{Amount\ plated}$$

2. **BSL-2** Determine whether catalase is produced by suspected streptococci. Using a toothpick, transfer some colonies to a slide, and perform the catalase test on the slide. Discard the toothpick in the biohazard container. The catalase test cannot be done directly on agar containing blood. Why not? _____

LABORATORY REPORT
Bacteria of the Mouth

PURPOSE _____

_____

_____

HYPOTHESIS

You will/will not culture *Streptococcus mutans* from your mouth.

RESULTS

MSB Agar

Suspected Species	Number	Description

Sucrose Blood Agar

	Colony Morphology		
	Suspected *S. salivarius*	Suspected *S. sanguinis*	Suspected *S. mutans*
Appearance of colonies			
Number of colonies			
CFU/ml			
BSL-2 Catalase production			

CONCLUSIONS

1. Do your results confirm your hypothesis? _____

2. Compare your data with those of your classmates, and draw a conclusion about your oral microbiota. _____

_____

3. Is MSB agar selective or differential? _____ Explain. _____

_____

4. Is sucrose blood agar selective or differential? _____ Explain. _____

_____

QUESTIONS

1. List three characteristics of streptococci found in the mouth. _____

_____

2. Studies have shown that both sucrose and bacteria are necessary for tooth decay. Why should this be true?

_____

_____

3. Even if a large number of *S. mutans* is in your saliva, how might you avoid tooth decay? _____

_____

_____

CRITICAL THINKING

1. *S. salivarius* forms large mucoid colonies on sucrose blood agar. Would you expect the same type of colonies on glucose blood agar? Briefly explain your answer.

2. Some dentists determine a patient's susceptibility to dental caries by measuring the pH of the saliva. What is the rationale for using this technique?

CLINICAL APPLICATION

One month after having her teeth cleaned by a dental hygienist, an 89-year-old woman with a hip replacement developed bacteremia. Gram-positive, catalase-negative cocci were isolated from her blood. The bacteria ferment inulin but not mannitol. They do not hydrolyze esculin and are V–P-negative. Use Appendix H, Figure H.4, to identify the species of this bacterium. What is the role of the artificial hip in this infection? _____

Bacteria of the Gastrointestinal Tract

OBJECTIVES

After completing this exercise, you should be able to:

1. List the bacteria commonly found in the gastrointestinal tract.
2. Define the following terms: *coliform, enteric*, and *enterococci*.
3. Interpret results from triple sugar iron (TSI) slants and phenylethyl alcohol (PEA)–blood agar.

BACKGROUND

The stomach and small intestine have relatively few microorganisms, as a result of the hydrochloric acid produced by the stomach and the rapid movement of food through the small intestine. But microbial populations in the large intestine are enormous, exceeding 10^{11} bacteria per gram of feces. Most of these intestinal organisms are commensals, and some are in mutualistic relationships with their human hosts. Some intestinal bacteria synthesize useful vitamins, such as folic acid and vitamin K. The normal intestinal microbiota prevent colonization of pathogenic species by producing antimicrobial substances and by competition.

The population of the large intestine consists primarily of anaerobes of the genera *Bacteroides, Bifidobacterium, Enterococcus*, and *Lactobacillus* and such facultative anaerobes as *Escherichia, Enterobacter, Citrobacter*, and *Proteus*.

Most diseases of the gastrointestinal system result from the ingestion of food or water that contains pathogenic microorganisms. Good sanitation practices, modern methods of sewage treatment, and the disinfection of drinking water help break the fecal–oral cycle of disease. A number of tests have been developed to identify facultative anaerobes associated with fecal contamination. Gram-negative, facultatively anaerobic rods are a large and diverse group of bacteria that includes the **enteric family** (Enterobacteriaceae).

Isolation of Enteric Bacteria

Media have been developed to differentiate lactose-fermenting enterics from non–lactose-fermenting enterics. The lactose fermenters are called **coliforms** and are generally not pathogenic. The non–lactose-

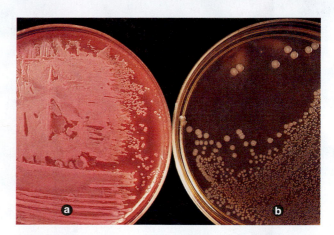

FIGURE 48.1 MacConkey agar. (a) Colonies of enteric bacteria that ferment lactose are pink to red. **(b)** Lactose-negative bacteria produce colorless or white colonies.

fermenting group includes such pathogens as *Salmonella* and *Shigella*. One of the most common media is MacConkey agar. **MacConkey agar** is selective in that bile salts are inhibitory to gram-positive organisms; thus, they allow the medium to selectively culture gram-negative organisms. MacConkey agar is differential in that lactose-fermenting organisms (coliforms) give red, opaque colonies, and non–lactose fermenters produce colorless, translucent colonies (**FIGURE 48.1**).

Identification of Enteric Bacteria

After isolation on MacConkey agar, differential screening media such as **triple sugar iron (TSI)** agar can be used to further characterize organisms. TSI contains the following:

0.1% glucose
1.0% lactose
1.0% sucrose
0.02% ferrous sulfate
Phenol red
Nutrient agar

As shown in **FIGURE 48.2**, if the organism ferments only glucose, the phenol red indicator will turn yellow in a few hours. The bacteria quickly exhaust the limited supply of glucose and start oxidizing amino acids for energy, giving off ammonia as an end-product.

(a) No sugar fermentation

Gas

Without gas production   With gas production
(b) Glucose fermented; lactose and sucrose not fermented

Gas

Without gas production   With gas production
(c) Glucose and lactose and/or sucrose fermented

(d) H_2S production can occur in addition to (a), (b), and (c).

FIGURE 48.2   TSI agar. Reactions in triple sugar iron (TSI) agar after incubation for 24 hours.

Oxidation of amino acids increases the pH, and the indicator in the slanted portion of the tube will turn back to red. The butt will remain yellow. If the organism in the TSI slant ferments lactose and/or sucrose, the butt and slant will turn yellow and remain yellow for days because of the increased level of acid production due to the adequate amount of sucrose and lactose. Gas production by an organism is indicated by the appearance of bubbles *in* the agar. TSI can also be used to ascertain whether hydrogen sulfide (H_2S) has been produced from desulfuration of sulfur-containing amino acids. H_2S reacts with ferrous sulfate in the medium, producing ferrous sulfide, a black precipitate. The key in Appendix H, Figure H.5, shows how enteric bacteria can be identified using MacConkey agar, TSI, and additional tests.

Enterococci

Enterococci are gram-positive, catalase-negative cocci. They are members of the normal intestinal microbiota of humans and other animals. The presence of enterococci can be used to indicate fecal contamination. *Enterococcus faecalis* and *E. faecium* are leading causes of hospital-acquired infections. Enterococci are the second most common cause of surgical wound infections, behind coagulase-negative staphylococci, and second to *E. coli* as the cause of healthcare-associated urinary tract infections. Enterococci can be isolated on an enriched medium (containing blood) and phenylethyl alcohol (PEA) that inhibits the growth of gram-negative bacteria. The key in Appendix H, Figure H.6, will help you identify *Enterococcus* species commonly found in the human gastrointestinal tract.

MATERIALS

FIRST PERIOD

Petri plate containing MacConkey agar
Petri plate containing PEA–blood agar
Tube containing sterile cotton swab
10-ml pipette
Sterile water
Fecal sample (animal or human feces; swab feces or the anus after a bowel movement with a sterile swab, and bring the contaminated swab to class in a sterile tube in a paper sack)

SECOND PERIOD

BSL-2 Gram stain reagents
BSL-2 TSI slant
BSL-2 Toothpicks
3% hydrogen peroxide

THIRD PERIOD   BSL-2

Fermentation tubes, as needed
BSL-1 Labs: TSI demonstration slants

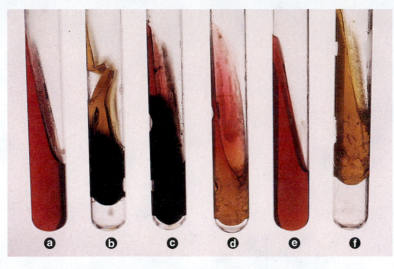

FIGURE 48.3 **Reactions in triple sugar iron (TSI) agar.** Tube **(a)** shows growth with no fermentation or hydrogen sulfide (H_2S). Tube **(b)** shows blackening due to H_2S, as well as acid and gas from fermentation of glucose and sucrose and/or lactose. Sucrose and lactose were not fermented in tube **(c)**; H_2S production masks the glucose fermentation reaction, although gas is produced. Acid and gas are produced in tube **(d)**. Tube **(e)** is uninoculated. Tube **(f)** shows acid and gas production from glucose and sucrose and/or lactose.

TECHNIQUES REQUIRED

Gram staining (Exercise 7)
Plate streaking (Exercise 11)
Fermentation tests (Exercise 14)
Catalase test (Exercise 17)
Blood agar (Exercise 46)

PROCEDURE First Period

1. If the feces are hard, break them apart in a sterile Petri dish and emulsify them in sterile water. Swab a small area on the MacConkey agar with an inoculated swab, and then streak for isolation with a loop (see Figure 45.2b on page 360).
2. Using the fecal swab, inoculate PEA–blood agar.
3. Seal the plate with tape. Incubate the plate, inverted, at 35°C for 24 to 48 hours.

Discard the test tube, tube of water, and other contaminated materials in the To Be Autoclaved area.

PROCEDURE Second Period

1. Observe your MacConkey plate and those of other students. Record the appearance and colors of the colonies (see Figure 48.1).
   BSL-2 Prepare a Gram stain from an isolated colony.

2. **BSL-2** Select one isolated colony using an inoculating needle, and inoculate the TSI slant. *Stab* into the butt of the agar, and then streak the surface of the slant. Incubate the tube at 35°C for 24 hours.
3. Observe the PEA–blood agar for the presence of hemolysis.
   BSL-2 Perform a Gram stain on isolated colonies. Transfer part of a colony to a slide for Gram staining. Transfer the remainder of the colony to another slide with a toothpick, and perform a catalase test. Discard the toothpick in the biohazard container. Why can't the catalase test be done on blood agar? _____ Is *Enterococcus* catalase-positive or catalase-negative? _____

PROCEDURE Third Period

1. **BSL-2** Observe and record results from the TSI slant inoculated (**FIGURE 48.3**). Observe other students' TSI slants.
   BSL-1 Labs: Observe the TSI demonstrations. Record the results (**FIGURE 48.3**).
2. Using the key in Appendix H, Figure H.5, identify your isolate. Inoculate the fermentation tubes. Incubate tubes at 35°C for 24 to 48 hours.

Name: _____ Date: _____ Lab Section: _____

LABORATORY REPORT
Bacteria of the Gastrointestinal Tract

PURPOSE _____

_____

_____

HYPOTHESIS

You will culture _____ on MacConkey agar and _____ on PEA–blood agar.

RESULTS

MacConkey Agar

Source of inoculum: _____

	Colony			
	1	**2**	**3**	**4**
Appearance of colonies				
Lactose fermentation				

TSI

Which colony was used? _____

	Appearance of Agar		
	Slant	**Butt**	**Results**
Acid from glucose			
Acid from lactose/sucrose			
Gas produced			
H_2S produced			

First PEA–Blood Agar Plate

	Colony		
	1	2	3
Colony description			
Hemolysis			

BSL-2 Which colony was isolated? _____

Catalase reaction: _____

Gram stain

Reaction: _____ Morphology: _____ Arrangement: _____

Additional biochemical tests

_____

_____

_____

CONCLUSIONS

1. Gram- _____ bacteria grew on MacConkey agar, and gram- _____ bacteria grew

   on PEA–blood agar. Do these data agree with your hypothesis? _____

2. What genus do you think you have in your TSI slant or the demonstration TSI slants? _____

   What additional information do you need to identify it? _____

   _____

3. Did you culture *Enterococcus* on the PEA–blood agar? _____

   How do you know? _____

4. What enterococci did you identify? _____

QUESTIONS

1. What tests, if any, would you need to perform to identify bacteria producing a red colony on MacConkey agar?

    _____

    _____

    _____

2. Differentiate among a coliform, an enteric that is not a coliform, and an *Enterococcus*. Give an example of an

    organism in each group. _____

    _____

    _____

3. Is MacConkey agar a selective medium? _____ A differential medium? _____

    How can you tell? _____

4. Is PEA–blood agar a selective medium? _____ A differential medium? _____

    How can you tell? _____

CRITICAL THINKING

1. How would you determine whether a colorless colony on MacConkey agar is *Salmonella* or *Shigella*? Why would you want to identify a colorless colony?

2. Tests to determine the presence of *Bacteroides* in clams and oysters are being developed. If these bacteria do not present a health hazard, why are these tests being developed?

CLINICAL APPLICATION

You are called to investigate an outbreak of diarrheal disease in a child care center. The symptoms include vomiting, fever, nausea, and cramps. You culture fecal samples from children and find a gram-negative, lactose-negative rod. The bacterium does not produce gas from glucose and makes colorless colonies on MacConkey agar. Use Appendix H

to identify the genus of this bacterium. _____

Bacteria of the Genitourinary Tract

OBJECTIVES

After completing this exercise, you should be able to:

1. List bacteria found in urine from a healthy individual.
2. Identify, through biochemical testing, bacteria commonly associated with urinary tract infections.
3. Determine the presence or absence of *Neisseria gonorrhoeae* in a gonococci (GC) smear.

BACKGROUND

The urinary and genital systems are closely related anatomically, and some diseases that affect one system also affect the other system, especially in the female. The upper urinary tract and urinary bladder are usually sterile. The urethra does contain resident bacteria, including *Streptococcus*, *Bacteroides*, *Mycobacterium*, *Neisseria*, and enterics. Most bacteria in urine are the result of contamination by skin microbiota during passage. The presence of bacteria in urine is not considered an indication of urinary tract infection unless there are more than 10^5 CFU per milliliter or 100 coliforms or more per milliliter of urine.

Many infections of the urinary tract, such as cystitis (inflammation of the urinary bladder) and pyelonephritis (inflammation of the kidney), are caused by opportunistic pathogens and are related to fecal contamination of the urethra and to medical procedures, such as catheterization.

Examination of Urine

Standard examination of urine consists of a plate count on blood agar for the total number of organisms, coupled with a streak plate of undiluted urine on MacConkey agar. Why? _____

_____

The patient is given a sterile container and instructed to collect a midstream sample, which is obtained by voiding a small volume from the bladder before collection. This washes away skin microbiota.

In the first part of this exercise, you will examine normal urine. In the second part, in BSL-2 labs, three gram-negative rods that commonly cause cystitis will be provided to you. *E. coli* and *Proteus* are enterics.

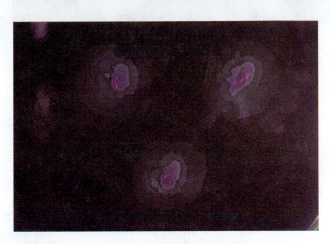

FIGURE 49.1 Cells of *Proteus mirabilis*. These bacteria are highly motile. *P. mirabilis* colonies exhibit swarming over EMB agar as young cells move away from the colony's center.

E. coli is one of the coliforms. What is a coliform? _____

_____

Proteus is actively motile and exhibits "swarming" on solid media (**FIGURE 49.1**), where the cells at the periphery move away from the main colony. *Pseudomonas* is a gram-negative aerobic rod. *Pseudomonas aeruginosa* is commonly found in the soil and other environments. Under the right conditions, particularly in weakened hosts, this organism can cause urinary tract infections, burn and wound infections, and abscesses. *P. aeruginosa* infections are characterized by blue-green pus. This bacterium produces an extracellular, water-soluble pigment called **pyocyanin** ("blue pus") that diffuses into its growth medium (**FIGURE 49.2**).

Sexually Transmitted Microbes

Most infections of the genital system are transmitted by sexual activity and are therefore called **sexually transmitted infections (STIs).** Most of the bacterial infections can be readily cured with antibiotics if treated early and can largely be prevented by the use of condoms. Nevertheless, STIs are a major U.S. public health problem.

The most common reportable communicable disease in the United States is chlamydial urethritis, an STI caused by *Chlamydia trachomatis*. Most testing for *Chlamydia* involves amplification of DNA in a urine

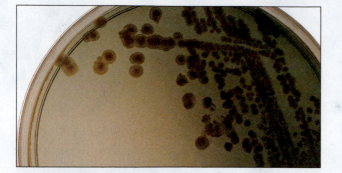

(a) Under white light

(b) Under UV light

FIGURE 49.2 *Pseudomonas aeruginosa.* *P. aeruginosa* produces a water-soluble blue pigment on Pseudomonas agar P in **(a)**, and the pigment fluoresces under ultraviolet light in **(b)**.

sample for DNA sequences specific for *C. trachomatis.* The second most often reported STI is gonorrhea, an STI caused by the gram-negative diplococcus *Neisseria gonorrhoeae*, also called gonococcus or GC. Gonorrhea is diagnosed by identifying the organism in the pus-filled discharges of patients, as demonstrated with GC smears in the third part of this exercise. Cultures from patients' discharges can be made on Thayer-Martin medium and incubated in a 3–10% CO_2 environment (see Figure 47.1 on page 374). An oxidase test is performed on characteristic colonies for confirmation.

Materials Urine Culture

First Period

Sterile widemouthed jar or plastic specimen cup, approximately 50 to 250 ml, and paper sack
Petri plate containing blood agar
Petri plate containing MacConkey agar
Sterile 10-µl calibrated loops (2)

Materials Cystitis

First Period BSL-2
Petri plate containing MacConkey agar
Petri plate containing Pseudomonas agar P

Second Period BSL-2
Tubes containing OF-glucose medium (4)
Urea agar slants (2)
Mineral oil
Oxidase reagent
Gram-stain reagents

Cultures BSL-2

Tube containing *Escherichia coli*, *Proteus vulgaris*, and *Pseudomonas aeruginosa*

Gc Smears

GC smears
One unknown smear # _____

Techniques Required

Gram staining (Exercise 7)
Plate streaking (Exercise 11)
OF test (Exercise 13)
Urea hydrolysis (Exercise 15)
Oxidase test (Exercise 17)
MacConkey agar (Exercise 48)

PROCEDURE Urine Culture

First Period

Work only with urine collected from your own body.

1. Collect a "clean-catch" urine specimen, as described by your instructor, using the sterile jars available in the lab. Place the jar in a paper sack, and refrigerate the specimen until you are ready to perform step 2. Why? _____
    _____

2. Gently shake the urine to suspend the bacteria. Using a 10-µl calibrated loop, streak one line down the center of the blood agar. Then streak back and forth, perpendicular to that line (see Figure 47.3 on page 375). *Discard the loop in disinfectant.* What volume of urine did you put on the agar? _____

3. Using another 10-µl loop, repeat step 2 to inoculate the MacConkey agar. *Discard the loop in disinfectant.*

Discard the jar and loops in the appropriate biohazard container.

4. Seal with tape, and incubate both plates, inverted, at 35°C for 24 to 48 hours.

Second Period

1. Count the colonies on the blood agar plate, and determine the number of colony-forming units (CFUs) per milliliter of urine:

$$\text{CFU/ml} = \frac{\text{Number of colonies}}{\text{Amount plated}}$$

2. Repeat step 1 to determine the CFU/ml on the MacConkey agar plate. Examine the MacConkey plate for the presence of possible coliforms. Consult your instructor if you are alarmed by your results.

PROCEDURE   Cystitis   BSL-2

First Period

1. Inoculate a MacConkey plate and a Pseudomonas agar P plate with the mixed culture of bacteria.
2. Incubate the plates, inverted, at 35°C for 24 to 48 hours.

Second Period

1. Examine the MacConkey plate for lactose-fermenting colonies. Which of the three organisms ferments lactose? _____

   Look for swarming (see Figure 49.1). Which organism is actively motile on solid media? _____

   Are any small, non–lactose-fermenting colonies present? _____

2. Prepare a Gram stain from each different colony type.
3. Examine the Pseudomonas agar P plate (see Figure 49.2). Can you identify *Pseudomonas*? _____
4. Perform an oxidase test on each different organism.
5. Inoculate four tubes of OF-glucose medium: two with *Proteus* and two with *Pseudomonas*. Plug one tube of each organism with mineral oil. Inoculate a urea agar slant with each organism. Incubate the tubes at 35°C for 24 to 48 hours.

Third Period

Record the results of the OF-glucose and urease production tests. Why isn't it necessary to perform these two biochemical tests on *E. coli*? _____

_____

PROCEDURE   GC Smears

1. Examine the GC smears provided.
2. Determine whether *N. gonorrhoeae* could be present in your unknown GC smear.

LABORATORY REPORT
Bacteria of the Genitourinary Tract

PURPOSE _____

_____

_____

HYPOTHESIS

Your urine plate count will indicate that a urinary tract infection is/is not present.

RESULTS

Urine Culture

	Blood Agar	MacConkey Agar
Number of colonies		
Total count (CFU/ml)		
Hemolysis		
Possible coliforms present		

Cystitis BSL-2

	Organism		
	E. coli	P. vulgaris	P. aeruginosa
MacConkey agar			
Appearance of colonies	_____	_____	_____
Lactose fermentation	_____	_____	_____
Swarming	_____	_____	_____
Gram stain	_____	_____	_____
Pseudomonas agar P			
Appearance of colonies	_____	_____	_____
Pigmentation	_____	_____	_____
Oxidase reaction	_____	_____	_____
OF-glucose (fermentative or oxidative)	_____	_____	_____
Urease production	_____	_____	_____

GC Smears

Sketch the appearance of each of the slides.

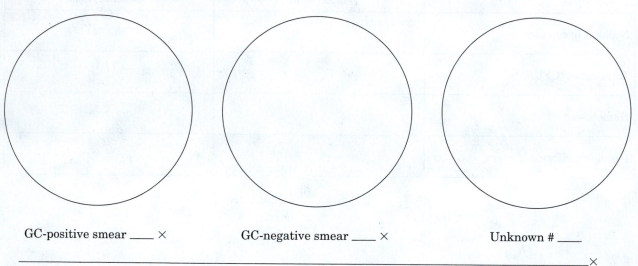

GC-positive smear ____ ×   GC-negative smear ____ ×   Unknown # ____

_____ ×

CONCLUSIONS

1. With the results obtained on the blood agar and MacConkey agar plates, is a urinary tract disease possible?

_____

2. Were there any coliforms in your urine sample? _____ How can you tell? _____

_____

3. Could *N. gonorrhoeae* be present in the unknown smear? _____

QUESTIONS

1. Why is MacConkey agar inoculated with a urine specimen? _____

   _____

   _____

2. The enterics and pseudomonads look alike microscopically. How can you easily distinguish between these two

   groups of bacteria? _____

   _____

   _____

3. Why are females more prone than males to urinary tract infections?_____

   _____

   _____

CRITICAL THINKING

1. Differentiate the pigment of *P. aeruginosa* on Pseudomonas agar P from the "pigment" of *E. coli* on MacConkey agar without referring to the colors.

2. What role does antibiotic treatment play in yeast infections of the urinary tract?

CLINICAL APPLICATION

An otherwise healthy 22-year-old woman was seen at a hospital emergency room because of frequent and painful urination. A urine culture reveals a gram-negative, lactose-positive rod that produces indole but no H_2S. Use

Appendix H to identify the genus of this bacterium. _____

Identification of an Unknown from a Clinical Sample

Science is built up of FACTS, as a house is built of STONES; but an accumulation of facts is no more a SCIENCE than a heap of stones is a house. – JULES-HENRI POINCARE

OBJECTIVES

After completing this exercise, you should be able to:

1. Separate and identify two bacteria from a sample.
2. Determine the sensitivity of the unknown bacteria to chemotherapeutic agents.

BACKGROUND

Clinical samples often contain many microorganisms, and a pathogen must be separated from resident normal microbiota. Differential and selective media are used to facilitate isolation of the pathogen, after which it must be identified. Also, antimicrobial sensitivity tests are performed to aid the physician in prescribing treatment. Effective treatment depends on identifying the pathogen; therefore, a clinical laboratory must provide results as quickly and accurately as possible. Microbiology laboratory results are usually available within 48 hours.

Identification

In this exercise, you will be provided with a simulated clinical sample containing two bacteria. The procedure will differ from that used in a clinical laboratory in that you will be asked to isolate and identify *both* organisms in the sample. In a clinical laboratory, selective and enrichment techniques are used to identify the pathogen only. After you have separated the two bacteria in pure culture, prepare stock and working cultures. To obtain pure cultures, inoculate a trypticase soy agar plate *and* a differential medium. Select the differential medium using the type of sample as a clue. For instance, a fecal sample should probably be inoculated onto _____ agar.

Considering that differential media should give the information needed to isolate the organisms, why is a trypticase soy agar inoculated also? _____

_____

MATERIALS

FIRST PERIOD

Petri plates containing trypticase soy agar
Gram stain reagents

SECOND PERIOD

Petri plate containing trypticase soy agar
All stains, reagents, and media previously used
Trypticase soy agar slants

THIRD PERIOD

All stains, reagents, and media previously used
Mueller-Hinton agar
Antimicrobial disks

CULTURE

Unknown sample # _____

TECHNIQUES REQUIRED

Gram staining (Exercise 7)
Plate streaking (Exercise 11)
Differential and selective media (Exercises 12–17)
Kirby-Bauer technique (Exercise 25)
Use of *Bergey's Manual* (Exercise 18)

PROCEDURE First Period

1. Review general information regarding identification of an unknown. (Study the identification schemes in Appendix H, Figure H.1 through H.6.) Try to identify the bacterial species in **FIGURE 50.1**.
2. You may be able to prepare a Gram stain directly from the unknown if it does not contain a high concentration of organic matter to interfere with the stain results.
3. Streak the unknown onto a trypticase soy agar plate and an appropriate differential medium. *Do not be wasteful.* Inoculate only the necessary media. Incubate the plate at 35°C for 24 to 48 hours.

PROCEDURE Second Period

1. Examine the plates, and select colonies that differ from each other in appearance and that are separated for easy isolation (**FIGURE 50.1A**).

2. Streak each organism for isolation onto half of a trypticase soy agar plate and an appropriate differential medium. *Do not be wasteful.* Inoculate only the necessary media. Incubate the plate at 35°C for 24 to 48 hours. The trypticase soy agar plate can be your first working culture. Prepare a stock culture of each organism on separate trypticase soy agar slants (one culture on each slant).

PROCEDURE Third Period

1. Prepare a Gram stain of each organism. What should you do if you think you do not have a pure culture? _____
_____

2. After determining the staining and morphologic characteristics of the two organisms, decide which biochemical tests you will need. Plan your work carefully. The same tests need not be performed on both bacteria.

3. Record which tests were performed.

PROCEDURE Next Period

1. Record the results of your biochemical tests.

2. Select six antimicrobials from Table 25.1 to test against your unknowns. Use the Mueller-Hinton agar and antimicrobial disks to determine whether your unknowns are resistant to any antimicrobial agents (see Figure 25.1 on page 200).

(a) Colonies after incubation at 25°C

FIGURE 50.1 Unknown identification. This is an example of how to identify an unknown. The uninoculated control tubes are on the left in each photograph. Use these results and Appendix H to identify the bacterial species shown in the photographs.

(b) Glucose fermentation

(d) Peptone iron agar

(c) Lactose fermentation

(e) MIO. Kovacs reagent was added to the test (right).

(f) Urea agar

LABORATORY REPORT
Identification of an Unknown
from a Clinical Sample

PURPOSE _____

_____

_____

RESULTS

Unknown # _____

Source: _____

Preliminary Gram stain results: _____

	Bacterium	
	A	B
Appearance of colonies on trypticase soy agar		
Gram reaction		
Cell morphology and arrangement		

On a separate piece of paper, make a flowchart of your procedure. Remember, the *same* tests may *not* be required for each bacterium. The following will help you get started:

TSA plate

Bacterium A   Bacterium B

Gram reaction: _____   _____

Medium: _____   _____

	Zone of Inhibition (mm)	
	Bacterium A	Bacterium B
Antimicrobial sensitivity to		
_____	_____	_____
_____	_____	_____
_____	_____	_____
_____	_____	_____
_____	_____	_____
_____	_____	_____

CONCLUSIONS

1. What two bacterial species were in unknown # _____ ? _____

    _____

2. Is either of your unknowns resistant to any antimicrobials? _____

    If so, which antimicrobials? _____

    _____

QUESTIONS

1. What organism is shown in Figure 50.1? _____

2. Why would a clinical microbiology laboratory perform antimicrobial susceptibility tests as well as identify the

    unknown? _____

    _____

3. Explain your rationale for arriving at the identities of your unknowns.

Rapid Identification Methods

OBJECTIVES

After completing this exercise, you should be able to:

1. Evaluate three methods of identifying enterics.
2. Name three advantages of the "systems approach" over conventional tube methods.

BACKGROUND

The microbiology laboratory must identify bacteria quickly and accurately. Using a series of standardized tests increases accuracy. The IMViC tests were developed as a means of separating members of the Enterobacteriaceae (enterics),* particularly the coliforms, to determine whether water was contaminated with sewage. The IMViC uses a standard combination of four tests, where each capital letter in **IMViC** represents a test; the *i* is added for easier pronunciation. The tests are as follows:

I for indole production from tryptophan
M for methyl red test for acid production from glucose
V for the Voges–Proskauer test for production of acetoin from glucose
C for the use of citrate as the sole carbon source

Although variation among strains does exist, IMViC reactions for selected species of the Enterobacteriaceae are given in **TABLE 51.1**.

Rapid identification methods have been developed that provide a large number of results from one inoculation. Examples are EnteroPluri-Test and API 20E for identifying oxidase-negative, gram-negative bacteria belonging to the family Enterobacteriaceae. EnteroPluri-Test (**FIGURE 51.1**) is divided into 12 compartments, each containing a different substrate in agar. A similar tube called Oxi/Ferm is used to identify oxidase-positive, gram-negative rods. API 20E

---

*Enterobacteriaceae are aerobic or facultatively anaerobic, gram-negative, non–endospore-forming, glucose-fermenting, rod-shaped bacteria. Coliforms are Enterobacteriaceae that ferment lactose with acid and gas formation within 48 hours at 35°C.

TABLE 51.1 IMViC REACTIONS FOR SELECTED SPECIES OF THE ENTEROBACTERIACEAE

Species	Indole	Methyl Red	Voges–Proskauer	Citrate
Escherichia coli	+(v)	+	–	–
Citrobacter freundii	–	+	–	+
Enterobacter aerogenes	–	–	+	+
Enterobacter cloacae	–	–	+	+
Serratia marcescens	–	+ or –*	+	+
Proteus vulgaris	+	+	–	–(v)
Proteus mirabilis	–	+	– or +†	+(v)

v = variable
*Majority of strains give + results.
†Majority of strains give – results.

(**FIGURE 51.2**) consists of 20 microtubes containing dehydrated substrates. The substrates are rehydrated by adding a bacterial suspension. No culturing beyond the initial isolation is necessary with these systems. Comparisons between these rapid identification methods and conventional culture methods show that they are as accurate as conventional test-tube methods.

Computerized analysis of test results increases accuracy because each test is given a point value. Tests that are more important than others get more points. IMViC uses only four tests of equal value.

As commercial identification systems are developed, they can provide greater standardization in identification because they overcome the limitations of hunting through a key, preparing media, and evaluating tests within a laboratory or between different laboratories. They also save time, money, and labor.

FIGURE 51.1 **EnteroPluri-Test.** Fifteen biochemical tests are performed in an EnteroPluri-Test. The top tube is uninoculated. The bottom tube is inoculated with *Citrobacter freundii*.

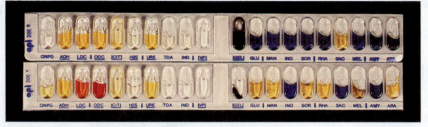

FIGURE 51.2 **API 20E strip.** Twenty biochemical tests are performed in an API 20E strip. The top strip is inoculated with *Proteus vulgaris*, and the bottom strip is inoculated with *Escherichia coli*.

MATERIALS

FIRST PERIOD

Petri plate containing nutrient agar

SECOND PERIOD

Oxidase reagent
IMViC tests and reagents
EnteroPluri-Test and reagents
API 20E tray and reagents
    5.0 ml sterile saline
    5-ml pipette
    Mineral oil
    Sterile Pasteur pipette

CULTURE

Unknown enteric # _____

TECHNIQUES REQUIRED

Inoculating loop and needle techniques (Exercise 4)
Aseptic technique (Exercise 4)
Plate streaking (Exercise 11)
MRVP tests (Exercise 14)
Fermentation tests (Exercise 14)
Protein catabolism (Exercises 15 and 16)
Respiration (Exercise 17)

PROCEDURE First Period

Isolation

Streak the nutrient agar plate with your unknown for isolation and to determine purity of the culture. Incubate the plate, inverted, at 35°C for 24 to 48 hours.

PROCEDURE Second Period

IMViC Tests

1. Inoculate tubes of tryptone broth (indole test), MRVP broths, and Simmons citrate agar with your unknown.
2. Incubate the tubes at 35°C for 48 hours or longer.

Isolation

1. Record the appearance of the colonies.
2. Determine the oxidase reaction of one of the colonies on the plate. Why? _____

   _____

   How will you determine the oxidase reaction? _____

   _____

   _____

IMViC Tests

Perform the appropriate tests. Add 4 to 5 drops of Kovacs reagent to the tryptone broth to test for indole. Mix the tube gently. A cherry red color indicates a positive test. (See Figure 16.4 on page 131.) Record your results.

EnteroPluri-Test (Figure 51.3)

1. Remove both caps from the EnteroPluri-Test. The straight end of the wire is used to pick up the inoculum; the bent end is the handle. Holding the EnteroPluri-Test, pick a well-isolated colony with the inoculating end of the wire (**FIGURE 51.3a**). Avoid touching the agar with the needle.

2. Inoculate the EnteroPluri-Test by holding the bent end of the wire and twisting; the tip of the wire should be visible in the citrate compartment. Withdraw the needle through all 12 compartments using a turning motion (**FIGURE 51.3b**).

3. Reinsert the needle into the EnteroPluri-Test, using a turning motion, through all 12 compartments until the notch on the wire is aligned with the opening of the tube. Break the wire at the notch by bending it (**FIGURE 51.3c**). The portion of the needle remaining in the tube maintains anaerobic conditions necessary for fermentation, production of gas, and decarboxylation.

4. Punch holes with the broken-off wire through the foil covering the air inlets of the last eight compartments (adonitol through citrate) to provide aerobic conditions (**FIGURE 51.3d**). Replace the caps on both ends of the tube.

 Discard the wire in disinfectant.

5. Incubate the tube lying on its flat surface at 35°C for 24 hours.

API 20E (Figure 51.4)

1. Prepare a bacterial suspension by touching the center of a well-isolated colony with a sterile loop and thoroughly mixing the inoculum in 5 ml of sterile saline (**FIGURE 51.4a** on page 405).

2. Place 5 ml of tap water into the corrugated incubation tray to provide a humid atmosphere during incubation.

3. Using a sterile Pasteur pipette, tilt the API 20E tray and fill the tube section of the microtubes with the bacterial suspension. Fill the tube *and* cupule sections of the CIT, VP, and GEL tubes (**FIGURE 51.4b**). Place the tip of the pipette against the side wall of the cupule and carefully inoculate without introducing bubbles into the cupule.

4. After inoculation, completely fill the cupule section of the ADH, LDC, ODC, H_2S, and URE tubes with mineral oil to create anaerobic conditions (**FIGURE 51.4c**).

5. Place the plastic lid on the tray and incubate the strip for 24 hours at 35°C (**FIGURE 51.4d**). If the

(a) Pick a well-isolated colony with the inoculating end of the wire.

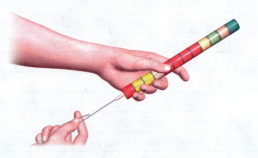

(b) Hold the bent end of the wire, and withdraw the needle through all 12 compartments with a turning motion.

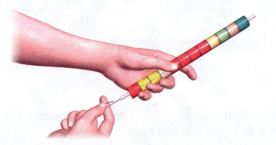

(c) Reinsert the wire through all 12 compartments. Then withdraw it to the notch on the wire. Break the wire at the notch.

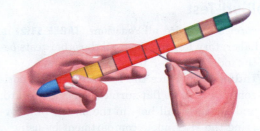

(d) Using the broken wire, punch holes through the foil covering the air inlets in the last eight compartments. Replace the caps loosely.

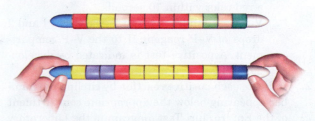

(e) After incubation, compare the tube to an uninoculated one to record results.

FIGURE 51.3 Inoculating an EnteroPluri-Test.

TABLE 51.2   ENTEROPLURI-TEST BIOCHEMICAL REACTIONS

Test	Comments	Indicator Changed	
		From	To
GLU	Acid from glucose	Red	Yellow
GAS	Gas produced from fermentation of glucose trapped in this compartment, causing separation of the wax		
LYS	Lysine decarboxylase	Yellow	Purple
ORN	Ornithine decarboxylase	Yellow	Purple
H$_2$S	Ferrous ion reacts with sulfide ions, forming a black precipitate		
IND	Kovacs reagent is added to the H$_2$S/IND compartment to detect indole	Beige	Red
ADON	Adonitol fermentation	Red	Yellow
LAC	Lactose fermentation	Red	Yellow
ARAB	Arabinose fermentation	Red	Yellow
SORB	Sorbitol fermentation	Red	Yellow
V–P	Voges–Proskauer reagents detect acetoin	Beige	Red
DUL	Dulcitol fermentation	Green	Yellow
PA	Phenylpyruvic acid released from phenylalanine after its deamination combines with iron salts to form a dark brown precipitate		
UREA	Ammonia changes the pH of the medium	Yellow	Pink
CIT	Citric acid used as a carbon source	Green	Blue

Source: Becton, Dickinson and Company, Sparks, MD 21152.

strip cannot be read after 24 hours, remove the strip from the incubator and refrigerate it.

PROCEDURE   Third Period

EnteroPluri-Test

1. Interpret and record all reactions (**TABLE 51.2**) in the Laboratory Report. Read all the other tests *before* the indole and V–P tests, which follow.
   a. **Indole test.** Place the EnteroPluri-Test horizontally, with its flat surface pointing upward, and melt a small hole in the plastic film covering the H$_2$S/indole compartment by using a warm inoculating loop. Add 3 to 4 drops of Kovacs reagent, and allow the reagent to contact the agar surface. A positive test is indicated by a red color within 10 seconds.
   b. **V–P test.** Add 3 drops of V–P reagent I and 2 drops of V–P reagent II to the V–P compartment. A positive test is indicated by development of a red color within 20 minutes.
2. Indicate each positive reaction by circling the number appearing below the appropriate compartment of the EnteroPluri-Test diagram in the Laboratory Report. Add the circled numbers only within each bracketed section, and enter this sum in the space provided below the arrow. Read the five numbers

in these spaces across as a five-digit number in the EnteroPluri-Test Codebook.*
3. Dispose of the EnteroPluri-Test by placing it in the autoclave bag.

API 20E

1. Interpret and record all reactions (**TABLE 51.3**) in the Laboratory Report. Read all the other tests before the TDA, VP, and IND tests, which follow.
   a. **TDA test.** Add 1 drop of 10% ferric chloride. A positive test is brownish-red. Indole-positive organisms may produce an orange color; this is a negative TDA reaction.
   b. **V–P test.** Add 1 drop of V–P reagent I and 1 drop of V–P reagent II. A positive reaction produces a red color (not pale pink) after 10 minutes.
   c. **Indole test.** Add 1 drop of Kovacs reagent. A red ring after 2 minutes indicates a positive reaction.
   d. **Nitrate reduction.** Before adding reagents, look for bubbles in the GLU tube. Bubbles indicate reduction of nitrate to N$_2$. Add 2 drops of nitrate reagent A and 2 drops of nitrate

*BD Bioscience, www.bd.com.

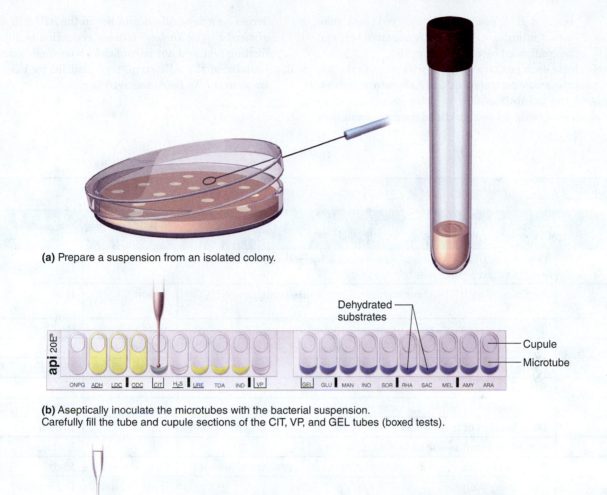

(a) Prepare a suspension from an isolated colony.

(b) Aseptically inoculate the microtubes with the bacterial suspension.
Carefully fill the tube and cupule sections of the CIT, VP, and GEL tubes (boxed tests).

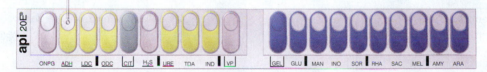

(c) Add mineral oil to the ADH, LDC, ODC, H₂S, and URE cupules (underlined tests).

(d) Incubate the strip in its plastic tray.

FIGURE 51.4 Inoculating the API 20E system.

reagent B. A positive reaction (red) may take 2 to 3 minutes to develop. A negative test can be confirmed by adding zinc dust.

2. Indicate each positive reaction with a (+) in the appropriate compartment of the Laboratory Report. Add the points for each positive reaction within each bold-outlined section. Read the seven numbers across as a seven-digit number in the *API 20E Analytical Profile Index*.* Nitrate reduction is a confirming test and not part of the seven-digit code.

3. Dispose of the API strip, tray, and lid by placing them in the To Be Autoclaved bag.

TABLE 51.3   API 20E BIOCHEMICAL REACTIONS

Test	Comments	Indicator Changed	
		From	To
ONPG	O-nitrophenyl-β-D-galactopyranoside is hydrolyzed by the enzyme that hydrolyzes lactose	Colorless	Yellow
ADH	Arginine dihydrolase transforms arginine into ornithine, NH_3, and CO_2	Yellow	Red
LDC	Decarboxylation of lysine liberates cadaverine	Yellow	Red
ODC	Decarboxylation of ornithine produces putrescine	Yellow	Red
CIT	Citric acid used as sole carbon source	Light green	Dark blue
H2S	Blackening indicates reduction of thiosulfate to H_2S		
URE	Urea is hydrolyzed by the enzyme urease to NH_3 and CO_2	Yellow	Red
TDA	Deamination of tryptophan produces indole, NH_3, and pyruvic acid	Yellow	Brown
IND	Kovacs reagent is added to detect indole	Yellow	Red ring
VP	Addition of KOH and a-naphthol detects the presence of acetoin	Colorless	Red
GEL	Gelatin hydrolysis	No diffusion	Diffusion of pigment
GLU MAN INO SOR RHA SAC MEL AMY ARA	Fermentation of glucose Fermentation of mannitol Fermentation of inositol Fermentation of sorbitol Fermentation of rhamnose Fermentation of sucrose Fermentation of melibiose Fermentation of amygdalin Fermentation of arabinose	Blue or green	Yellow or yellow-green
NO_2, N_2 gas	Nitrate reduction Red after addition of reagents indicates nitrite. Bubbles in the GLU indicate N_2.		

Source: bioMérieux, www.biomerieux.com.

LABORATORY REPORT
Rapid Identification Methods

PURPOSE _____

_____

_____

RESULTS

Unknown # _____

Appearance on nutrient agar: _____

Oxidase reaction: _____

IMViC

Indicate positive (+) and negative (−) results for each test.

Indole: _____

Methyl red: _____

V–P: _____

Citrate: _____

EnteroPluri-Test

Circle the number corresponding to each positive reaction below the appropriate compartment. Then determine the five-digit code.

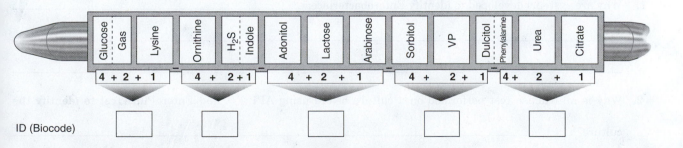

ID (Biocode)

API 20E

Indicate positive (+) and negative (−) results in the Results line. Then determine the seven-digit code.

	ONPG 1	ADH 2	LDC 4	ODC 1	CIT 2	H₂S 4	URE 1	TDA 2	IND 4	VP 1	GEL 2	GLU 4

Results

Profile number

	MAN 1	INO 2	SOR 4	RHA 1	SAC 2	MEL 4	AMY 1	ARA 2	Oxi-dase 4	NO₂	N₂ gas

Results

Profile number

CONCLUSIONS

1. What species was identified in unknown # _____ by the IMViC tests? _____

    By the EnteroPluri-Test? _____

    By the API 20E? _____

2. Did all methods you used agree? _____ If not, explain any discrepancies. _____

    _____

    _____

3. Which method did you prefer? _____ Why? _____

    _____

    _____

QUESTIONS

1. Why are systems developed to identify Enterobacteriaceae? _____

    _____

    _____

2. Why is an oxidase test performed on a culture before using API 20E and EnteroPluri-Test to identify the

    culture? _____

    _____

CRITICAL THINKING

1. Why should the first digit in the five-digit EnteroPluri-Test ID value be equal to or greater than 4, and the fourth digit in the API 20E profile number be equal to or greater than 4?

2. Why can one species have two or more numbers in a rapid identification system? (For example, *E. coli* is EnteroPluri-Test numbers 51210 and 51240; *Citrobacter braakii* is API 20E numbers 3504552 and 3504553.)

3. Use Table 51.1 to give an example of a limitation of the IMViC tests. Note that not all the enterics are listed in this table.

EXERCISES

Microorganisms are omnipresent in our environment; however, the presence of some microorganisms in certain places, such as food or water, may be undesirable. In Exercises 52 and 53, we will use the presence of nonpathogenic bacteria to indicate water pollution. The presence and number of bacteria in foods can indicate that foods were contaminated during processing; these bacteria may result in food spoilage (Exercise 54).

Harmful microorganisms are only a small fraction of the total microbial population. The activities of most microbes are in fact beneficial. In Exercise 55, we will use selected microorganisms to produce desired flavors in foods and to preserve foods.

The activities of soil bacteria are essential to maintaining life on Earth. Nitrogen fixation (Exercise 56) was first described in 1893 by Sergei Winogradsky, who entered into the study of biogeochemical cycles because he was

> ... impressed by the incomparable glitter of Pasteur's discoveries. I started to investigate the great problem of fixation of atmospheric nitrogen. I succeeded without too much difficulty in isolating an anaerobic rod called Clostridium that would perform this function.*

*Quoted in H. A. Lechevalier and M. Solotorovsky. *Three Centuries of Microbiology*. New York: Dover Publications, 1974, p. 263.

Winogradsky named the organism *Clostridium pasteurianum* in honor of Pasteur. In addition to fixing nitrogen and recycling other elements, some microbes are able to degrade dangerous pollutants in the environment (Exercise 57).

CASE STUDY: *Exxon Valdez* Oil Spill

In 1989, the oil tanker *Exxon Valdez* slammed into the Bligh Reef in Alaska's Prince William Sound, spilling more than 11 million gallons of crude oil, which spread over 1300 miles of pristine shoreline. The oil slick killed thousands of sea birds, otters, and *Orca* whales, along with billons of fish eggs. In an attempt to clean up some of the spilled oil as quickly as possible, high-pressure hot water and steam were used to clean the oil-saturated rocks and sand. Biodegradation of oil by microbes is a slow process and may not occur rapidly enough to prevent ecological damage. In some areas, this slow biodegradation was enhanced by biostimulation via the addition of garden fertilizer to increase the rate of microbial metabolism of the oil. Today, over 25 years later, some species of sea otters and ducks have recovered, but some *Orca* whale pods appear to be on their way to being extirpated (locally extinct). Oil is still trapped between buried rocks and sand. The continuing effects of the chronic exposure of birds, otters, and other wildlife to the remaining oil are unknown. Ironically, areas that were not treated with hot water and steam have much lower amounts of oil remaining.

Questions

1. Why does the shoreline that was cleaned with hot water and steam have the highest level of oil remaining today?
2. Why did addition of fertilizer increase the rate of oil degradation?
3. Bioaugmentation, wherein microbes are added to the oil spill, was not shown to be beneficial in the Alaska oil spill. Why?
4. The presence of oil-degrading microbes in an environment can be determined by adding the indigenous microbes to a Petri plate of minimal salts containing opaque emulsified oil. If degradation of the oil has occurred, what result will be observed?

Microbes in Water: Multiple-Tube Technique

OBJECTIVES

After completing this exercise, you should be able to:

1. Define the term *coliform*.
2. Provide the rationale for determining the presence of coliforms as an indication of fecal contamination.
3. List and explain each step in the multiple-tube technique.

BACKGROUND

Coliforms

Tests that determine the bacteriological quality of water have been developed to prevent transmission of water-borne diseases of fecal origin. However, it is not practical to look for pathogens in water supplies because pathogens occur in such small numbers that they might be missed by sampling. Moreover, when pathogens are detected, it is usually too late to prevent occurrence of the disease. Rather, the presence of **indicator organisms** is used to detect fecal contamination of water. An indicator organism must be present in human feces in large numbers and must be easy to detect. The most frequently used indicator organisms are the coliform bacteria. **Coliforms** are aerobic or facultatively anaerobic, gram-negative, non–endospore-forming, rod-shaped bacteria that ferment lactose with acid and gas formation within 48 hours at 35°C. Coliforms are not usually pathogenic, although they can cause opportunistic infections. The IMViC tests (Exercise 51) historically were used to identify coliforms, such as *Escherichia coli*.

Coliforms are not restricted to the human gastrointestinal tract and are found in other animals and in the soil. Tests that determine the presence of **fecal coliforms** (found in birds and mammals) have been developed. Appearance of coliform colonies on differential media incubated at 44.5°C is now considered positive for fecal coliforms. One of these differential media, MUG agar, will be used in this exercise and is discussed below.

Established public health standards specify the maximum number of fecal coliforms allowable in each 100 ml of water, depending on the intended use of the water (EPA Standards,* for example, allow 0 per 100 ml in water for drinking 200 per 100 ml for water-contact sports, or 800 per 100 ml for landscape irrigation).

MPN Test

The most accurate number of coliform bacteria is obtained by testing a large sample of water. Total coliforms can be detected and enumerated in the **multiple-tube technique** (**FIGURE 52.1**). In this method, coliforms are detected in two stages. In the **presumptive test**, various amounts of the water sample are added to lactose fermentation tubes. The lactose broth can be made selective for gram-negative bacteria by adding lauryl sulfate or brilliant green and bile. Fermentation of lactose to gas is a positive reaction.

Samples from the positive presumptive tube at the highest dilution are examined for coliforms by inoculating a differential medium in the **confirmed test**. A confirmed test can be done on **MUG agar.** Almost all strains of *E. coli* produce the enzyme β-glucuronidase (GUD). If *E. coli* is added to a nutrient medium containing 4-methylumbelliferone glucuronide (MUG), GUD converts MUG to a fluorescent compound that is visible with an ultraviolet lamp (**FIGURE 52.2**).

The number of coliforms is determined by a statistical estimation called the **most probable number (MPN)** method. In the presumptive test, tubes of lactose broth are inoculated with samples of the water being tested. The tubes showing acid and gas are counted, and the count then is compared to statistical tables such as the one shown in **TABLE 52.1** on page 416. The MPN number is *the most probable number* of coliforms per 100 ml of water.

MATERIALS

FIRST PERIOD

Water sample, 50 ml or BSL-2 bring your own from a pond or stream.
9-ml, single-strength lactose fermentation tubes (6)
20-ml, 1.5-strength lactose fermentation tubes (3)
Sterile 10-ml pipette
Sterile 1-ml pipette

SECOND PERIOD

Petri plate containing MUG agar

---

*EPA, *Federal Register* 54: 27544, June 29, 1989.

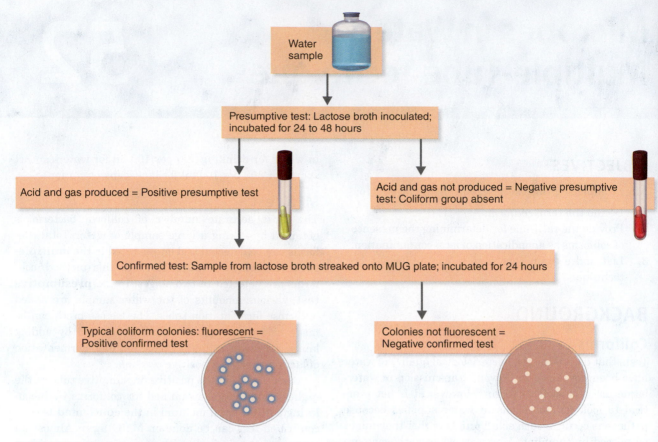

FIGURE 52.1 Analysis of drinking water for coliforms by the multiple-tube technique.

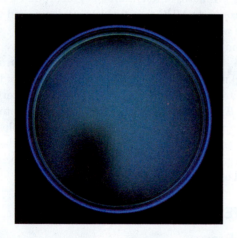

FIGURE 52.2 *E. coli.* E. coli is identified by its ability to convert MUG to a fluorescent compound that is visible with an ultraviolet lamp.

TECHNIQUES REQUIRED

Aseptic technique (Exercise 4)
Plate streaking (Exercise 11)
Fermentation tests (Exercise 14)
Pipetting (Appendix A)

PROCEDURE First Period

Using **FIGURE 52.3** as a reference, complete the following presumptive test.

1. Label three single-strength lactose broth tubes "0.1," label another three tubes "1," and label the three 1.5-strength broth tubes "10."

 Wear safety goggles when pipetting liquids.

2. Inoculate each 0.1 tube with 0.1 ml of your water sample.
3. Inoculate each 1 tube with 1.0 ml of your water sample.
4. Inoculate each 10 tube with 10 ml of your water sample. Why is 1.5-strength lactose broth used for this step? _____ _____
5. Incubate the tubes for 24 to 48 hours at 35°C.

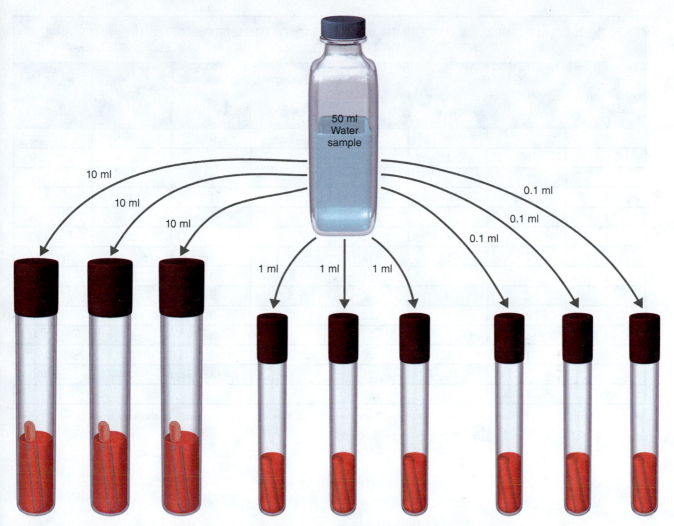

FIGURE 52.3 **Most probable number (MPN) dilution series.** There are three sets of tubes and three tubes in each set. Each tube in the first set of three tubes receives 10 ml of the inoculum. Each tube in the second set of three tubes receives 1 ml of the sample, and the third set, 0.1 ml each.

PROCEDURE Second Period

1. Record the results of your presumptive test (see Figure 52.1). Which tube has the highest dilution of the water sample? _____

2. Determine the number of coliforms per 100 ml of the original sample (see Table 52.1). If a tube has gas, streak the MUG agar with the positive broth. Incubate the plate, inverted, for 24 to 48 hours at 44.5°C.

PROCEDURE Third Period

1. Examine the plate using an ultraviolet lamp.

 Do not look directly at the ultraviolet light, and do not leave your hand exposed to it.

2. Record the results of your confirmed test (see Figure 52.1 and Figure 52.2). How can you tell whether fecal coliform colonies are present? _____
_____

TABLE 52.1 MOST PROBABLE NUMBER (MPN) INDEX FOR VARIOUS COMBINATIONS OF POSITIVE AND NEGATIVE RESULTS WHEN THREE 10-ml PORTIONS, THREE 1-ml PORTIONS, AND THREE 0.1-ml PORTIONS ARE USED

Number of Tubes Giving Positive Reaction Out of			MPN Index per 100 ml	Number of Tubes Giving Positive Reaction Out of			MPN Index per 100 ml
3 of 10 ml Each	3 of 1 ml Each	3 of 0.1 ml Each		3 of 10 ml Each	3 of 1 ml Each	3 of 0.1 ml Each	
0	0	0	<3	3	0	0	23
0	0	1	3	3	0	1	39
0	1	0	3	3	0	2	64
1	0	0	4	3	1	0	43
1	0	1	7	3	1	1	75
1	1	0	7	3	1	2	120
1	1	1	11	3	2	0	93
1	2	0	11	3	2	1	150
2	0	0	9	3	2	2	210
2	0	1	14	3	3	0	240
2	1	0	15	3	3	1	460
2	1	1	20	3	3	2	1,100
2	2	0	21	3	3	3	≥2,400
2	2	1	28				

Source: *Standard Methods for the Examination of Water and Wastewater,* 13th ed. New York: American Public Health Association, 1971.

LABORATORY REPORT
Microbes in Water: Multiple-Tube Technique

PURPOSE _____
_____
_____

HYPOTHESIS

You will/will not be able to determine whether the water sample is fecally contaminated by the multiple-tube technique.

RESULTS

Water sample source: _____

Presumptive Test

Results	Number of Tubes with a Positive Result		
	Inoculum		
	10 ml	1 ml	0.1 ml
Growth			
Acid			
Gas			
Number of tubes positive for all 3			

Confirmed Test

Tube: _____

Appearance of colonies with white light: _____
_____

Appearance of colonies with UV light: _____

Data from water samples tested by other students:

Sample	Coliforms Present?	MPN	Fecal Coliforms?

CONCLUSIONS

1. What is the MPN of your water sample? _____ per 100 ml

2. Are fecal coliforms present? _____

    Did your results confirm your hypothesis? _____

QUESTIONS

1. Could the water have a high concentration of the pathogenic bacterium *Vibrio cholerae* and give negative

    results in the multiple-tube technique? Briefly explain. _____

    _____

    _____

2. Why are coliforms used as indicator organisms if they are not usually pathogens? _____

    _____

3. Why isn't a pH indicator needed in the lactose broth fermentation tubes? _____

    _____

4. If coliforms are found in a water sample, the IMViC tests will help determine whether the coliforms are of fecal

    origin and not from plants or soil. What IMViC results would indicate the presence of fecal coliforms?

    _____

    _____

    _____

CRITICAL THINKING

1. Why didn't we inoculate MUG agar directly and bypass lactose broth?

2. Use the data shown below to explain how waterborne diseases can best be eliminated.

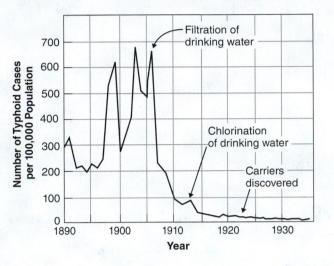

CLINICAL APPLICATION

The following table lists the etiologies of illness associated with recreational waters in 2009–2010.

Microorganism	Number of Cases
Cryptosporidium spp.	422
Campylobacter spp.	166
Escherichia coli O157:H7	31
Pseudomonas aeruginosa	50

Source: *MMWR* 63(1): 6–10, January 10, 2014.

Which of the water samples would have shown a high coliform count?

Which diseases could have been prevented by chlorinating the water?

Microbes in Water: Membrane Filter Technique

EXERCISE

53

OBJECTIVES

After completing this exercise, you should be able to:

1. Explain the principle of the membrane filter technique.
2. Use the membrane filter technique to determine the bacteriologic quality of water.

BACKGROUND

Fecal contamination of water can be determined by the presence of fecal coliforms in a water sample, via the multiple-tube technique (Exercise 52). Fecal coliforms, as well as enterococci, can also be detected by the membrane filter technique.

In the **membrane filter technique,** water is drawn through a thin filter (see Appendix F). Filters with a variety of pore sizes are available. Pores of 0.45 μm are used for filtering out most bacteria. Bacteria are retained on the filter, which is then placed on a suitable nutrient medium. In field situations, nutrients are added to a thick absorbent pad on which the filter is placed. Nutrients that diffuse through the filter can be metabolized by bacteria trapped on the filter. Each bacterium that is trapped on the filter will develop into a colony. Bacterial colonies growing on the medium can then be counted. When a selective or differential medium is used, desired colonies will have a distinctive appearance.

Selective Media

Endo agar is frequently used as a selective and differential medium with the membrane filter technique. The composition of Endo agar is shown in **TABLE 53.1**. Endo agar is selective because desoxycholate and sodium lauryl sulfate inhibit gram-positive bacteria. Endo agar is differential in that the colonies of lactose-fermenting bacteria have pink to red colonies with a metallic green sheen (**FIGURE 53.1**).

The enterococci are fecal streptococci that include *Enterococcus faecalis* and *E. faecium*. *Enterococcus* spp. differ from streptococci by their ability to grow in 6.5% NaCl, at pH 9.6, and at 10°C and 45°C. **Enterococcus**

TABLE 53.1	CHEMICAL COMPOSITION OF ENDO AGAR
Tryptose	0.75%
Yeast extract	0.12%
Lactose	0.94%
Dipotassium phosphate	0.4%
Sodium desoxycholate	0.01%
Sodium lauryl sulfate	0.005%
Basic fuchsin	0.008%
Agar	1.5%

Source: Difco Manual.

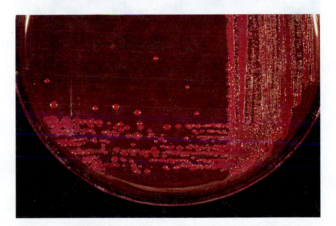

FIGURE 53.1 Endo agar. Coliforms produce rose-red colonies with a metallic green sheen within 24 hours.

agar is a selective and differential medium for enterococci (**TABLE 53.2**). Azide inhibits growth of gram-negative bacteria, and the dye TTC is picked up by growing cells to produce red colonies (**FIGURE 53.2**).

421

TABLE 53.2	CHEMICAL COMPOSITION OF ENTEROCOCCUS AGAR
Tryptose	2.0%
Yeast extract	0.5%
Dextrose	0.2%
Dipotassium phosphate	0.4%
Sodium azide	0.04%
Agar	1.0%
Triphenyl tetrazolium chloride (TTC)	0.01%

Source: Difco Manual.

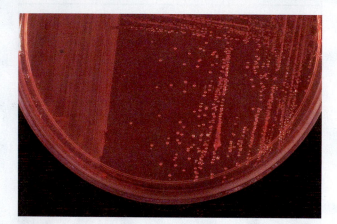

FIGURE 53.2 Enterococcus agar. Enterococci produce pink to red colonies; gram-negative bacteria are inhibited by azide in the agar.

MATERIALS

FIRST PERIOD

Water sample, 100 ml or **BSL-2** bring your own from a pond or stream
47-mm Petri plate containing Endo agar
47-mm Petri plate containing Enterococcus agar
Sterile membrane filter apparatus
Sterile 0.45-μm filters (2)
Blunt-tip forceps
Beaker of alcohol
Sterile pipette or graduated cylinder, as needed
Sterile rinse water

SECOND PERIOD

Dissecting microscope

TECHNIQUES REQUIRED

Special media (Exercise 12)
Pipetting (Appendix A)
Dissecting microscope (Appendix E)
Membrane filtration (Appendix F)

PROCEDURE FIRST PERIOD

1. Set up the filtration equipment (see Appendix F). Remove wrappers as each piece is fitted into place. Why shouldn't all the wrappers be removed at once? _____

   _____

   a. Attach the filter trap to the vacuum source. What is the purpose of the filter trap?_____

   _____

   b. Place the filter holder base (with stopper) on the filtering flask. Attach the flask to the filter trap.

Disinfect the forceps by dipping in alcohol and burning off the alcohol. Keep the beaker of alcohol away from the flame.

   c. Using the sterile forceps, place a filter on the filter holder. Why must the filter be centered exactly on the filter holder? _____

   _____

   d. Set the funnel on the filter holder, and fasten it in place.

Wear safety goggles when pipetting liquids.

2. Filter the sample.
   a. Shake the water sample well to resuspend all material, and pour or pipette a measured volume into the funnel. Your instructor will help you determine the volume.* (*For samples of 10 ml or less, pour 20 ml of sterile water into the funnel first.*)
   b. Turn on the vacuum, and allow the sample to pass into the filtering flask. Leave the vacuum on.
   c. Pour sterile rinse water into the funnel. Rotate the funnel while pouring to wash bacteria from the sides of the funnel. (*Use the same volume as the sample.*) Allow the rinse water to go through the filter. Turn off the vacuum.
3. Inoculate the filter (**FIGURE 53.3**).
   a. Carefully remove the filter from the filter holder using sterile forceps. Why does the filter have to be "peeled" off? _____

   _____

Suggested sample volumes: lakes and wells, 50–100 ml; treated sewage, 0.1–10 ml; rivers and storm water runoff, 0.01–1 ml.

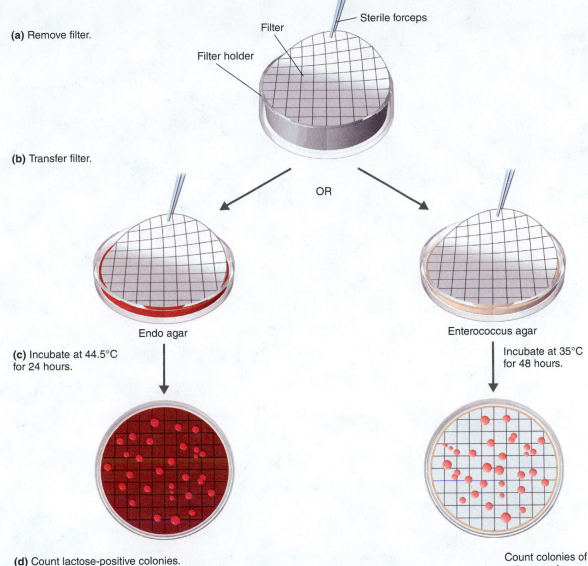

(a) Remove filter.

Sterile forceps

Filter

Filter holder

(b) Transfer filter.

OR

Endo agar

Enterococcus agar

(c) Incubate at 44.5°C for 24 hours.

Incubate at 35°C for 48 hours.

(d) Count lactose-positive colonies.

Count colonies of enterococci.

FIGURE 53.3 Inoculation. Using sterile forceps, remove the filter from the filter holder. Place the filter on the culture medium, gradually laying it down from one edge to the other.

b. Carefully place the filter on the Endo agar. Do not bend the filter; place one edge down first, then carefully set the remainder down. Do not leave air spaces between the filter and agar. Place the filter on the agar as it was in the filter holder.

4. Invert the plate and incubate it for 24 hours at 44.5°C.

5. Repeat steps 1, 2, and 3 using the same water source. Place the filter on Enterococcus agar. Invert the plate, and incubate it for 48 hours at 35°C.

PROCEDURE Second Period

1. Examine the Endo plate using a dissecting microscope. On Endo agar, coliforms will form red colonies with a green metallic sheen. Count plates that

have 20 to 80 coliform colonies and not more than 200 colonies of all types.

2. Examine the Enterococcus plate using a dissecting microscope. On Enterococcus agar, enterococci will form pink to red colonies. Count plates that have 20 to 60 enterococcus colonies and not more than 200 colonies of all types.

3. Calculate the bacteria in the original water sample:

Number of fecal coliforms (or enterococci) per 100 ml of water =

$$100 \times \frac{\text{Number of fecal coliform (or enterococci) colonies}}{\text{Volume of sample filtered}}$$

LABORATORY REPORT
Microbes in Water: Membrane Filter Technique

PURPOSE _____

_____

_____

HYPOTHESES

1. Coliforms will/will not be present in the water sample, as evidenced by red colonies on the Endo agar.

2. Enterococci from the water sample will/will not be found on the Enterococcus agar.

RESULTS

Sample tested: _____

Fecal Coliforms	Enterococci
Amount of water tested: _____	Amount of water tested: _____
General appearance of the colonies on Endo agar:_____ _____ _____	General appearance of the colonies on Enterococcus agar:_____ _____ _____
Number of fecal coliform colonies: _____	Number of *Enterococcus* colonies: _____
Number of fecal coliforms per 100 ml of water: _____	Number of enterococci/100 ml of water: _____
Use the space below for your calculations.	Use the space below for your calculations.

Data from water samples tested by other students:

Sample	Fecal Coliforms/100 ml	Enterococci/100 ml

CONCLUSIONS

1. Which water sample(s) is (are) potable? _____

2. Which water sample(s) is (are) contaminated with fecal material? _____

QUESTIONS

1. What basic assumption is made in this technique if the number of bacteria is determined from the number of

   colonies? _____

   _____

2. Will *E. coli* grow on Enterococcus agar, and will *Enterococcus* grow on Endo agar? Briefly explain. _____

   _____

   _____

3. Why can filtration be used to sterilize culture media? _____

   _____

4. If you performed the multiple-tube technique (Exercise 52), list one advantage and one disadvantage of each

   method of detecting coliforms. _____

   _____

   _____

   _____

5. Why is the membrane filter technique useful for a sanitarian working in the field? _____

   _____

CRITICAL THINKING

1. Design an experiment to detect the presence and number of *Acetobacter* in wine. Why would you want to perform such a test?

2. A water sample could be negative for *Enterococcus* and coliforms and still be a major public health threat. Why?

CLINICAL APPLICATION

The following table lists the etiologies of illness associated with drinking water in 2009–2010.

Microorganism	Number of Cases
Campylobacter spp.	829
Legionella pneumophila	63
Norovirus	47
Hepatitis A virus	2
Escherichia coli O157:H7	39
Giardia intestinalis	14

Source: *MMWR* 62(35): 714–720, September 6, 2013.

Which of the water samples would definitely have shown a high coliform count?

Which etiologies could have been detected by membrane filtration?

Which diseases could have been prevented by membrane filtration?

Microbes in Food: Contamination

OBJECTIVES

After completing this exercise, you should be able to:

1. Determine the approximate number of bacteria in a food sample using a heterotrophic plate count.
2. Provide reasons for monitoring the bacteriologic quality of foods.
3. Explain why the heterotrophic plate count is used in food quality control.

BACKGROUND

Illness and food spoilage can result from microbial growth in foods. The sanitary control of food quality is concerned with testing foods for the presence of microbes. During processing (grinding, washing, and packaging), food may be contaminated with soil microbes and microbiota from animals, food handlers, and machinery.

Foods are the primary vehicle responsible for transmitting diseases of the digestive system. For this reason, they are examined for the presence of coliforms because the presence of coliforms usually indicates fecal contamination.

Plate counts are routinely performed on food and milk by food-processing companies and public health agencies. The **heterotrophic plate count** is used to determine the total number of viable bacteria in a food sample. The presence of large numbers of bacteria is undesirable in most foods because it increases the likelihood that pathogens will be present, and it increases the potential for food spoilage.

Calculating Numbers of Bacteria

In a heterotrophic plate count, the number of **colony-forming units (CFUs)** is determined. Each colony may arise from a group of cells rather than from one individual cell. The initial sample is diluted through serial dilutions (Appendix B) to obtain a small number of colonies on each plate. A known volume of the diluted sample is placed in a sterile Petri dish, and melted cooled nutrient agar is poured over the inoculum. After incubation, the colonies are counted. Plates with between 25 and 250 colonies are suitable for counting. A plate with fewer than 25 colonies is unsuitable for

counting because a single contaminant could influence the results. A plate with more than 250 colonies is extremely difficult to count. The microbial population in the original food sample can then be calculated using the following equation:

Colony-forming units per gram or ml of sample =

$$\frac{\text{Number of colonies}}{\text{Amount plated} \times \text{dilution*}}$$

A limitation of the heterotrophic plate count is that only bacteria capable of growing in the culture medium and environmental conditions provided will be counted. A medium that supports the growth of most heterotrophic bacteria is used.

MATERIALS

Melted plate count or nutrient agar, cooled to 45°C
Sterile 1-ml pipettes (8)
Sterile Petri dishes (10)
Sterile weighing dish
Sterile spatula
Sterile 99-ml dilution blanks (3)
Sterile 9-ml dilution blanks (2)
Food samples
Vortex mixer

TECHNIQUES REQUIRED

Aseptic technique (Exercise 4)
Pour plate technique (Exercise 11)
Pipetting (Appendix A)
Serial dilution technique (Appendix B)

---

*Dilution *refers to the tube prepared by serial dilutions (Appendix B). For example, if 250 colonies were present on the $1{:}10^6$ plate, the calculation would be as follows:*

$$\text{Colony-forming units per gram} = \frac{250 \text{ colonies}}{0.1 \text{ ml} \times 10^{-5}}$$
$$= 250 \times 10^5 \times 10^1$$
$$= 250{,}000{,}000$$
$$= 2.5 \times 10^8$$

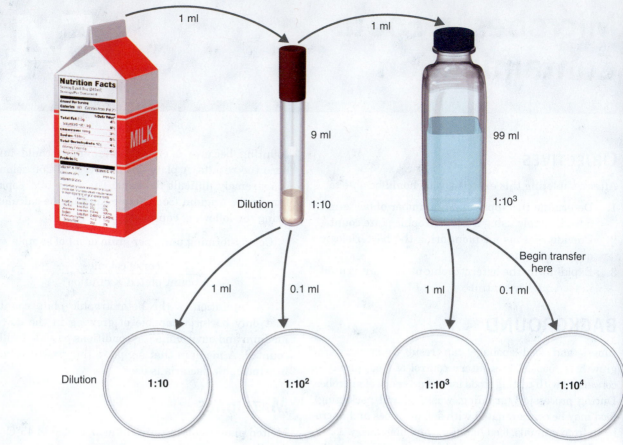

(a)

(b)

FIGURE 54.1 **Heterotrophic plate count of milk. (a)** Make serial dilutions of a milk sample. **(b)** Mark four Petri dishes with the dilutions.

⚠️ **Wear safety goggles when pipetting liquids.**

PROCEDURE First Period

Bacteriologic Examination of Milk

1. Obtain a sample of either raw or pasteurized milk.
2. Using a sterile 1-ml pipette, aseptically transfer 1 ml of the milk into a 9-ml dilution blank; label the tube "1:10," and discard the pipette (**FIGURE 54.1a**). Mix the contents of the tube on a vortex mixer.
3. Using a sterile 1-ml pipette, aseptically transfer 1 ml of the 1:10 dilution into a 99-ml dilution blank; label the bottle "1:10³" and discard the pipette. Shake the bottle 20 times, with your elbow resting on the table, as shown in **FIGURE 54.2**.
4. Label the bottoms of four sterile Petri dishes with the dilutions: "1:10," "1:10²," "1:10³," and "1:10⁴" (**FIGURE 54.1b**).

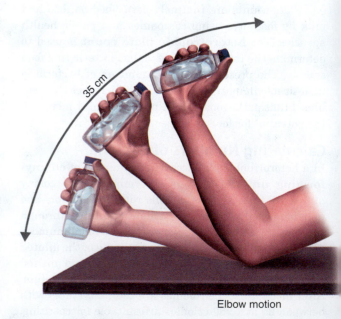

FIGURE 54.2 **Shaking the dilution bottle 20 times through a 35-cm arc.**

Elbow motion

Nutrient agar

Mix

(b)

(a)

Inoculum

(c)

FIGURE 54.3 Pour plate. (a) Pipette inoculum into a Petri dish. **(b)** Add liquefied nutrient agar, and **(c)** mix the agar with the inoculum by gently swirling the plate.

5. Using a 1-ml pipette, aseptically transfer 0.1 ml of the $1:10^3$ dilution into the bottom of the $1:10^4$ dish. *Note:* 0.1 ml of the $1:10^3$ dilution results in a $1:10^4$ dilution of the original sample. Using the same pipette, transfer 1.0 ml of the $1:10^3$ dilution into the dish labeled $1:10^3$. Pipette 0.1 ml and 1.0 ml from the 1:10 dilution into the $1:10^2$ and 1:10 dishes, respectively, with the same pipette (**FIGURE 54.3a**). Why is it important to proceed from the highest to the lowest dilution? _____
_____

6. Check the temperature of the water bath containing the nutrient agar. Test the temperature of the outside of the agar container with your hand. It should be warm, but not too hot for your bare hand. Why? _____
_____

7. Pour the melted nutrient agar into one of the dishes (to about one-third full) (**FIGURE 54.3b**). Cover the plate, and swirl it gently (**FIGURE 54.3c**) to distribute the milk sample evenly through the agar. Continue until all the plates are poured.

8. When each plate has solidified, seal with tape, invert it, and incubate all plates at 35°C for 24 to 48 hours.

Bacteriologic Examination of Hamburger, Hot Dog, or Frozen Vegetables

1. Weigh 1 g of raw hamburger (or hot dog) or frozen, thawed vegetables by using the sterile spatula and weighing dish.

2. Transfer the 1 g of food into a 9-ml dilution blank; label the tube "1:10" (**FIGURE 54.4a**). Mix the contents on a vortex mixer. Mixing will shake bacteria off the food and put them in suspension.

3. Using a sterile 1-ml pipette, aseptically transfer 1 ml of the 1:10 dilution into a 99-ml dilution blank; label the bottle "$1:10^3$" and discard the pipette. Shake the bottle 20 times, with your elbow resting on the table (as shown in **FIGURE 54.2**). Make a $1:10^5$ dilution using another 99-ml dilution blank. Shake the bottle as before.

4. Label the bottoms of six sterile Petri dishes with the dilutions: "1:10," "$1:10^2$," "$1:10^3$," "$1:10^4$," "$1:10^5$," and "$1:10^6$" (**FIGURE 54.4b**).

5. Using a 1-ml pipette, aseptically transfer 0.1 ml of the $1:10^5$ dilution into the $1:10^6$ dish. *Note:* 0.1 ml of a $1:10^5$ dilution results in a $1:10^6$ dilution of the original sample. Using the same pipette, repeat

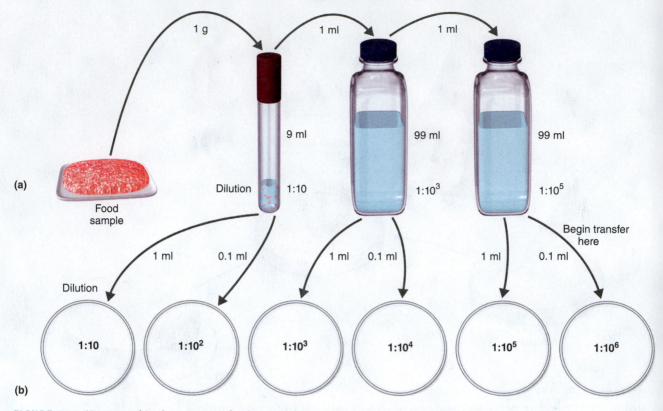

FIGURE 54.4 Heterotrophic plate count on food. (a) Prepare serial dilutions of a food sample. **(b)** Label six Petri dishes with the dilutions to be plated.

this procedure with the 1:10³ dilution and the 1:10 dilution until all the dishes have been inoculated (see **FIGURE 54.3a**). Why can the same pipette be used for each transfer? _____

_____

_____

6. Check the temperature of the water bath containing the nutrient agar. Test the temperature of the outside of the agar container with your hand. It should be warm, but not too hot to hold. Why? ____

_____

7. Pour the melted nutrient agar into one of the dishes (to about one-third full) (see **FIGURE 54.3b**). Cover the plate, and swirl it gently (see **FIGURE 54.3c**) to distribute the sample through the agar evenly. Continue until all the plates are poured.

8. When each plate has solidified, seal with tape, invert it, and incubate all plates at 35°C for 24 to 48 hours.

PROCEDURE Second Period

1. Arrange each plate in order from lowest to highest dilution.

2. Select the plate with 25 to 250 colonies. Record data for plates with fewer than 25 colonies as *too few to count (TFTC)* and those with more than 250 colonies, as *too numerous to count (TNTC)*.

3. Count the colonies on the plate selected.

4. Calculate the number of colony-forming units in the original food. For example, if 129 colonies were counted on a 1:10³ dilution, calculate as follows:

$$\frac{129 \text{ colonies}}{1 \text{ ml} \times 10^{-3}} = 129,000$$
$$= 1.29 \times 10^5 \text{ CFU/ml or}$$
$$\text{gram of milk or food}$$

LABORATORY REPORT
Microbes in Food: Contamination

PURPOSE _____

_____

_____

HYPOTHESIS

You will be able to determine the number of colony-forming units (CFUs) in food by counting the colonies on the nutrient agar. Agree/disagree

RESULTS

Milk sample: _____

Dilution	Colonies per Plate
1:	
1:	
1:	
1:	

Food sample: _____

Dilution	Colonies per Plate
1:	
1:	
1:	
1:	
1:	
1:	

CONCLUSIONS

1. Did your results agree with your hypothesis? _____

2. What did you test? _____

   Number of colony-forming units per ml (or gram) of original sample: _____

   Make your calculations in the space below.

3. In a quality-control laboratory, each dilution is plated in duplicate or triplicate. Why would this increase the accuracy of a standard plate count? _____

_____

4. Results from your classmates who tested the same product:

Student Data	CFU/ml (or gram)
1	
2	
3	
Average	

Record data for other foods tested by other students.

Food	CFU/ml (or gram)

How do you account for the range of numbers? _____

_____

_____

QUESTIONS

1. Why are plates with 25 to 250 colonies used for calculations? _____

_____

_____

2. There are other techniques for counting bacteria, such as a direct microscopic count and turbidity. Why is the heterotrophic plate count preferred for food? _____

_____

_____

3. Why is ground beef a better bacterial growth medium than a steak or roast? _____

_____

4. Why does repeated freezing and thawing increase bacterial growth in meat? _____

_____

_____

CRITICAL THINKING

1. Assume that a heterotrophic plate count indicates that substantial numbers of microbes are present in a food sample. What should be done next to determine whether the food sample is a danger to the consumer?

2. Food and Drug Administration guidelines state that pasteurized milk is allowed 20,000 CFU/ml. How many colonies would be present if 1 ml of a 10^{-3} dilution was plated? _____ Is 20,000 CFU/ml a health hazard? _____

CLINICAL APPLICATIONS

1. An outbreak of botulism associated with home-canned chili peppers killed 16 members of one family. Would a heterotrophic plate count have detected the etiologic agent in the canned chili peppers? Briefly explain.

2. An outbreak of enteropathogenic *Escherichia coli* has occurred. The source of the epidemic is thought to be alfalfa sprouts. A heterotrophic plate count of the sprouts indicates low levels of bacteria. Can you assume that the sprouts are not the source of the outbreak?

Microbes Used in the Production of Foods

OBJECTIVES

After completing this exercise, you should be able to:

1. Explain how the activities of microorganisms are used to preserve food.
2. Define the term *fermentation*.
3. Produce an enjoyable product!

BACKGROUND

Microbial fermentations are used to produce a wide variety of foods. **Fermentation** means different things to different people. In industrial usage, it is any large-scale microbial process occurring with or without air. To a biochemist, it is the group of metabolic processes that release energy from a sugar or other organic molecule, do not require oxygen or an electron transport system, and use an organic molecule as a final electron acceptor.

In dairy fermentations, such as yogurt production, microorganisms use lactose and produce lactic acid without using oxygen. In nondairy fermentations, such as wine production, yeast use sucrose to produce ethanol and carbon dioxide under anaerobic conditions. If oxygen is available, the yeast will grow aerobically, liberating carbon dioxide and water as metabolic end-products. In this exercise, we will examine a lactic acid fermentation used in the production of food.

Historically, nearly every human population used milk that had been fermented by selected microbes. The acids and bacteriocins (antibiotics) produced during fermentation prevented the growth of spoilage or pathogenic bacteria. These "sour" milks have varied from country to country depending on the source of milk, conditions of culture, and microbial "starter" used. Milk from donkeys to zebras has been used, with the Russian *kumiss* (horse milk), containing 2% alcohol, and Swedish *surmjölk* (reindeer milk) being unusual examples. The bacteria yield lactic acid, and the yeast produce ethanol. Currently, two fermented cows' milk products, buttermilk and yogurt, are widely used.

Buttermilk is the fluid left after cream is churned into butter. Today, buttermilk is actually prepared by souring true buttermilk or by adding bacteria to skim milk and then flavoring it with butterflake. *Lactococcus lactis* ferments the milk, producing lactic acid (sour); neutral fermentation products (diacetyls) are produced by *Leuconostoc*. Yogurt originated in the Balkan countries, with goat milk as the primary source. Yogurt is milk in which the milk solids have been concentrated by heating. Then the milk is fermented at elevated temperatures. *Streptococcus* produces lactic acid, and *Lactobacillus* produces the flavors and aroma of yogurt.

MATERIALS

FIRST PERIOD

Homogenized milk
Nonfat dry milk
Sterile water
Large sterile beaker
Sterilized stirring rod
Thermometer
Sterile glass test tube containing sterile water
Hot plate or ring stand and asbestos pad
Sterile 5-ml pipette
Disposable cups with lids

SECOND PERIOD

Plastic spoons
Petri plate containing trypticase soy agar
pH paper
Gram stain reagents (for Third Period, too)
Optional: jam, jelly, honey, and so on

CULTURES

Commercial yogurt *or*
Streptococcus salivarius subsp. *thermophilus* and
    Lactobacillus delbrueckii

TECHNIQUES REQUIRED

Gram staining (Exercise 7)
Plate streaking (Exercise 11)
Pipetting (Appendix A)

PROCEDURE First Period

Be sure all glassware is clean.

1. Add 100 ml of milk per person (in your group) to a wet beaker (wash out the beaker first with sterile water to decrease the sticking of the milk). Put a

437

thermometer in a sterile glass test tube containing sterile water before placing it in the beaker.

2. Heat the milk on a hot plate or over a Bunsen burner on wire gauze placed on a ring stand, to about 80°C for 10 to 20 minutes. Stir occasionally. Do not let it boil. Why is the milk heated? _____

_____

3. Cover and cool the milk to about 65°C. Add 3 g of nonfat dry milk per person. Stir to dissolve. Why is dry milk added? _____

_____

4. Rapidly cool the milk to about 45°C. Pour the milk equally into the cups.

5. Inoculate each cup with 1 to 2 teaspoonfuls of commercial yogurt, or 2.5 ml of *S. salivarius* subsp. *thermophilus* and 2.5 ml of *L. delbrueckii*. Cover and label the cups. Make a smear with the inoculum on a slide, heat-fix it, and save it for the second period.

6. Incubate the cups at 45°C for 4 to 18 hours or until they are firm (custard-like).

PROCEDURE   Second Period

1. Cool the yogurt to about 5°C. Save a small amount for steps 2 through 4. Then, outside of the laboratory, taste it with a clean spoon. Add jam or some other flavor if you desire. Eat and enjoy!

Do not work with other bacteria or perform other exercises while eating the yogurt.

2. Determine the pH of the yogurt.

3. Make a smear next to the smear prepared previously, and, after heat-fixing it, prepare a Gram stain. Observe the slides with the oil immersion objective, and record your results.

4. Streak for isolation on trypticase soy agar. Incubate the plate, inverted, at 45°C.

PROCEDURE   Third Period

After distinct colonies are visible, record your observations. Prepare and observe Gram stains from each different colony.

LABORATORY REPORT

Microbes Used in the Production of Foods

PURPOSE _____

_____

_____

HYPOTHESIS

You will produce an edible yogurt with gram-positive streptococci and lactobacilli. Agree/disagree

RESULTS

Yogurt characteristics:

Taste: _____

Consistency: _____

Odor: _____

pH: _____

Gram stain results:

Inoculum: _____

Yogurt: _____

Streak plate results: _____

_____

_____

Gram stain of isolated colonies: _____

_____

CONCLUSIONS

1. Do your results support your hypothesis? _____
_____
_____

2. Compare your yogurt to commercial yogurt. How do you account for any differences? _____
_____
_____

3. What Gram stain result would indicate that the yogurt is contaminated? _____
_____

QUESTIONS

1. How can pathogens enter yogurt, and how can this be prevented? _____
_____
_____

2. What could cause an inferior product in a microbial fermentation process? _____
_____

3. How are microbial fermentations used to preserve foods? _____
_____
_____

CRITICAL THINKING

1. What was the source of the bacteria and yeast originally used in dairy product fermentations and breads?

2. Describe how you would show that *Streptococcus* and *Lactobacillus* are in commercial yogurt.

CLINICAL APPLICATION

Yogurt has been used as a probiotic for the gastrointestinal tract after extensive antibiotic therapy or surgery. What is the rationale for this use of yogurt?

Microbes in Soil: The Nitrogen and Sulfur Cycles

Soil is the **PLACENTA** *of life.* – P E T E R   F A R B

OBJECTIVES

After completing this exercise, you should be able to:

1. Diagram the nitrogen cycle, showing the chemical changes that occur at each step.
2. Differentiate symbiotic from nonsymbiotic nitrogen fixation.
3. Describe various aspects of the sulfur cycle as it occurs in a Winogradsky column.
4. Explain the importance of the nitrogen and sulfur cycles.

BACKGROUND

The Nitrogen Cycle

One aspect of soil microbiology that has been studied extensively is the nitrogen cycle. All organisms need nitrogen for the synthesis of proteins, nucleic acids, and other nitrogen-containing compounds. The recycling of nitrogen by different organisms is called the **nitrogen cycle** (**FIGURE 56.1**). Microbes play a fundamental, irreplaceable role in the nitrogen cycle by participating in many different metabolic reactions involving nitrogen-containing compounds. When plants, animals, and microorganisms die, microbes decompose them by proteolysis and ammonification.

Proteolysis is the hydrolysis of proteins to form amino acids. **Ammonification** liberates ammonia by deamination of amino acids or catabolism of urea to ammonia (see Figure 15.4 on page 124). In most soil, ammonia dissolves in water to form ammonium ions:

$$NH_3 + H_2O \rightarrow NH_4OH \rightarrow NH_4^+ + OH^-$$
Ammonia Water Ammonium   Ammonium Hydroxide
                hydroxide   ion   ion

Some of the ammonium ions are used directly by plants and bacteria for the synthesis of amino acids.

The next sequence of the nitrogen cycle is the oxidation of ammonium ions in **nitrification.** Two genera of soil bacteria are capable of oxidizing ammonium ions in two successive stages as follows:

$$2NH_4^+ + 2O_2 \xrightarrow{Nitrosomonas} 2NO_2^-$$
Ammonium Oxygen   Nitrite ions
ions

$$2NO_2^- + O_2 \xrightarrow{Nitrobacter} 2NO_3^-$$
Nitrite ions Oxygen   Nitrate ions

These reactions are used to generate energy (ATP) for the cells. Nitrifying bacteria are chemoautotrophs, and many are inhibited by organic matter. Nitrates are an important source of nitrogen for plants.

Denitrifying bacteria reduce nitrates and remove them from the nitrogen cycle. **Denitrification** is the reduction of nitrates to nitrogen gas. This conversion may be represented as follows:

$$NO_3^- \rightarrow NO_2^- \rightarrow N_2O \rightarrow N_2$$
Nitrate   Nitrite   Nitrous   Nitrogen
ion   ion   oxide   gas

Denitrification is **anaerobic respiration.** Many genera of bacteria, including *Pseudomonas* and *Bacillus*, are capable of denitrification under anaerobic conditions.

Atmospheric nitrogen can be returned to the soil by the conversion of nitrogen gas into ammonia, a process called **nitrogen fixation.** Microbes possessing the nitrogenase enzyme can fix nitrogen under anaerobic conditions as follows:

$$N_2 + 6H^+ + 6e^- \rightarrow 2NH_3$$
Nitrogen Hydrogen   Electrons Ammonia
gas   ions

Some free-living prokaryotic organisms, such as *Azotobacter*, clostridia, and cyanobacteria, can fix nitrogen. However, many of the nitrogen-fixing bacteria live in close association with the roots of grasses in the **rhizosphere**, where root hairs contact the soil.

Symbiotic bacteria serve a more important role in nitrogen fixation. One such symbiotic relationship is a

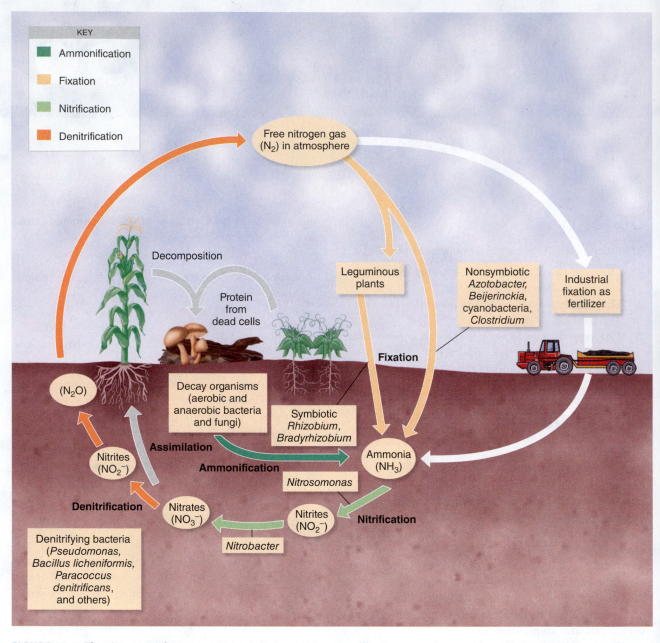

FIGURE 56.1 The nitrogen cycle.

mutualistic relationship between *Rhizobium* and the roots of legumes (such as soybeans, beans, peas, alfalfa, and clover) (**FIGURE 56.2**). There are thousands of legumes. Farmers grow soybeans and alfalfa to replace nitrogen in their fields. Many wild legumes are able to grow in the poor soils found in tropical rainforests or arid deserts. *Rhizobium* species are specific for the host legume that they infect. When a root hair and rhizobia make contact in the soil, a root nodule forms on the plant. The nodule provides the anaerobic environment necessary for nitrogen fixation.

Symbiotic nitrogen fixation also occurs in the roots of some nonleguminous plants. The actinomycete *Frankia* forms root nodules in alders.

Any break in the nitrogen cycle could be catastrophic for the survival of all life.

The Sulfur Cycle

Purple and green bacteria are involved in another biogeochemical cycle, the **sulfur cycle** (**FIGURE 56.3**).

Green photosynthetic bacteria are colored by bacteriochlorophylls, although they may appear brown because of the presence of red accessory photosynthetic pigments called **carotenoids. Purple photosynthetic bacteria** appear purple or red because of large amounts of carotenoids. Purple bacteria also have bacteriochlorophylls.

Photosynthetic bacteria use **bacteriochlorophylls** to generate electrons for ATP synthesis and use

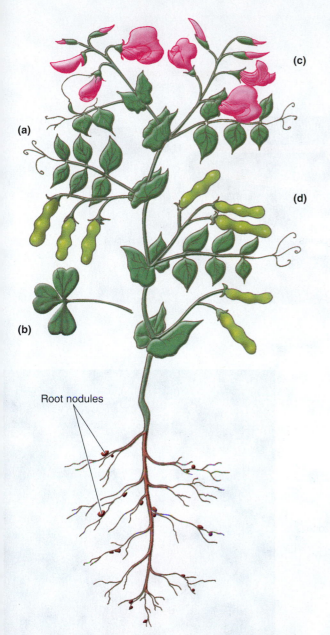

FIGURE 56.2 Some characteristic features of legumes. The leaves are opposite and may be **(a)** pinnately or **(b)** palmately compound. The flowers **(c)** have five asymmetrical petals, and the fruits **(d)** are pealike.

sulfur, sulfur-containing compounds, hydrogen gas, or organic molecules as electron donors. The generalized equation for anoxygenic bacterial photosynthesis is as follows:

$$6CO_2 \quad + \quad 12H_2S \xrightarrow{\text{Bacteriochlorophyll}}$$

Carbon dioxide   Hydrogen sulfide

$$C_6H_{12}O_6 \quad + \quad 12S^0 \quad + \quad 6H_2O$$

Sugar   Sulfur   Water

Some photosynthetic bacteria store sulfur granules in or on their cells. The stored sulfur can be used as an electron donor in photosynthesis, resulting in the production of sulfates.

In nature, hydrogen sulfide is produced from the reduction of sulfates in anaerobic respiration and the degradation of sulfur-containing amino acids. Sulfates can be reduced to hydrogen sulfide by several genera of **sulfate-reducing** bacteria (the best known of which is *Desulfovibrio*). Carbon dioxide used by photosynthetic bacteria is provided by the fermentation of carbohydrates in an anaerobic environment.

In this exercise, we will use an enrichment culture technique involving a habitat-simulating device called a *Winogradsky column* to enhance the growth of bacteria involved in the anaerobic sulfur cycle. A variety of organisms will be cultured depending on their exposure to light and the availability of oxygen (**FIGURE 56.4**). We will also investigate the steps in the nitrogen cycle.

MATERIALS

Ammonification
Tube containing peptone broth
Soil (Bring your own.)

SECOND PERIOD
Ammonia test reagent
Tube containing peptone broth
Ammonium hydroxide
Spot plate

Denitrification
Tubes containing nitrate-salts broth (2)
Soil (Bring your own.)

SECOND PERIOD
Nitrate reagents A and B
Zinc dust

Nitrogen Fixation
Petri plate containing mannitol–yeast extract agar
Methylene blue
Sterile razor blade
Legumes (Bring your own; see Figure 56.2.)

Winogradsky Column
Mud mixture (mud, $CaCO_3$ hay or paper, and $CaSO_4$)
Large test tube or graduated cylinder
Glass rod
Plain mud
Winogradsky buffer ($CaCO_3$, $CaSO_4$) NH_4Cl, $Na_2S\cdot9$
    H_2O, KH_2PO_4, K_2HPO_4)
Aluminum foil
Opaque tape
Light source

CULTURES

Pseudomonas aeruginosa `BSL-2`
Bacillus licheniformis

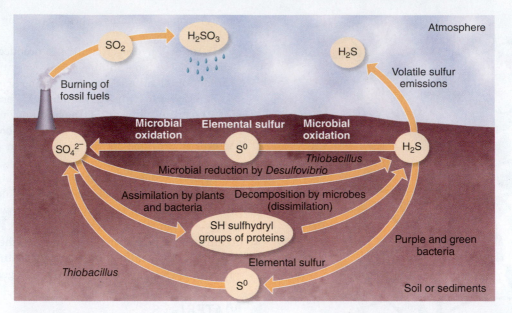

FIGURE 56.3 The sulfur cycle.

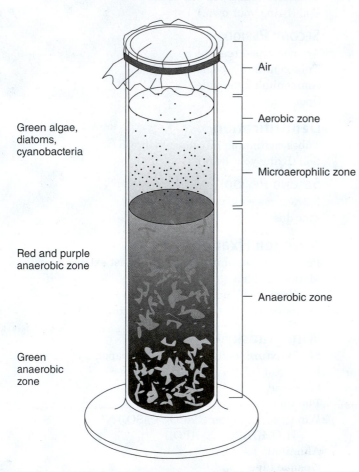

(a) Diagram of a Winogradsky column.

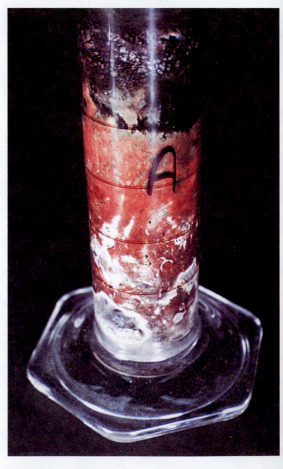

(b) Colonization of photosynthetic bacteria in a Winogradsky column.

FIGURE 56.4 Winogradsky column.

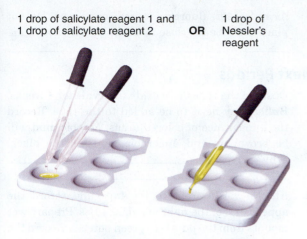

1 drop of salicylate reagent 1 and
1 drop of salicylate reagent 2 **OR**

1 drop of Nessler's reagent

(a) Place test reagent in one well of a spot plate.

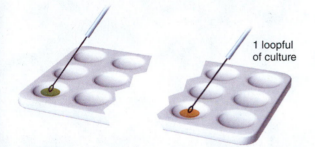

1 loopful of culture

(b) Add one loopful of culture to the reagent in the well.

FIGURE 56.5 Testing for ammonia. Your instructor will tell you which ammonia test reagent is available. **(a)** Place ammonia test reagent in a well of a spot plate. **(b)** Add a loopful of the sample to be tested, and mix. Observe for a color change.

DEMONSTRATIONS

Rhizobium inoculum used by farmers
Slides of stained root nodules

TECHNIQUES REQUIRED

Compound light microscopy (Exercise 1)
Wet-mount technique (Exercise 2)
Simple staining (Exercise 5)
Aseptic technique (Exercise 4)
Nitrate reduction test (Exercise 17)

PROCEDURE Ammonification

First Period

1. Obtain an inoculum of soil on a wetted loop, and inoculate the peptone broth. What is peptone? ____
    _____

2. Incubate the peptone broth at room temperature, and test for ammonia at 2 days and at 7 days.

Second Period

1. Test for ammonia by placing ammonia test reagent in a spot plate. Add a loopful of peptone broth, and mix (**FIGURE 56.5**). A yellow to brown color indicates the presence of ammonia. Compare your results to a spot to which a drop of ammonium hydroxide and ammonia test reagent have been added. Use an uninoculated tube of peptone broth as a control. Why is a control necessary? _____
    _____

2. Discard the used reagents as instructed.

PROCEDURE Denitrification

First Period

1. Inoculate one nitrate-salts broth with soil by using a wetted loop to pick up the soil. Inoculate the other tube with *Bacilus licheniformis* or (in BSL-2 labs) with *P. aeruginosa*.

2. Incubate both tubes at room temperature for 1 week.

Second Period

Test for nitrate reduction. Remember how? Describe.
_____
_____

What result indicates that denitrification has occurred?_____
Record your results.

PROCEDURE Nitrogen Fixation

First Period

1. Cut off a nodule from a legume, and wash it under tap water (see Figure 56.2). Observe.
2. Cut the nodule in half with a sterile razor blade. Observe. Crush the nodule between two slides, and make a smear by rotating the slides together.
3. Streak a loopful of the crushed nodule on the mannitol–yeast extract agar. Incubate the plate at room temperature for 7 days.
4. Air-dry the slide, and heat-fix it. Stain the slide for 1 minute with methylene blue. Rinse the slide, and observe it under oil immersion.
5. Observe the demonstrations. What is in the farmer's inoculum? _____

Second Period

Observe the growth on the plate, and make a simple stain of the growth with methylene blue. Compare this stain to the stain prepared from the nodule.

PROCEDURE Winogradsky Column

First Period

1. Pack a large test tube two-thirds full with the mud mixture (see Figure 56.4a). Pack it with the glass stirring rod to eliminate air bubbles. Why? _____

   _____

2. Carefully pack a narrow layer of plain mud on top of the first layer.

3. Gently pour the buffer down the side of the tube, taking care not to disturb the mud surface. Fill the tube as full as possible.

4. Cover the top of the tube with foil, and place it in front of a light source. Cover the back of the cylinder with opaque tape, such as duct tape, so one portion of the mud will not be exposed to light. The instructor may assign different light sources (e.g., incandescent, fluorescent, red, or green). Do not place the tube too close to the bulb. Why not? _____

   _____

Next Periods

1. Observe the tubes at weekly intervals for 4 weeks. Buffer may need to be added to the tube. Record the appearance of colored areas. Aerobic mud will be brownish, and anaerobic mud will be black. Why? _____

   _____

2. After 4 weeks, remove the tape and record the appearance of the light and dark sides. Prepare wet mounts from the purple or green patches. Record the microscopic appearance of the bacteria and whether sulfur granules are present (see Figure 56.4b).

LABORATORY REPORT

Microbes in Soil:
The Nitrogen and Sulfur Cycles

EXERCISE

56

PURPOSE _____

_____

_____

EXPECTED RESULTS

1. Ammonification of peptones will result in a(n) _____ color when ammonia test reagent is added.

2. Nitrate reduction to nitrite will give a(n) _____ color when nitrate A and nitrate B are added to the

   nitrate-salts tube inoculated with soil.

3. The Winogradsky column will demonstrate the sulfur cycle by the production of _____ when exposed to light.

RESULTS

Ammonification

| | | Ammonia Test Reagent | | | |
Incubation	Growth	Color of Inoculated Peptone	Color of NH_4OH	Color of Control	Ammonia Present
2 days					
7 days					

Denitrification

Inoculum	Growth	Gas	Nitrate Reduction	Denitrification
Soil				
Pseudomonas or *Bacillus*				

Nitrogen Fixation

Diagram of a root nodule:

Diagram microscopic appearance of bacteria and cells in a root nodule.

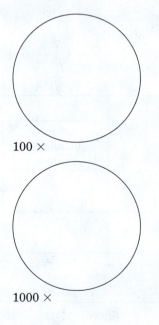

_____×

Morphology: _____

Describe colonies on the mannitol–yeast extract plate. _____

_____

_____

Sketch the microscopic appearance of bacteria cultured from the inoculum.

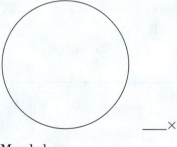

_____×

Morphology: _____

Demonstration slides:
Label root, nodule, and bacteria.

100 ×

1000 ×

Winogradsky Column

Diagram the appearance of your enrichment at weekly intervals. Label patches of photosynthetic bacteria. Indicate the aerobic and anaerobic regions on the diagrams of your enrichment column.

Light source: _____

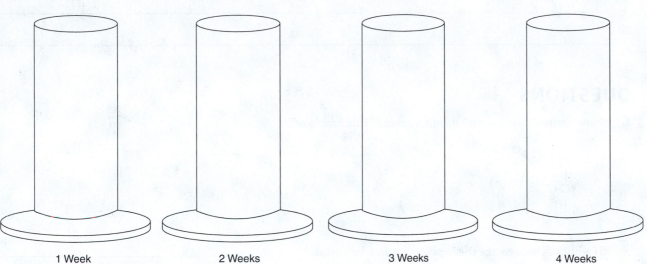

| 1 Week | 2 Weeks | 3 Weeks | 4 Weeks |

Microscopic observations:

Source	Morphology	Motility	Sulfur Granules

If other light sources were used, compare the appearance of the enrichments after 4 weeks. _____

_____

_____

_____

CONCLUSIONS

1. Did your expected results match your actual results?

    a. Ammonification _____

    b. Nitrate reduction _____

    c. Winogradsky column _____

2. Did ammonification occur in your peptone broth inoculated with soil? _____

   In your uninoculated peptone broth? _____

3. How did the illuminated side of the Winogradsky column differ from the covered side? _____

   _____

   _____

QUESTIONS

1. Show ammonification and desulfuration of this amino acid.

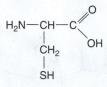

   Of what advantage are these processes to a bacterium? _____

   _____

   _____

2. Why can't you smell ammonia in the peptone broth tubes? _____

   _____

   _____

3. Why is denitrification a problem for farmers? _____

   _____

   _____

4. What gas is in the gas trap of a positive denitrification test? _____

   _____

5. Which process in the nitrogen cycle requires oxygen? _____

   _____

   _____

6. Why is the nitrogen cycle important to all forms of life? _____

_____

_____

_____

7. What was the purpose of each of the following chemicals in the Winogradsky enrichment?

Hay or paper _____ $Na_2S \cdot 9H_2O$ _____

$CaSO_4$ _____ KH_2PO_4 and K_2HPO_4 _____

$CaCO_3$ _____ NH_4Cl _____

CRITICAL THINKING

1. What genetic engineering project would you propose concerning nitrogen fixation?

2. Why do the bacteria in the root nodule look different from the bacteria cultured from the nodule?

3. Design an experiment that would determine the effect of heat produced by the lights on the growth of bacteria in the Winogradsky column.

CLINICAL APPLICATION

Newborn babies' wet diapers do not smell, but after several weeks a wet diaper will have a strong ammonia smell. What has changed?

Microbes in Soil: Bioremediation

OBJECTIVES

After completing this exercise, you should be able to:

1. Define the following terms: *bioremediation, bioenhancement*, and *hydrocarbon*.
2. Demonstrate enrichment of oil-degrading bacteria.
3. Describe the degradation or decomposition of lipids.

BACKGROUND

The use of bacteria to eliminate pollutants is called **bioremediation.** Unlike some forms of environmental cleanup, in which dangerous substances are removed from one place only to be dumped in another, bacterial cleanup eliminates the toxic substance and often returns a harmless or useful substance to the environment (**FIGURE 57.1** and **FIGURE 57.2**).

Many bacteria use exoenzymes such as lipase that hydrolytically decompose lipids. The lipid molecule is first broken down into glycerol and three fatty acids (**FIGURE 57.3**). Some bacteria ferment the glycerol, while others oxidize the fatty acids. In the presence of oxygen, the bacteria remove two carbons at a time in a process called *beta-oxidation*. When lipid hydrolysis occurs in decomposition of foods such as butter, it results in a rancid flavor and aroma caused by the fatty acids. Bacterial cultures are being used to degrade grease from restaurants and meat processors before disposal. On their own, bacteria degrade large, complex molecules in petroleum too slowly to be helpful in cleaning up an oil spill. However, scientists have hit upon a simple way to speed them up. Nitrogen and phosphorus plant fertilizers can be added in a process called *bioenhancement*. This approach was used on an Alaskan beach affected by the 1989 *Exxon Valdez* oil spill; as a result, the beach was free of oil after 1 week.

In this exercise, we will enrich for hydrocarbon-degrading bacteria. Degradation of the hydrocarbon should result in the opaque, emulsified oil becoming small, soluble molecules, such as fatty acids and acetyl groups (**FIGURE 57.4**; see also Figure 18.1e on page 147).

MATERIALS

FIRST PERIOD

Soil slurry (1 g of soil in 9 ml of water)
Enrichment (hydrocarbon–minimal salts broth inoculated with soil slurry and incubated for 1 week)
Sterile 1-ml pipettes (6)
Propipette or pipette bulb
Sterile 99-ml dilution blanks (4)
Petri plates containing hydrocarbon–minimal salts (4)
Spreading rod
Alcohol
Gram-staining reagents (second period)

FIGURE 57.1 Activated sludge in secondary sewage treatment. Air is added to sewage, and bacteria degrade soluble organic matter.

FIGURE 57.2 Land bioremediation. Soil is excavated and layered with pipes at a former truck stop. The pipes are connected to supplies of nitrogen, phosphate, and air to provide additional nutrients for bacteria that use hydrocarbons as their carbon source.

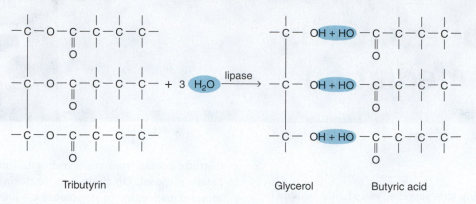

FIGURE 57.3 Tributyrin digestion by hydrolysis. Lipids are composed of a glycerol and three fatty acid molecules.

TECHNIQUES REQUIRED

Compound light microscopy (Exercise 1)
Gram staining (Exercise 7)
Spreading rod (Exercise 27)
Pipetting (Appendix A)
Serial dilution technique (Appendix B)

PROCEDURE First Period

1. Obtain a soil slurry.

 Wear safety goggles when pipetting!

2. Using a sterile 1-ml pipette, aseptically transfer 1 ml of the soil slurry to a 99-ml dilution blank; label the bottle "$1:10^2$" and discard the pipette. Mix the contents of the bottle by shaking it 20 times (see Figure 54.2).

3. Make a $1:10^4$ dilution by transferring 1 ml of the $1:10^2$ dilution to another 99-ml dilution blank. Thoroughly mix the contents of the bottle as before.

4. Label the bottoms of two hydrocarbon agar plates "10^{-4}" and "10^{-5}." Which hydrocarbon is in the plates?

_____

5. Using a 1-ml pipette, aseptically transfer 0.1 ml of the $1:10^4$ dilution onto the agar of the "10^{-5}" plate. Using the same pipette, transfer 1 ml of the $1:10^4$ dilution onto the agar of the "10^{-4}" plate. Spread the inoculum over the agar using the rod. Disinfect the spreading rod. (See Figure 27.4 on page 216.)

 Disinfect the spreading rod by dipping it in alcohol. Keep the flame away from the beaker.

Seal the plates and incubate at room temperature for 24 to 48 hours.

6. Obtain an enrichment of the same hydrocarbon used in the plates in step 4. What is the hydrocarbon?

_____

7. Repeat steps 2 through 5 to inoculate these plates from the enrichment.

PROCEDURE Second Period

1. Observe and describe the growth on the plates. Count the number of hydrocarbon-degrading colonies (see Figure 57.4). How will you identify them?

_____

Calculate the number of hydrocarbon-degrading bacteria using the following formula:

$$\text{CFU/gram of soil} = \frac{\text{Number of colonies}}{\text{Amount plated} \times \text{dilution}}$$

2. **BSL-2** Prepare a Gram stain of some of the oil-degrading colonies.

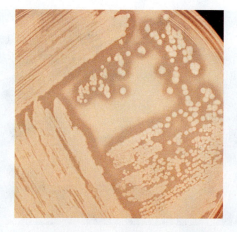

FIGURE 57.4 Hydrocarbon degradation. Bacteria that degrade the hydrocarbon in the agar produce water-soluble products, resulting in clearing around the colonies.

LABORATORY REPORT
Microbes in Soil: Bioremediation

PURPOSE _____

_____

_____

HYPOTHESIS

Microbes capable of bioremediation will/will not be present in the soil slurry.

RESULTS

Before enrichment (soil slurry):

Colony Color Description	Oil Degrader	Morphology and Gram Reaction	Number of This Type

Number of hydrocarbon-degrading (CFU/gram of soil): _____

After enrichment:

Colony Color Description	Oil Degrader	Morphology and Gram Reaction	Number of This Type

Number of hydrocarbon-degrading (CFU/gram of soil): _____

CONCLUSIONS

1. Did you isolate any oil-degrading bacteria? How can you tell? _____

_____

_____

2. Compare the results before and after enrichment. _____

_____

_____

QUESTIONS

1. What is an enrichment? _____

2. Besides watching for clearing, how else could you determine whether lipids had been hydrolyzed? _____

_____

_____

3. The enrichment broth consisted of a hydrocarbon and minimal salts. What was the purpose of the hydrocarbon?

_____

_____

What was the purpose of the minimal salts? _____

_____

CRITICAL THINKING

1. Show beta-oxidation of the following hydrocarbon:

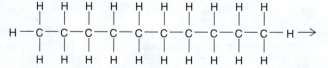

2. Here are the partial formulas of two detergents that have been manufactured. Which of these would be readily degraded by bacteria? Why?

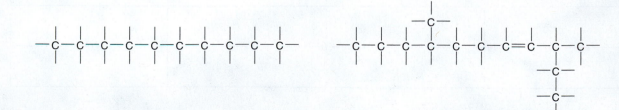

3. How could the oil-degrading bacteria you grew be used commercially in a detergent or drain cleaner?

4. Design an experiment to isolate mercury-utilizing bacteria from soil.

CLINICAL APPLICATION

Propionibacterium acnes metabolizes sebum, a natural oil on skin, forming fatty acids. Why would that be harmful to the skin?

Pipetting

Pipettes are used for measuring small volumes of fluid. They are usually coded according to the total volume and graduation units (**FIGURE A.1**). To fill a pipette, use a bulb or other mechanical device, as shown in **FIGURE A.2**.

Never use your mouth to fill a pipette. Wear safety goggles when handling liquids.

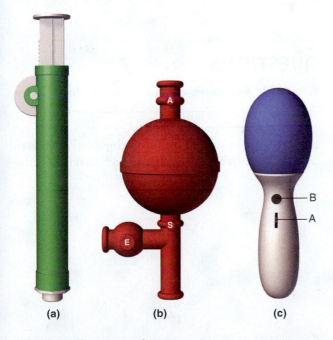

FIGURE A.1 **Pipette.** This pipette holds a total volume of 1 ml when filled to the zero mark. It is graduated in 0.01-ml units.

FIGURE A.2 **Three types of pipette aspirators. (a)** Attach this plastic pump to the pipette, and turn the wheel to draw fluid into the pipette. Turning the wheel in the other direction will release the fluid. **(b)** Insert a pipette into this bulb. While pressing the A valve, squeeze the bulb, and it will remain collapsed. To draw fluid into the pipette, press the S valve; to release fluid, press the E valve. **(c)** Insert a pipette into the stem. Raise the lever (A) to draw fluid up; lower the lever to release the fluid. Pushing the button (B) will release the last drop.

Draw the desired amount of fluid into the pipette. Read the volume at the bottom of the meniscus (**FIGURE A.3**). Fill the pipette to above the zero mark, and then allow it to drain just to the zero. The desired amount can then be dispensed.

Microbiologists use two types of pipettes. The **serological pipette** is meant to be emptied to deliver the total volume (**FIGURE A.4a**). Note that the graduations stop above the tip. The **measuring pipette** delivers the volume read on the graduations (**FIGURE A.4b**). This pipette is not emptied, but the flow must be stopped when the meniscus reaches the desired level.

Aseptic use of a pipette is often required in microbiology. Bring the entire closed pipette container to your work area. Lay down the canister, as shown in **FIGURE A.5**. If the pipettes are wrapped in paper, open the wrapper at the end opposite the delivery end; in a

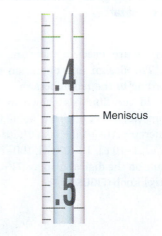

FIGURE A.3 **The meniscus.** Read the fluid volume at the *lowest* level of the meniscus.

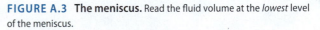

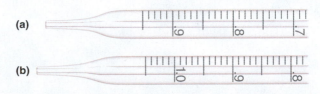

FIGURE A.4 **Types of pipette. (a)** A serological pipette. **(b)** A measuring pipette.

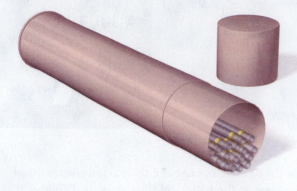

FIGURE A.5 Proper placement of a pipette canister.

canister, the delivery end will be at the bottom of the canister. Do not touch the delivery end of a sterile pipette. Remove a pipette, attach the pipette aspirator, and with your other hand, pick up the sample to be pipetted. Remove the cap from the sample with the little finger of the hand holding the pipette. Fill the pipette, and replace the cap on the sample.

 After pipetting, place the contaminated pipette in the appropriate container of disinfectant. If it is a disposable pipette or micropipette tip, discard it in a biohazard bag or any container designated for biohazards.

Micropipettes are used to measure volumes of less than 1 ml. The design and operation of micropipettes vary according to the manufacturer. A disposable tip is used to hold the fluid, as shown in **FIGURE A.6**. To use a micropipette, select one that holds the volume you need. The micropipette is labeled with its range—for example, 0.5–10 μl, 1–10 μl, 10–100 μl, 100–1000 μl. Set the desired volume on the digital display (**FIGURE A.7a**) by turning the control knob (**FIGURE A.7b**).

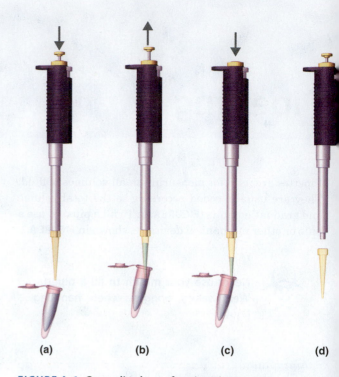

(a) (b) (c) (d)

FIGURE A.6 Generalized use of a micropipette. (a) Depress the control button, and insert the tip into the liquid. **(b)** Smoothly release the button to allow the liquid to enter the tip. **(c)** Place the tip against the inside of the receiving tube, and depress the button. **(d)** Eject the used tip by pressing the eject button or by pressing the control button to the final stop.

QUESTIONS

1. Which micropipette would you use to measure 15 μl? _____
   _____

2. If this is the display for the micropipette you chose for question 1, write in the 1 and 5 to show 15 μl.

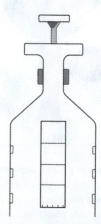

3. How do you eject the tip on your micropipette?
   _____

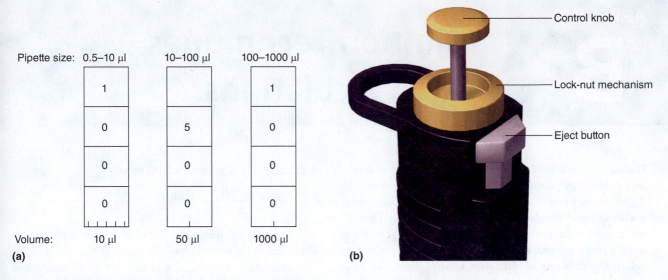

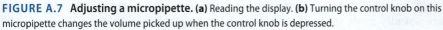

FIGURE A.7 Adjusting a micropipette. (a) Reading the display. **(b)** Turning the control knob on this micropipette changes the volume picked up when the control knob is depressed.

Bacteria, under good growing conditions, will multiply into such large populations that it is often necessary to dilute them to isolate single colonies or to obtain estimates of their numbers. This requires mixing a small, accurately measured sample of the bacterial culture with a large volume of sterile water or saline, which is called the **diluent** or **dilution blank.** Accurate dilutions of a sample are obtained through the use of pipettes. For convenience, dilutions are usually made in multiples of 10.

A single dilution is calculated as follows:

$$Dilution = \frac{Volume \ of \ the \ sample}{Total \ (volume \ of \ the \ sample + the \ diluent)}$$

For example, the dilution of 1 ml into 9 ml equals

$$\frac{1}{1+9} \ , which \ is \ \frac{1}{10} \ and \ is \ written \ 1:10$$

The same formula applies for all dilutions, regardless of the volumes. A dilution of 0.1 ml into 0.9 ml equals

$$\frac{0.1}{0.1+0.9} \ , which \ is \ \frac{0.1}{1} = \frac{1}{10}$$

*Adapted from C. W. Brady. "Dilutions and Dilution Calculations." Unpublished paper. Whitewater: University of Wisconsin, n.d.

A dilution of 0.5 ml into 4.5 ml equals

$$\frac{0.5}{0.5+4.5} \ , which \ is \ \frac{0.5}{5.0} = \frac{1}{10}$$

Experience has shown that better accuracy is obtained with very large dilutions if the total dilution is made out of a series of smaller dilutions rather than one large dilution. This series is called a **serial dilution,** and the total dilution is the product of each dilution in the series. For example, if 1 ml is diluted with 9 ml, and then 1 ml of that dilution is put into a second 9-ml diluent, the final dilution will be

$$\frac{1}{10} \times \frac{1}{10} = \frac{1}{100} \ or \ 1:100$$

To facilitate calculations, the dilution is written in exponential notation. In the example above, the final dilution 1:100 would be written 10^{-2}. Remember,

$$1:100 = \frac{1}{100} = 0.01 = 10^{-2}$$

(See the section Exponents, Exponential Notation, and Logarithms, on page 466.) A serial dilution is illustrated in **FIGURE B.1**.

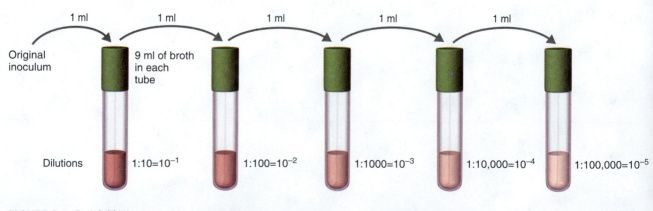

1 ml   1 ml   1 ml   1 ml   1 ml

Original inoculum

9 ml of broth in each tube

Dilutions   1:10=10^{-1} 1:100=10^{-2} 1:1000=10^{-3} 1:10,000=10^{-4} 1:100,000=10^{-5}

FIGURE B.1 **Serial dilution.** A 1-ml sample from the first tube will contain 1/10 the number of cells present in 1 ml of the original sample. A 1-ml sample from the last tube will contain 1/100,000 the number of cells present in 1 ml of the original sample.

Twofold dilutions are commonly used to dilute patient's serum to measure antibodies. The same formula applies: a dilution of 100 μl of sample into 100 μl of saline equals

$$\frac{100}{100 + 100} \text{, which is } \frac{100}{200} = \frac{1}{2}$$

If 100 μl of this 1:2 dilution is put in 100 μl of saline, the final dilution is

$$\frac{1}{2} \times \frac{1}{2} = \frac{1}{4}$$

PROCEDURE

1. Aseptically pipette 1 ml of sample into a dilution blank.
   a. If the dilution is into a tube, mix the contents on a vortex mixer or by rolling the tube back and forth between your hands.
   b. If the dilution is into a 99-ml blank, hold the cap in place with your index finger and shake the bottle up and down through a 35-cm arc (see Figure 54.2 on page 430).
2. It is necessary to use a fresh pipette for each dilution in a series, but it is permissible to use the same pipette to remove several samples from the same bottle, as when plating out samples from a series of dilutions.

PROBLEMS

Practice calculating serial dilutions using the following problems.

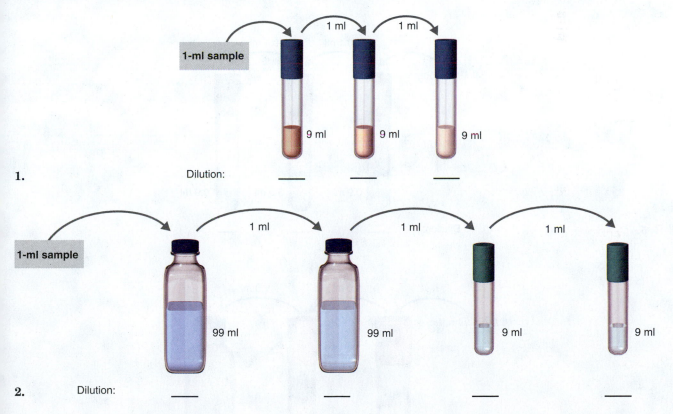

1. Dilution: ___ ___ ___

2. Dilution: ___ ___ ___ ___

3. Design a serial dilution to achieve a final dilution of 10^{-8}.

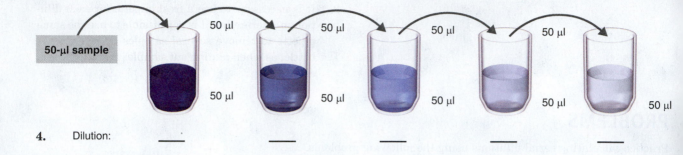

50-µl sample   50 µl   50 µl   50 µl   50 µl

50 µl   50 µl   50 µl   50 µl   50 µl

4. Dilution: ____ · ____ ____ ____

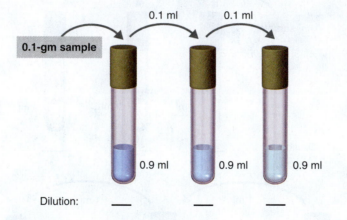

0.1 ml   0.1 ml

0.1-gm sample

0.9 ml   0.9 ml   0.9 ml

5. Dilution: ____ ____ ____

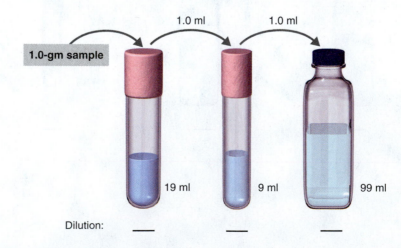

1.0 ml   1.0 ml

1.0-gm sample

19 ml   9 ml   99 ml

6. Dilution: ____ ____ ____

EXPONENTS, EXPONENTIAL NOTATION, AND LOGARITHMS*

Very large and very small numbers—such as 4,650,000,000 and 0.00000032—are cumbersome to work with. It is more convenient to express such numbers in exponential notation—that is, as a power of 10. For example, 4.65×10^9 is written in **standard exponential notation,** or **scientific notation;** 4.65 is the **coefficient,** and 9 is the power, or **exponent.** In standard exponential notation, the coefficient is a number between 1 and 10, and the exponent is a positive or negative number.

To change a number into exponential notation, follow two steps. First, determine the coefficient by moving the decimal point so you leave only one nonzero digit to the left of it. For example:

$$0.00000032$$

The coefficient is 3.2. Second, determine the exponent by counting the number of places you moved the decimal point. If you moved it to the left, the exponent is a positive number. If you moved it to the right, the exponent is negative. In the example, you moved the decimal point 7 places to the right, so the exponent is −7. Thus

$$0.00000032 = 3.2 \times 10^{-7}$$

Now suppose we are working with a very large number instead of a very small number. The same rules apply, but our exponential value will be positive rather than negative. For example:

$$4,650,000,000 = 4.65 \times 10^{+9}$$
$$= 4.65 \times 10^9$$

To multiply numbers written in exponential notation, multiply the coefficients and *add* the exponents. For example:

$$(3 \times 10^4) \times (2 \times 10^3) =$$
$$(3 \times 2) \times 10^{4+3} = 6 \times 10^7$$

To divide numbers written in exponential notation, divide the coefficients and *subtract* the exponents. For example:

$$\frac{3 \times 10^4}{2 \times 10^3} = \frac{3}{2} \times 10^{4-3} = 1.5 \times 10^1$$

Microbiologists use exponential notation in many kinds of situations. For instance, exponential notation is used to describe the number of microorganisms in a population. Such numbers are often very large. Another application of exponential notation is to express concentrations of chemicals in a solution—chemicals such as media components, disinfectants, or antibiotics. Such numbers are often very small. Converting from one unit of measurement to another in the metric system requires multiplying or dividing by a power of 10, which is easiest to carry out in exponential notation.

A **logarithm** is the power to which a base number is raised to produce a given number. Usually we work with logarithms to the base 10, abbreviated **log₁₀.** The first step in finding the $\log_{10}$ of a number is to write the number in standard exponential notation. If the coefficient is exactly 1, the $\log_{10}$ is simply equal to the exponent. For example:

$$\log_{10} 0.00001 = \log_{10}(1 \times 10^{-5})$$
$$= -5$$

If the coefficient is not 1, as is often the case, a calculator must be used to determine the logarithm.

Microbiologists use logs for pH calculations and for graphing the growth of microbial populations in culture.

---

*Source: G. J. Tortora, B. R. Funke, and C. L. Case. *Microbiology: An Introduction*, 12th ed. San Francisco, CA: Pearson Education, 2016, Appendix B.

Use of the Spectrophotometer

In a **spectrophotometer,** a beam of light is transmitted through a bacterial suspension to a photoelectric cell (**FIGURE C.1**). As bacterial numbers increase, the broth becomes more turbid, causing the light to scatter and allowing less light to reach the photoelectric cell. The change in light is registered on the instrument as **percentage of transmission,** or **%T** (the amount of light getting through the suspension) and **absorbance (Abs.)** (a value derived from the percentage of transmission). Absorbance is a logarithmic value and can be used to plot bacterial growth on a graph.

Several brands of spectrophotometers are available; however, all follow the same principles of operation. You will use one of the following (analog or digital) procedures, regardless of the make and model of your spectrophotometer.

> ⚠ Always wipe the surface of the spectrophotometer tube with a low-lint, non-abrasive paper such as a Kimwipe before placing the tube in the spectrophotometer.

OPERATION OF THE ANALOG SPECTRONIC 20

1. Turn on the power, and allow the instrument (**FIGURE C.2**) to warm up for 15 minutes.
2. Set the wavelength for maximum absorption of the bacteria and minimal absorption of the culture medium.
3. For the Spectronic 20, *zero* the instrument by turning the zero control until the needle measures 0% transmission.

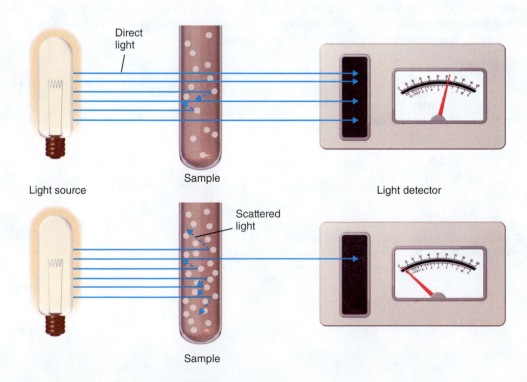

Direct light

Sample

Light source

Light detector

Scattered light

Sample

FIGURE C.1 Estimation of bacterial numbers by turbidity. The amount of light picked up by the light detector is proportional to the number of bacteria. The less light transmitted, the more bacteria in the sample.

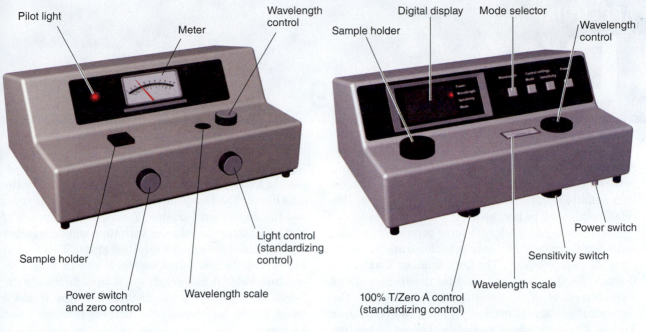

FIGURE C.2 Bausch and Lomb Spectronic 20.

FIGURE C.3 Bausch and Lomb Spectronic 21, digital model.

4. To read a metered scale, look directly at the meter so the needle is superimposed on its reflection in the mirror.
5. Place an uninoculated tube of culture medium (*control*) in the sample holder. To *standardize* the instrument, turn the light control until the needle registers 100% transmission.
6. To take a sample measurement, place an inoculated tube of culture medium in the sample holder, wait about 45 seconds, and record a reading. There will be some fluctuation in turbidity depending on the alignment of particles. Waiting longer than 45 seconds is not recommended because the tube will heat up, creating convection currents that cause rod-shaped cells to line up.
7. %T is usually read on the Spectronic 20 because it is the more accurate scale. Calculate the absorbance as follows:

$$\text{Absorbance} = -\log\left(\frac{\%\text{T}}{100}\right)$$

OPERATION OF THE DIGITAL SPECTRONIC 21

1. Turn on the power, and allow the instrument (**FIGURE C.3**) to warm up for 15 minutes.
2. Set the wavelength for maximum absorption of the bacteria and minimal absorption of the culture medium.

3. Select the operating mode: A (absorbance) or %T (transmission). Absorbance data are more linear and usually preferable.
4. Set the sensitivity switch to LO.
5. Place an uninoculated tube of culture medium (*control*) in the sample holder. To *standardize* the instrument, turn the 100% T/Zero A control until the display registers 000 A. If 000 A can't be reached, turn the sensitivity to M or HI. A U in the upper-left corner of the display indicates that you are under the range of the light detector, and an O in the display indicates that you are over the range. In either case, use the sensitivity switch and the standardizing control to make the necessary adjustments.
6. To take a sample measurement, place an inoculated tube of culture medium in the sample holder, wait about 45 seconds, and record a reading. There will be some fluctuation in turbidity depending on the alignment of particles. Waiting longer than 45 seconds is not recommended because the tube will heat up, creating convection currents that cause rod-shaped cells to line up.

Graphing

A **graph** is a visual representation of the relationship between two variables. Whenever one variable changes in a definite way in relation to another variable, this relationship may be graphed.

Microbiologists work with large populations of bacteria and frequently use graphs to illustrate the activity of these populations. The horizontal, or **X-axis,** is a linear scale. The X-axis is used for the **independent variable**—that is, the variable not being tested. The dependent variable is marked off along the vertical, or **Y-axis.** The **dependent variable** changes in relation to the independent variable. The intersection point of the X-axis and Y-axis is called the **origin,** and all units are marked off from this point.

Use a computer-graphing application to graph the numbers of bacteria at each temperature, as shown in **TABLE D.1**. First enter the data into a spreadsheet.

Next, convert the numbers of bacteria to their log values. At times when no readings were taken, leave the cells blank. The independent variable (X-axis) is time, and the dependent variable (Y-axis) is the number of bacteria. Select the columns with time and log number of cells, and choose an XY (scatter) graph. This format will distribute the points on the X-axis at their correct intervals. (A line graph will simply distribute the X-axis data evenly so the distance between 0 and 4 hours will be the same as the distance between 4 and 5 hours.)

A graph of the numbers of bacteria in the culture incubated at 20°C is shown in **FIGURE D.1**. A best-fit line is drawn through the points. Use the log trend line in your graphing application or draw the line by hand. The line should be straight and does not have to connect all the points. When the points do not fall in a

TABLE D.1 NUMBERS OF *E. coli* IN THREE CULTURES INCUBATED AT DIFFERENT TEMPERATURES			
	Number of Bacteria per ml at:		
Time (hr)	15°C	20°C	35°C
0	1.50×10^5	5.00×10^5	5.40×10^5
4	1.50×10^5	*N*	*N*
5	*N*	4.60×10^5	*N*
6	*N*	*N*	5.20×10^5
10	5.62×10^5	5.78×10^5	6.00×10^5
11	*N*	8.80×10^5	1.23×10^6
12	*N*	1.05×10^6	2.58×10^6
13	*N*	1.16×10^6	4.93×10^6
14	*N*	2.32×10^6	9.00×10^6
15	1.62×10^6	3.80×10^6	*N*
16	*N*	7.00×10^6	*N*
17	*N*	8.10×10^6	*N*
18	*N*	9.60×10^6	*N*
19	*N*	1.95×10^7	*N*
20	2.70×10^6	*N*	*N*

N = No reading at that time.

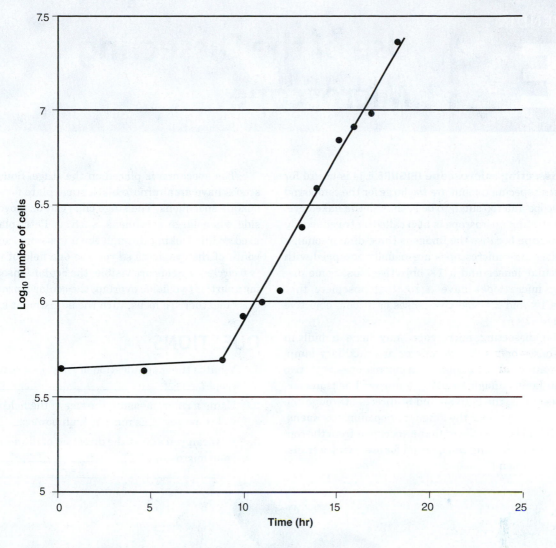

FIGURE D.1 Growth of *E. coli* at 20°C.

straight line, draw a line between the points, leaving an equal number of points above and below the line.

When you draw graphs that compare the same variables under different conditions (for example, bacterial growth rates at different temperatures), it is best to draw all the graphs on the same chart so comparisons are obvious and easily made. Title graphs to explain the data presented.

The data for cultures incubated at 15°C and 35°C (see Table D.1) can be plotted on the same graph. It is not necessary to have bacterial numbers for the same "time points" in each graph. The missing points will be

filled in by the line. This is possible because the line between each point is an *interpolation* of what was happening between measurements.

ACTIVITY

Using the data provided (see Table D.1), graph the growth curves on graph paper or using a computer graphing application, and compare the effect of temperature on the rate of growth. Because *rate* is a change over time, the rate can be interpreted by the slope of the line. What will you title this graph? _____

_____

Use of the Dissecting Microscope

The **dissecting microscope** (FIGURE E.1) is useful for observing specimens that are too large for the compound microscope and too small to be seen with the naked eye. The dissecting microscope is also called a **stereoscopic microscope** because the image is three-dimensional.

Dissecting microscopes are usually equipped with 10× ocular lenses and a 1× objective lens. Some dissecting microscopes have a rotating nosepiece that houses 1× and 2× objective lenses or a zoom objective lens (1×−7×).

The dissecting microscope may have a built-in light source or a substage mirror and auxiliary lamp. Illumination can be adjusted on microscopes with two built-in lamps and/or by using a mirror. For transparent specimens, the light should be directed through the specimen from under the stage. For opaque specimens, light should be directed onto the specimen from the top. Both types of lighting can be used for observing a translucent specimen.

The specimen is placed on the stage. Some microscopes have an alternate black stage plate for use with opaque specimens. The stage clips can be moved to the side when large specimens, such as Petri plates, are used. While looking through the microscope, adjust the width of the eyepieces so you see one field of vision. If two circles or fields are visible, the ocular lenses are too far apart; if two fields overlap, the ocular lenses are too close together. To focus, turn the adjustment knob.

QUESTIONS

1. What is the magnification of your dissecting microscope? _____

2. Using a ruler, measure the size of the field of vision at low power: ___ mm. At high power: ___ mm.

3. How can you adjust the direction of illumination on your microscope? _____
_____

Ocular lenses Provides primary magnification, usually 10 times.

Nosepiece Houses objective lenses. May be rotated on some microscopes for additional magnification.

Arm

Adjustment knob Moves lenses for focusing.

Objective lens Provides additional magnification.

Lamp Provides illumination.

Stage clips May be used to hold specimen in place.

Mirror Directs light through specimen.

Stage Provides platform to hold specimen.

Stage plate Provides contrasting or transparent background for specimen.

FIGURE E.1 The dissecting microscope: principal parts and their functions.

Use of the Membrane Filter

Membrane filtration can be used to separate bacteria, algae, yeasts, and molds from solutions. Nitrocellulose or polyvinyl membrane filters with 0.45-μm pores are commonly used to trap microorganisms and generate sterile solutions. Membrane filtration is used to separate viruses from their host cells because the viruses will pass through the filter. Membrane filters are used to trap bacteria found in water in order to count the bacteria.

Membrane filters have many uses in industrial microbiology. They are used to filter wine, soft drinks, air, and water to detect or remove microorganisms and particulate matter. In addition, membrane filters can be used to separate large molecules, such as DNA, from solutions.

The receiving flask, filter base, and filter support can be wrapped in paper and sterilized by autoclaving (**FIGURE F.1**). Membrane filters can also be sterilized by autoclaving. For use, the filter base must be unwrapped aseptically. With sterile, flat forceps, a sterile filter is placed on the filter support, and the filter funnel is clamped or screwed into place. Presterilized, disposable polystyrene membrane filtration units are available (**FIGURE F.2**).

Gravity alone will not pull a sample through the small filter pores, so the filtering flask is attached to a vacuum source (see Figure F.1). Because air pressure is then greater outside the flask, the sample is pushed into the vacuum inside the flask. A vacuum pump or vacuum line is usually used to provide the necessary vacuum. An aspirator trap is placed between the vacuum source and the receiving flask to prevent the flow of solution

FIGURE F.1 Membrane filtration setup.

into the pump or vacuum line. With small filters and small volumes of fluid, a syringe can be used to supply the needed vacuum or to push fluid through the filter.

Following filtration, the filtrate in the flask is free of all microorganisms larger than viruses. For observation of cells trapped on the filter, the dried filter can be dipped into immersion oil to make it translucent and then mounted on a microscope slide. Microorganisms can be cultured from the filter by placing the filter on a solid nutrient medium.

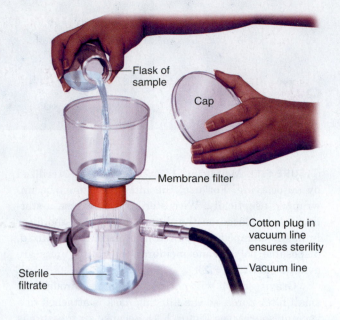

Flask of sample

Cap

Membrane filter

Cotton plug in vacuum line ensures sterility

Vacuum line

Sterile filtrate

FIGURE F.2 Disposable presterilized plastic filtration unit.

Electrophoresis

Charged macromolecules can be separated according to size and charge by a process called **electrophoresis.** The material to be tested is placed at one end of a gel matrix. Nucleic acids and most proteins are colorless, so a tracking dye is usually added to the sample to enable you to track progress through the gel. The dye molecules will migrate faster than the sample because they are smaller than the sample molecules. Electrophoresis is stopped when the dye reaches the anodic end of the gel. Additionally, tracking dyes are in glycerol, which is denser than water and makes loading the wells easier. Polyacrylamide gel or agarose is used to separate proteins and small pieces of DNA and RNA. Agarose is the usual matrix for separation of nucleic acids. A **buffer** (salt solution) is used to conduct electric current through the matrix. When an electric current is applied, the molecules start to move. Molecules with different charges and sizes will migrate at different rates.

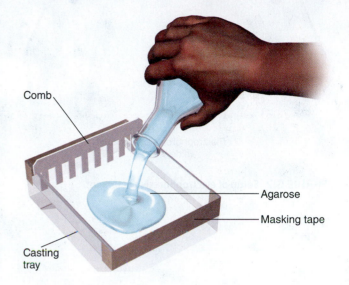

Comb

Agarose

Masking tape

Casting tray

FIGURE G.1 Preparation of an agarose minigel.

PROCEDURE

1. Place the well-forming comb in position on a casting tray (**FIGURE G.1**). Seal the open ends of the casting tray with masking tape. Or, if the tray has movable gates, raise the gate at each end. Pour melted agarose onto the casting tray until the agar surrounds the teeth. The agar should be about 5 mm deep and have a flat surface. Let the gel harden for about 20 minutes. When the gel is solid, carefully remove the comb, and then remove the tape or lower the end gates.
2. Fill the buffer chamber with the appropriate buffer. Add just enough buffer to cover the gel. Place the agar in the buffer chamber with or without the casting tray.
3. Fill the wells with samples to be tested. Close the lid and apply current (**FIGURE G.2**).

4. During electrophoresis, a visible marker dye will migrate in front of the sample. If the dye isn't moving in the right direction, you have accidentally reversed the polarity of the electric field. If the marker doesn't move and if there is little current, you have an open circuit or have not added the correct buffer. In this case, you must start over.
5. The two components of the tracking dye migrate at different rates. Why?

Turn off the power when the two dyes are 4 or 5 cm apart. Remove the gel. The gel can now be stained to locate the macromolecules. Store stained or unstained gels in sealed plastic bags in the refrigerator.

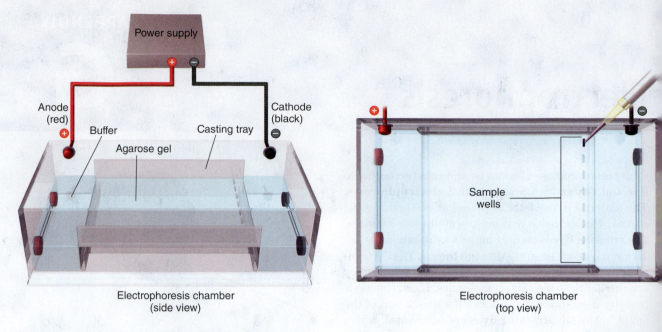

Electrophoresis chamber
(side view)

Electrophoresis chamber
(top view)

FIGURE G.2 Setup of electrophoresis equipment.

Keys to Bacteria

Appendix H deals with the identification of bacteria. The information is arranged in **dichotomous keys.** In a dichotomous key, identification is based on successive questions, each of which has two possible answers (*di-* means two). After answering each question, the investigator is directed to another question until an organism is identified. There is no single "correct" way to write a key; the goal is to conclude with one unambiguous identification. These keys do not include all bacteria. You should refer to *Bergey's Manual* to verify your identification.

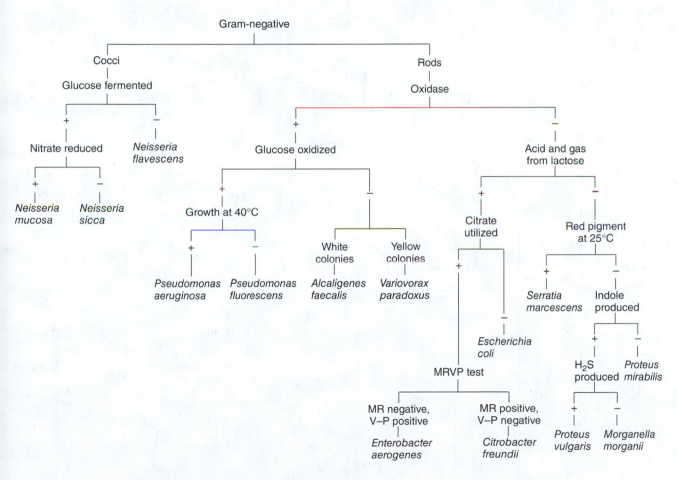

FIGURE H.1 **Key to selected gram-negative heterotrophic bacteria.**

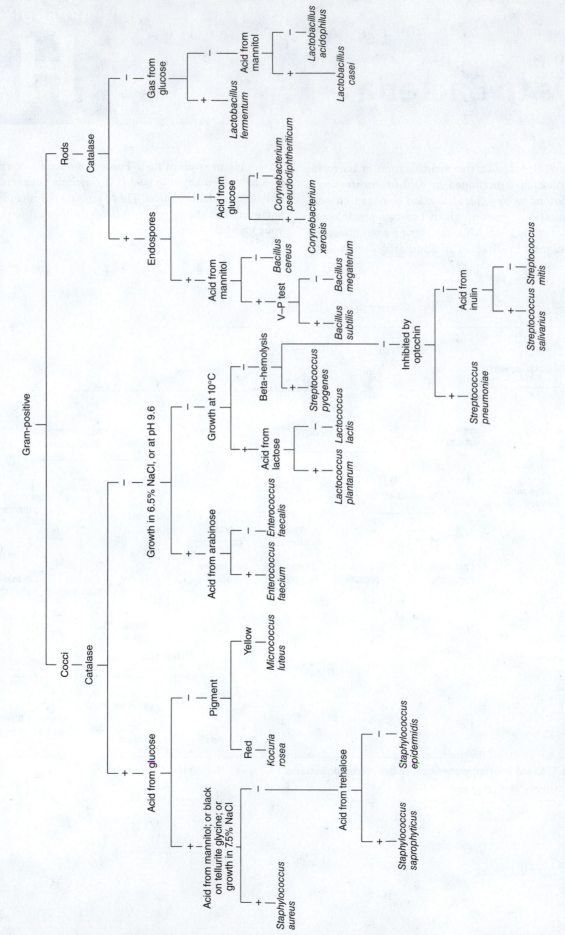

FIGURE H.2 Key to selected gram-positive heterotrophic bacteria.

Gram-positive cocci
↓
Catalase-positive
↓

I. Cells arranged in tetrads, glucose not fermented

    A. Colonies have yellow pigment   *Micrococcus luteus*

    B. Colonies have red pigment   *Kocuria rosea*

II. Cells arranged in grapelike clusters, glucose fermented   *Staphylococcus*

    A. Acid produced from mannitol

        1. Coagulase-positive   *S. aureus*

        2. Coagulase-negative

            a. Acid produced from trehalose

                1) Acid from xylose   *S. arlettae*

                2) No acid from xylose   *S. saprophyticus*

            b. No acid from trehalose   *S. capitis*

    B. No acid from mannitol

        1. Acid produced from trehalose   *S. saprophyticus*

        2. No acid from trehalose   *S. epidermidis*

FIGURE H.3 Key to gram-positive cocci commonly found on human skin.

Gram-positive cocci in chains
↓
Catalase-negative
↓
No growth in 6.5% NaCl broth, or at pH 9.6
↓
Alpha- or gamma-hemolysis

I. Inulin fermented

  A. Mannitol fermented

    1. Salicin fermented   *S. mutans*

    2. Salicin not fermented   *S. sobrinus*

  B. Mannitol not fermented

    1. Acetoin produced (V–P test)   *S. salivarius*

    2. Acetoin not produced (V–P test)   *S. sanguinis*

II. Inulin not fermented

  A. Mannitol fermented

    1. Esculin hydrolyzed   *S. anginosus*

    2. Esculin not hydrolyzed   *S. oralis*

  B. Mannitol not fermented   *S. mitis*

FIGURE H.4 Key to streptococci found in the mouth.

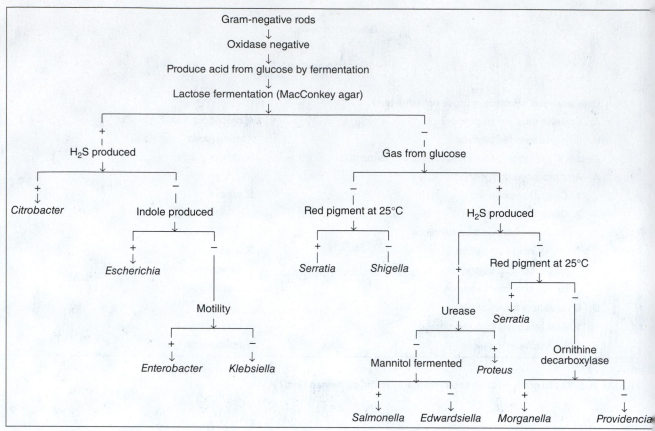

FIGURE H.5 Identification scheme for enteric genera primarily using MacConkey agar and triple sugar iron agar.

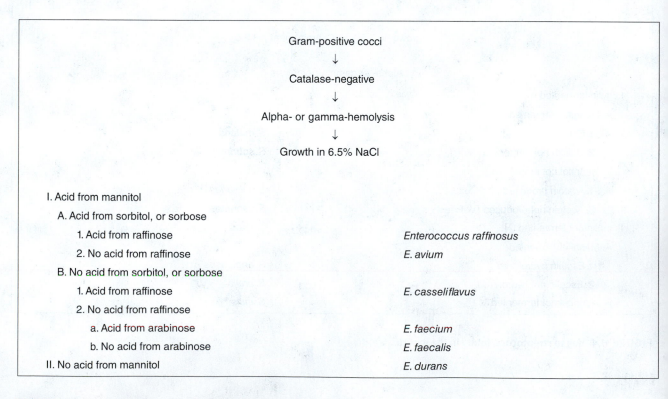

FIGURE H.6 Key to enterococci commonly found in the human intestine.

Art and Photography Credits

ART CREDITS

All illustrations by Precision Graphics unless otherwise noted.

The following figures are adapted from *Microbiology,* 12e, by Tortora, Funke, and Case, © 2016 Pearson Education, Inc.: Figures 1.1, 1.2, 19.2, 21.1, 22.1, 23.1, 35.1, 36.1, 36.2, 36.3, 37.1, 44.1, 47.1, 56.1, 56.3, F.2, and Part openers for Parts 6, 7, 8, 9, and 11.

PHOTOGRAPHY CREDITS

All photographs by Christine Case unless otherwise noted.

Figure 1.1: Leica Microsystems.

Figure 8.2: CDC.

Figure 9.4: CDC.

Figure 18.1: CDC.

Figure 29.UN.1: B. Manicassamy et al., "Analysis of in vivo dynamics of influenza virus infection in mice using a GFP reporter virus," Proceedings of National Academy of Sciences of the United States of America 107, no. 25 (June 22, 2010: 11531-11536).

Figure 36.UN.5: CDC.

Figure 36.UN.6a: Dr. Mae Melvin/CDC.

Figure 36.UN.6b: CDC.

Figure 57.2: Randall von Wedel/CytoCulture International.

Index

A *t* following a page number indicates tabular material and an *f* following a page number indicates a figure.